AF780624

Advanced Control and Operation of Microgrids and Power Distribution Systems

Advanced Control and Operation of Microgrids and Power Distribution Systems

Guest Editors

Il-Yop Chung
Seon-Ju Ahn

MDPI

Basel • Beijing • Wuhan • Barcelona • Belgrade • Novi Sad • Cluj • Manchester

Guest Editors
Il-Yop Chung
School of Electrical Engineering
Kookmin University
Seoul
Republic of Korea

Seon-Ju Ahn
Department of Electrical Engineering
Chonnam National University
Gwangju
Republic of Korea

Editorial Office
MDPI AG
Grosspeteranlage 5
4052 Basel, Switzerland

This is a reprint of the Special Issue, published open access by the journal *Energies* (ISSN 1996-1073), freely accessible at: https://www.mdpi.com/journal/energies/special_issues/DYH649I3O0.

For citation purposes, cite each article independently as indicated on the article page online and as indicated below:

Lastname, A.A.; Lastname, B.B. Article Title. *Journal Name* **Year**, *Volume Number*, Page Range.

ISBN 978-3-7258-5489-9 (Hbk)
ISBN 978-3-7258-5490-5 (PDF)
https://doi.org/10.3390/books978-3-7258-5490-5

© 2025 by the authors. Articles in this book are Open Access and distributed under the Creative Commons Attribution (CC BY) license. The book as a whole is distributed by MDPI under the terms and conditions of the Creative Commons Attribution-NonCommercial-NoDerivs (CC BY-NC-ND) license (https://creativecommons.org/licenses/by-nc-nd/4.0/).

Contents

energies

MDPI

Article

Research on the Hybrid Wind–Solar–Energy Storage AC/DC Microgrid System and Its Stability during Smooth State Transitions

Qiushuo Li *, Xinwei Dong, Mengru Yan, Zhao Cheng and Yu Wang

School of Electrical and Engineering, China University of Mining and Technology, Xuzhou 221116, China; 4614@cumt.edu.cn (X.D.); ts22230173p31@cumt.edu.cn (M.Y.); ts22230097p31@cumt.edu.cn (Z.C.); ts22230161p31@cumt.edu.cn (Y.W.)
* Correspondence: ts22230034a31@cumt.edu.cn; Tel.: +86-113852155620

Abstract: The hybrid AC/DC microgrid is an independent and controllable energy system that connects various types of distributed power sources, energy storage, and loads. It offers advantages such as a high power quality, flexibility, and cost effectiveness. The operation states of the microgrid primarily include grid-connected and islanded modes. The smooth switching between these two states is a key technology for ensuring the flexible and efficient operation of the microgrid. In this paper, the typical structure of an AC–DC hybrid microgrid and its coordination control strategy are introduced, and an improved microgrid model is proposed. Secondly, an adaptive current–voltage–frequency integrated control method based on signal compensation is proposed to solve the impulse current and voltage generated during the switching between a grid-connected state and an off-grid state. Finally, in response to unplanned grid-connected scenarios, an improved pre-synchronization control strategy based on BP neural networks is introduced to rapidly restore stable operation. The proposed control strategies enhanced the steady-state and transient stability of the hybrid wind–solar–energy storage AC/DC microgrid, achieving seamless grid-connected and islanded transitions without disturbances. The simulation and experimental results validated the correctness and effectiveness of the proposed theories.

Keywords: hybrid AC/DC microgrid; smooth switching between grid-connected and islanded modes; control strategies; BP neural networks algorithm; stability control

1. Introduction

In recent years, with the aggravation of resource scarcity and environmental pollution issues, the proportion of distributed generation (DG) represented by photovoltaic and wind power in the power grid is increasing, which brings many advantages such as protecting the environment, reducing power generation costs, and improving resource utilization. It also leads to intermittent output fluctuation, power quality decline, and many other hazards. Microgrids, especially hybrid AC/DC microgrids, have emerged as intelligent micro-power systems that maximize the advantages of DG. They integrate various types of distributed energy sources, energy storage systems, loads, controls, and various protection measures [1]. Its unique bidirectional AC/DC converter, which enables smooth power flow between AC and DC sub-microgrids, reduces the power conversion process and improves power supply reliability [2]. Presently, hybrid AC/DC microgrids are predominantly solar-storage-based and wind–solar-storage-based. However, at the same time, because of the intermittent photovoltaic and wind resources, they have a certain instability.

Hybrid AC/DC microgrids operate in two basic states: grid-connected and islanded. During state transitions, the system structure will change, leading to voltage and frequency variations, and various issues like current surges, voltage fluctuations, and frequency mutation. This can jeopardize the stable operation of the microgrid and even affect the safety of the main grid. Therefore, ensuring the smooth transition between grid-connected and islanded states is essential for improving the microgrid's safety and stability.

Currently, the control strategies for individual AC or DC microgrids during state transitions are relatively mature [3–8], but there are few studies involving both AC and DC microgrids [9]. In order to solve the problem of voltage shock during the switching process, the literature [3] proposed an improved voltage loop controller, which reduces the surge current, but lacks a solution for addressing current and frequency transients. Reference [4] put forward a hybrid control method based on master–slave control, combining droop control in the grid-connected mode and V/f control in the islanded mode. Although this improved the control accuracy, switching between different control methods can cause instability among the related parameters. In order to ensure smooth switching between the operating modes and reduce the impact of transient switching, the literature [5] proposed a new software phase-locked loop (PLL) control, including instantaneous grid phase detection, off-grid phase-locking technology, and a frequency compensation phase pre-synchronization algorithm. However, pure software control has high hardware stability requirements and cannot meet the needs of unplanned grid-connected and islanded transitions. In the literature [10], an improved hybrid voltage and current control mode was proposed to realize smooth switching between inverters and off-grid, but the control mode needed to be changed during the switching process, and there were deviations in the voltage and frequency under the off-grid operation. In addition to improving the voltage and current loop control, pre-synchronous control is also an effective control method for realizing the off-grid state to the grid-connected state. The literature [11] used pre-synchronization control to pre-adjust the output voltage, frequency, and power of the inverter before switching, effectively suppressing the problem of the voltage and current distortion in the traditional switching process. However, due to the functional limitations of the PI controller itself, the phase difference after several cycles was approximately zero after the output of the PI controller, resulting in an insufficient adjustment function.

Based on the above research results, it can be found that the current AC–DC hybrid microgrid lacks a microgrid model that can solve its steady-state instability in the process of stable operation. There is a lack of control methods that can solve the transient fluctuation of the AC–DC hybrid microgrid in the process of on-grid and off-grid state switching.

On this basis, this paper presents an improved model of a wind–solar storage hybrid AC–DC microgrid based on a doubly-fed induction generator (DFIG), along with control methods for smooth transitions between the grid-connected and islanded states, ensuring transient and steady-state stability. The structure of this paper is as follows. In Section 2, the typical structure and coordinated control strategy of an AC–DC hybrid microgrid are introduced, and an improved microgrid model is proposed. The active power control and doubly-fed technology of the DFIG are used to actively adjust the output of the generator to reduce the external impact. In Section 3, an adaptive integrated control method based on the signal compensation for the current, voltage, and frequency is introduced to mitigate the impact of transient currents and voltages during state transitions. Morever, this section also addresses unplanned grid-connected scenarios and presents an improved pre-synchronization control strategy based on BP neural networks to resolve the inadequacies of traditional pre-synchronization control. In Section 4 presents the improved wind–wind storage hybrid microgrid model built on the Matlab/Simulink (R2022b) simulation platform, and the accuracy and effectiveness of the impact-free smooth grid-connected and off-grid switching control strategy is verified. Finally, Section 5 provides a summary of the paper and future prospects.

2. Introduction of the Hybrid Wind–Solar–Energy Storage AC/DC Microgrid System

2.1. Typical Topology of the Hybrid Wind–Solar–Energy Storage AC/DC Microgrid System

The hybrid AC/DC microgrid is composed of an AC microgrid, DC microgrid, and bidirectional AC/DC converter. Compared with the traditional DC microgrid and AC microgrid, it has stronger flexibility and includes two sub-microgrids, an AC microgrid and a DC microgrid, which can not only accept a variety of units, but also provide energy for different types of loads. In the hybrid AC–DC microgrid, the DC power supply and load are connected to the DC bus, and the bidirectional AC/DC converter is used to exchange power with the AC bus, so that the operation control of the AC side can be better carried out and power can flow between the AC microgrid and the DC microgrid, providing power support for the other sub-microgrids. It ensures the power quality during the grid-connected operation, improves energy utilization, and ensures the safe and stable operation of the whole large system. The typical structure of the hybrid AC–DC microgrid is shown in Figure 1, which is also known as a Type I AC–DC hybrid microgrid structure.

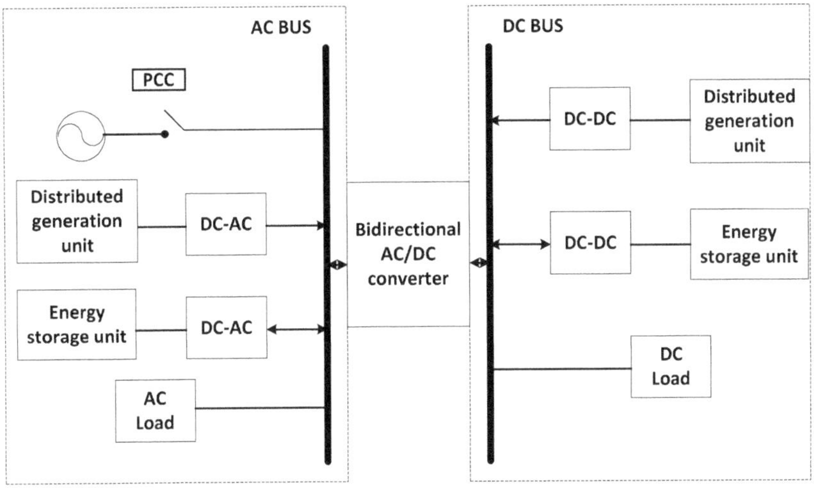

Figure 1. The topology diagram of a Type I hybrid AC/DC microgrid.

The hybrid AC/DC microgrid model used in this paper, namely, the Type I hybrid AC/DC microgrid, has a basic structure in which the point of common coupling (PCC) is connected to both the main grid and the AC/DC microgrid. The AC and DC sub-microgrids are interconnected through bidirectional AC/DC converters. The AC sub-microgrid consists of AC distributed generation units and AC loads, while the DC sub-microgrid comprises DC distributed generation units and DC loads. Due to its connection method and structural characteristics, the Type I hybrid AC/DC microgrid additionally benefits from the main grid's voltage and frequency support for the AC sub-microgrid. This configuration is suitable for microgrids with a significant proportion of AC renewable distributed generation units and AC loads, aligning with the increasing role of renewable energy sources in the power grid. Additionally, it enhances the stability of microgrid system operation [12].

2.2. The General Control Strategy for a Wind–Solar–Energy Storage Hybrid AC/DC Microgrid

The overall control strategy of the microgrid can be divided into master–slave control, peer-to-peer control, and hierarchical control.

(a) Master–Slave Control: The controller of each distributed power generation unit in the microgrid is set up in a subordinate relationship. Typically, the master controller is chosen to be a unit with high inertia and capacity, responsible for providing voltage magnitude and frequency support to the distributed generation units within the microgrid. The slave controllers, benefiting from this support, generally employ direct P/Q control to adjust the power output. This control strategy offers simplicity and flexibility during normal microgrid operation but may face control breakdown during abnormal microgrid operation.

(b) Peer-to-Peer Control: In contrast to master–slave control, peer-to-peer control treats all the units in the microgrid as equals and often employs droop control. All the micro-sources can adjust their power output by varying the frequency and amplitude of their output voltage. However, droop control also has the disadvantage of causing the bus voltage and frequency to drop.
(c) Hierarchical Control: Hierarchical control typically involves a coordinating controller responsible for harmonizing the operation of local distributed energy sources and loads, ensuring the secure and stable operation of the microgrid. To maximize the consumption of distributed energy and ensure the stable operation of the microgrid, a two-times non-differential voltage regulation and a two-times non-differential frequency regulation of the AC sub-microgrid in the local control layer can be achieved. Based on these advantages, the wind–solar–energy storage hybrid AC/DC microgrid proposed in this paper will employ a hierarchical control approach for overall microgrid control.

The bidirectional AC/DC converter plays a crucial role in connecting the AC subnet and the DC subnet and realizing the power flow between the subnets. When there is a power surplus or power deficit in the AC/DC subnetwork, the bidirectional AC/DC converter can work in the rectification or inverter mode to transfer or provide power to another subnetwork. Therefore, researching the control strategy of bidirectional AC/DC converters is of great significance in the steady-state and transient stability studies of hybrid AC–DC microgrids. The traditional control methods of a bidirectional AC/DC converter include V/F control, P/Q control, and droop control.

V/F control [13] means constant voltage and constant frequency control, as shown in Figure 2. When the AC/DC hybrid microgrid operates in the island mode, the AC sub-microgrid loses the support of the large power grid, and the bidirectional AC/DC converter provides constant voltage and frequency support for the bus of the AC sub-microgrid to maintain the stable operation of the AC–DC hybrid microgrid.

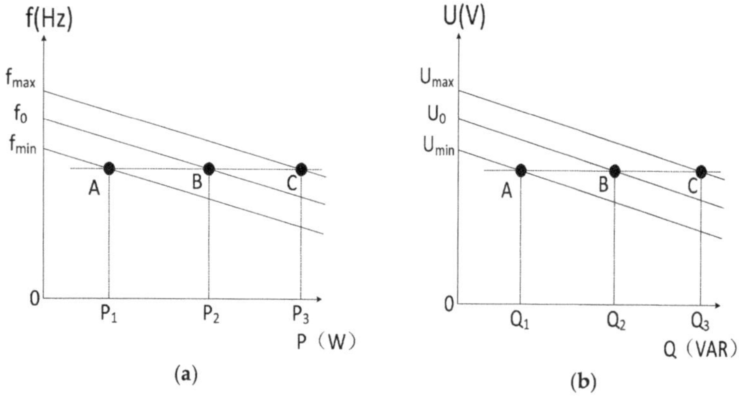

Figure 2. Schematic diagram of V/F control. (**a**) The P-f curve; (**b**) the Q-U curve.

The P/Q control, also known as power balance control [14], is depicted in Figure 3. The voltage of the AC and DC buses is normalized. By using the normalized deviation "n" between the AC and DC bus voltages, the power balance of the hybrid AC–DC microgrid is determined, and the PI controller is used to control $U_{ac} = U_{dc}$. When n > 0, the AC sub-grid provides power support to the DC sub-grid; when n < 0, the DC sub-grid provides power support to the AC sub-grid; and when n = 0, the power of the AC–DC microgrid reaches balance.

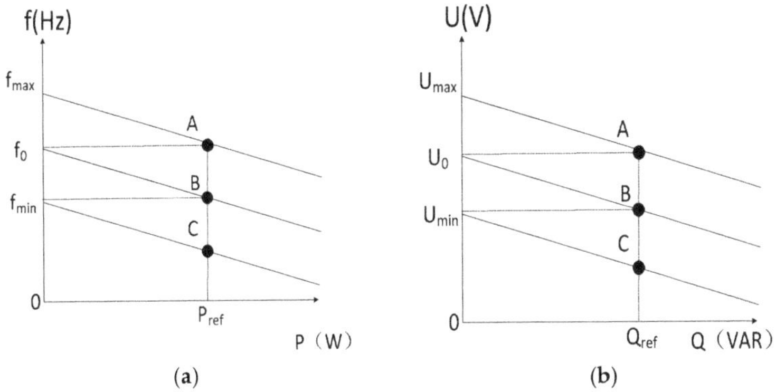

Figure 3. Schematic diagram of P/Q control. (**a**) The P-f curve; (**b**) the Q-U curve.

As shown in Figure 4, the droop control strategy employed in the hybrid AC–DC microgrid is actually a droop control strategy determined by two variables [15], which adjusts the power of distributed generation according to the droop characteristic P-f. Specifically, the transmission power of the bidirectional AC/DC converter is obtained according to the frequency–voltage deviation of the DC sub-grid named $P_{ac} - f_{ac}$ and the frequency–voltage deviation of the AC sub-grid named $P_{dc} - f_{dc}$. When the active power in the microgrid increases and the frequency decreases, the droop controller adjusts the active power of distributed generation based on the droop characteristics. When the reactive power in the microgrid increases and the voltage amplitude decreases, the droop controller adjusts the reactive power based on the droop characteristics. However, during grid-connected and off-grid switching, the frequency of the microgrid fluctuates seriously. Droop control, a control mode guided by frequency power control, has the disadvantage of causing a drop in the bus voltage and frequency, so it is not suitable for grid-connected and off-grid switching.

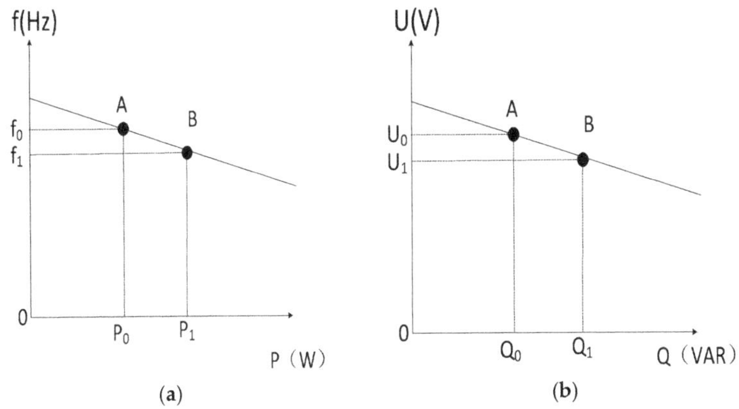

Figure 4. Schematic diagram of droop control. (**a**) The P-f curve; (**b**) the Q-U curve.

Based on the above analysis, the overall control of the wind–wind storage AC–DC hybrid microgrid proposed in this paper adopted the hierarchical control mode. The bidirectional AC/DC converter adopted the P/Q control mode and V/F control mode, and switches the control mode according to the grid-connected and off-grid real-time signals. Specifically, when the microgrid is grid-connected, the bidirectional AC/DC converter operates in P/Q control mode. When the microgrid is in an off-grid state, the bidirectional

AC/DC converter switches to the V/F control mode. The selection of the control mode maximizes the stability of the steady-state and transient operation of the microgrid.

2.3. *Improved Hybrid Wind–Solar–Energy Storage AC/DC Microgrid System Model*

In order to achieve a greater degree of clean energy utilization, this paper selected the Type I hybrid AC/DC microgrid and made improvements to its specific topology, as illustrated in Figure 5 The enhanced hybrid AC/DC microgrid comprises a DC sub-grid, an AC sub-grid, and a control center.

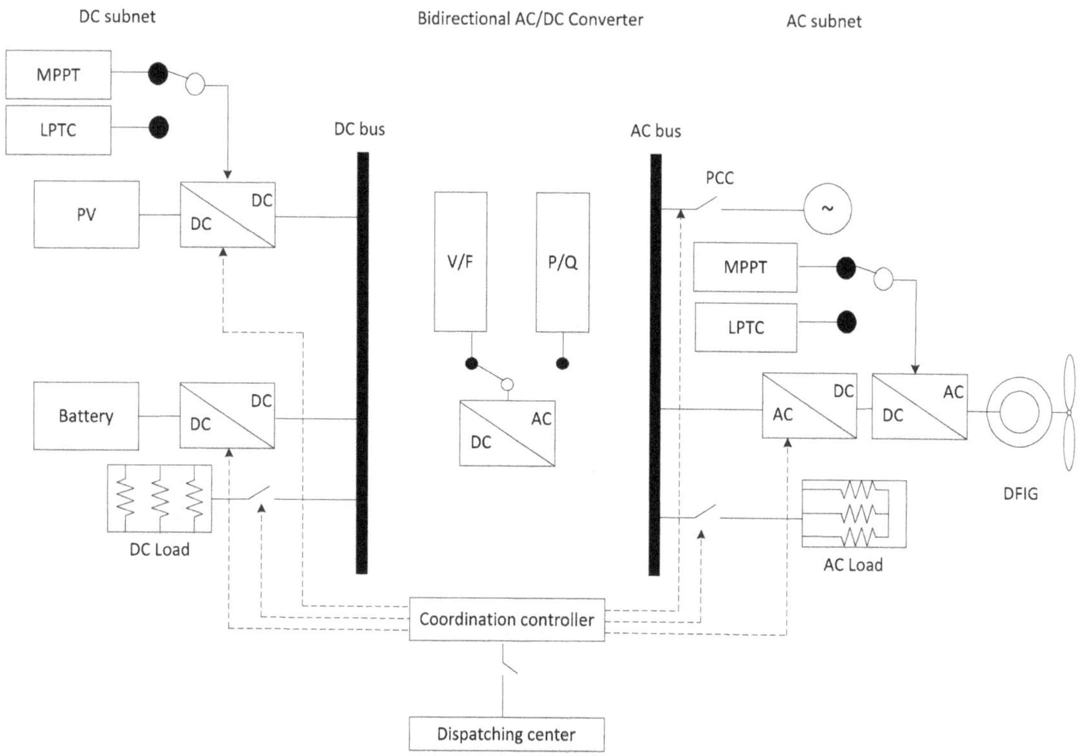

Figure 5. The topology diagram of the improved hybrid wind–solar–energy storage AC/DC microgrid system.

The DC sub-grid consists of photovoltaic generation units, a battery bank, DC loads, power converters, and a DC bus. The photovoltaic generation units and the battery bank are connected in series with power converters and then connected in parallel to the DC bus.

The AC sub-grid includes a DFIG, AC loads, inverters, rectifiers, an AC bus, and a PCC. The DFIG, inverters, and rectifiers are connected in series to the AC bus. The PCC is used to link the main grid and the hybrid AC/DC microgrid, while the AC loads and wind power generation units are connected in parallel to the AC bus.

The control system comprises bidirectional AC/DC converters, a coordinating controller, and a dispatch center. The coordinating controller includes a pre-synchronization controller, a phase-locked loop (PLL), and current–voltage–frequency control loops for stabilizing the relevant parameters of the microgrid.

In order to improve the steady-state stability of the system, the traditional simple wind power module was modified into a DFIG. As shown in Figure 6, the DFIG employs a vector control method, where the q-axis component (i_q) of the rotor current controls active power, and the d-axis component (i_d) controls reactive power, achieving decoupled control of active and reactive power. The DFIG utilizes fundamental control modes, including maximum power point tracking (MPPT) and load power tracking control (LPTC), and issues different current reference commands (i_{qref}, i_{dref}) to control the real-time active and reactive power output. Due to its advantages in the structure and control mode, the DFIG has partial power transfer capability. When the power grid fails or sudden load fluctuation occurs, the DFIG can provide additional reactive power by adjusting the rotor speed and magnetic field to reduce the risk of power failure [16].

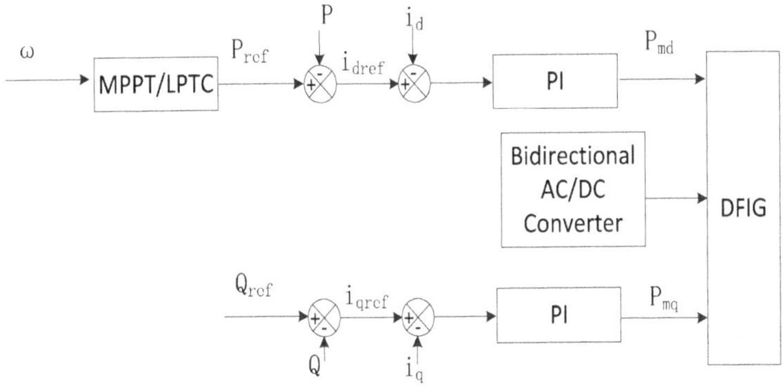

Figure 6. The block diagram of the DFIG.

3. Smooth Switching Control Strategy of the Hybrid Wind–Solar–Energy Storage AC/DC Microgrid System in Grid-Connected and Off-Grid States

In order to achieve smooth transitions between the grid-connected and islanded operation modes of the microgrid and enhance its transient stability, it was necessary to devise a rational control logic. Based on the hierarchical control strategy of microgrids, a set of state switch control logic was designed in this paper, which included two components:

(a) An adaptive integrated control method for the current, voltage, and frequency based on signal compensation.
(b) An improved pre-synchronization control strategy based on a BP neural network.

3.1. The Adaptive Integrated Control Method for the Current, Voltage, and Frequency Based on Signal Compensation

When the AC/DC hybrid microgrid switches between the on-grid state and off-grid state, the control mode of the bidirectional AC/DC converters toggles between the P/Q control and V/F control modes. However, with sudden changes in the outer loop, the voltage and current of the PCC will produce shock and large overshoot, and the frequency will have a large fluctuation, which pose a threat to the safe operation of other renewable energy units and the hybrid AC/DC microgrid.

The voltage loop of the microgrid is a control component within the microgrid system responsible for controlling and maintaining the internal voltage stability. It monitors the voltage level within the microgrid and takes necessary control measures to keep the voltage within an appropriate range. If the voltage inside the microgrid deviates from the set standards or safety levels, the voltage loop employs control strategies such as adjusting the output of the distributed energy resources or scheduling energy storage devices to ensure voltage restoration to normal levels. Similar to the voltage loop, the current loop and frequency loop are utilized to maintain the stability of the current and frequency within the microgrid system, respectively.

This paper adopted an adaptive integrated control method based on signal compensation, as depicted in Figure 7, which included a current loop, voltage loop, and frequency loop, all of which were interconnected and mutually compensated within the three loops.

In this figure, "dq" represents the coordinate transformation section, with "d" and "q" denoting the dq-axis component of the control quantity, respectively, and the subscript "ref" represents the reference value of the control quantity. The switches in the current loop and the voltage loop are configured similarly, where switch state 1 corresponds to the grid-connected state of the microgrid and switch state 2 corresponds to the off-grid state of the microgrid.

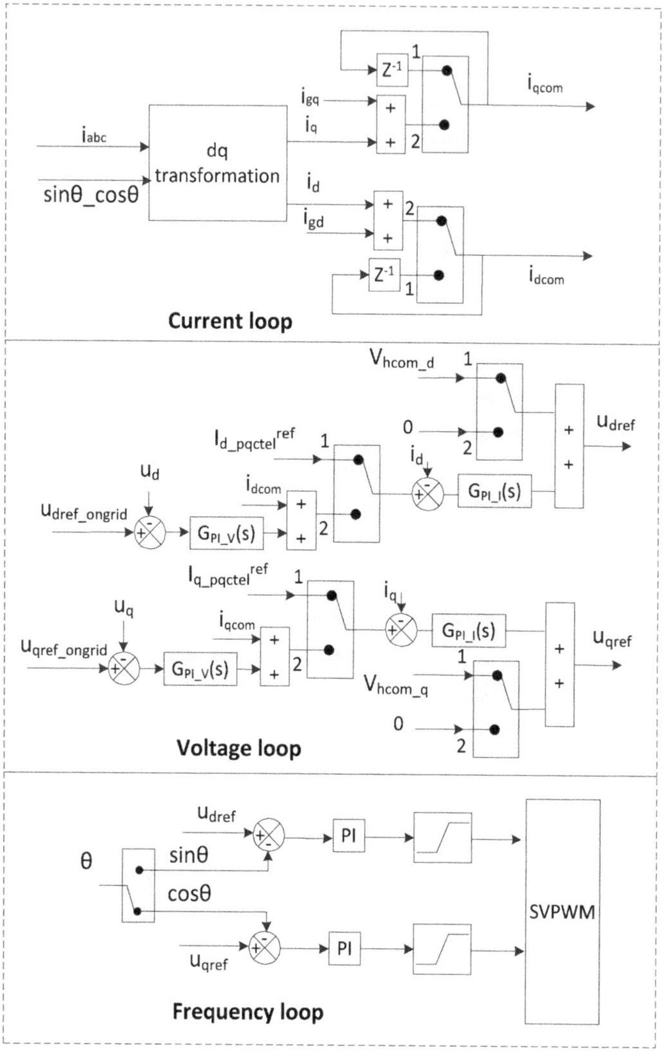

Figure 7. Structure diagram of the signal compensation-based adaptive integrated control method for the current, voltage, and frequency.

In the current loop, during the grid-connected state, the input was the three-phase current signal of the bus. After passing through the dq transformer, it produced separate current signals for the d-axis I_d and q-axis I_q. These signals, first added to the grid current signal I_{gd} and I_{gq}, and then changed to I_{dcom} and I_{qcom} after being adjusted by the PI controller, became the input signal for the voltage outer loop. In the off-grid state, in order to maintain the stable operation of the system, a delay was applied to I_{dcom} and I_{qcom}, so that the current signal in the off-grid state remains the current value in the grid-connected state until the system is stable.

In the voltage loop during the grid-connected state, the difference between the external reference currents $I_{d\_pqctel^{ref}}$ and $I_{q\_pqctel^{ref}}$ on the d and q axes and the internal current signals I_d and I_q on the d and q axes was taken. This difference was then transformed into voltage signals through the current transformation function $G_{PI\_I}(s)$. Finally, along with the constant voltage reference values V_{hcom_d} and V_{hcom_q} on the d and q axes, it served as the input signals to obtain the output values U_{dref} and U_{qref} of the voltage loop. During the islanded state, the voltage loop underwent a more complex signal processing procedure. The real-time voltage signals U_d and U_q on the d and q axes were compared to the reference voltages U_{dref_ongrid} and U_{qref_ongrid} during the grid-connected operation, resulting in differences. These differences were then transformed into current signals through the voltage transformation function $G_{PI\_V}(s)$. After combining with the current signals I_{dcom} and I_{qcom} from the current loop, the calculated current values were obtained. The differences between these calculated current values and the current signals I_d and I_q on the d and q axes were then transformed into voltage signals through the current transformation function $G_{PI\_I}(s)$. Simultaneously, the voltage reference values were set to zero, directly yielding the output values U_{dref} and U_{qref} of the voltage loop.

In the frequency loop, the input was the voltage phase angle signal, which was transformed into a frequency signal. This frequency signal was then combined and transformed with the output signal of the voltage outer loop, resulting in U_α and U_β. These, along with the DC voltage U_{dc} output from the DC subnet, served as the input signals for the space vector pulse width modulation (SVPWM) inverter. Through mutual signal compensation, this approach more accurately maintained the voltage, current, and frequency stability within the microgrid, enhancing the reliability and power quality of the microgrid.

In the voltage outer loop, when switch = 1, the microgrid was in the grid-connected mode, the output voltage values were V_{Hcom_d1} and V_{Hcom_q1}, and the output current values were i_{dref1} and i_{qref1}. When switch = 2, it was in the islanded mode, the output voltage values were V_{Hcom_d2} and V_{Hcom_q2}, and the output current values are i_{dref2} and i_{qref2}. To avoid voltage deviations and current impacts during the transition between the grid-connected and islanded states, the voltage loop was improved by incorporating current feedback information. In the grid-connected mode, the values were set to $i_{dref2} = 0$, $i_{qref2} = 0$, $V_{Hcom\_d2} = 0$, and $V_{Hcom\_q2} = 0$, and the integral initial value in the PI controller always tracked i_d. At the moment of switching to the islanded mode, the current reference value compensation was introduced and i_d was assigned to the integral initial value, causing it to start changing from the i_d point at the switching moment. This ensured that the output of the voltage loop remained unchanged at the switching moment. Finally, the compensated current reference value was transformed through the G(s) function, superimposed with V_{Hcom_d} and V_{Hcom_q}, and input into the dq transformation. This completed the current superimposition for the voltage loop input value, with the compensation equation as follows.

$$\begin{cases} i_{dref} = i_{dref1} + i_{dref2} \\ i_{qref} = i_{qref1} + i_{qref2} \end{cases} \quad (1)$$

It was assumed that the output power of the bidirectional AC/DC converter was P_{ref2} and Q_{ref2} during the off-grid operation and P_{ref1} and Q_{ref1} during the grid-connected operation, with the objective of making $P_{ref1} = P_{ref2}$ and $Q_{ref1} = Q_{ref2}$, resulting in $i_{dref1} = i_{dref2}$ and $i_{qref1} = i_{qref2}$. This approach followed the provided P_{ref} and Q_{ref}, thereby reducing current disturbances to the microgrid and the main grid.

Based on the analysis presented above, the signal compensation-based adaptive integrated control method proposed in this paper offers the following advantages. The AC/DC hybrid microgrid does not need to use any state following device when transferring from grid-connected to off-grid, and the input quantity of the outer loop of the microgrid does not change at the moment of off-grid. Moreover, the change in the control mode during the transition between the grid-connected and islanded operations only affects the outer loops, without impacting the inner loops. This greatly enhances the microgrid's transient stability and achieves a smooth, non-impactful transition between the grid-connected and islanded modes.

3.2. Improved Pre-Synchronization Control Method Based on a BP Neural Network

When the operation mode of the microgrid is switched directly from off-grid to grid-connected, due to the inconsistency between the PCC point voltage and the voltage amplitude, frequency, and phase angle of the grid, a large impulse current may be generated, which may damage the power electronic devices and cause the system to crash in serious cases. Therefore, pre-synchronization should be added during the transition from an off-grid state to a grid-connected state.

3.2.1. Details of BP Neural Network

This paper proposes an improved pre-synchronization control method based on BP neural networks, as illustrated in Figure 8. The control deviation of the PID controller is reduced by neural network algorithm, and it is applied to the pre-synchronization controller to reduce the difference in the reference voltage between the phase-locked loop and interconnecting converter in the pre-synchronization control method, so as to improve the control effect of the pre-synchronization control method. It can effectively solve the problem that the output of the phase difference after several cycles is approximately zero after the PID controller in the traditional pre-synchronization control method, resulting in an insufficient adjustment function, reducing the impact current caused by different phase and amplitude, and effectively ensuring the smooth switching of the microgrid off-grid connection.

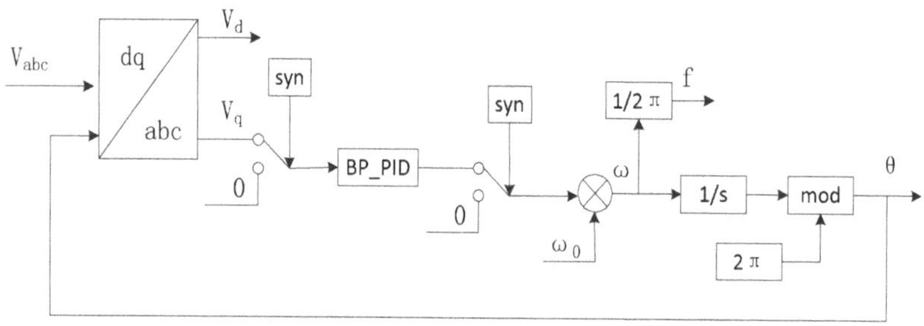

Figure 8. Pre-synchronization control structure diagram.

The specific control block diagram of the PID improved by the BP neural network algorithm proposed in this paper is shown in Figure 9, where Vq is the q-axis component of the real-time AC voltage; $Vqref$ is the q-axis reference voltage; e is the difference between the two; and kp, ki, and kd are the relevant control parameters of the PID controller. The BP neural network algorithm was used to adjust the PID parameters in the PID controller in the pre-synchronization control method in real time. The proportion, integral, and differential gain of the adjusted PID controller can be expressed as follows.

$$\begin{cases} kp = kp0 + \Delta kp \\ ki = ki0 + \Delta ki \\ kd = kd0 + \Delta kd \end{cases} \tag{2}$$

where $kp0$, $ki0$, and $kd0$ are the initial values of the PID parameters, Δkp, Δki, and Δkd are the adjustment quantities generated by the BP neural network algorithm, and kp, ki, and kd are the adjusted PID parameters.

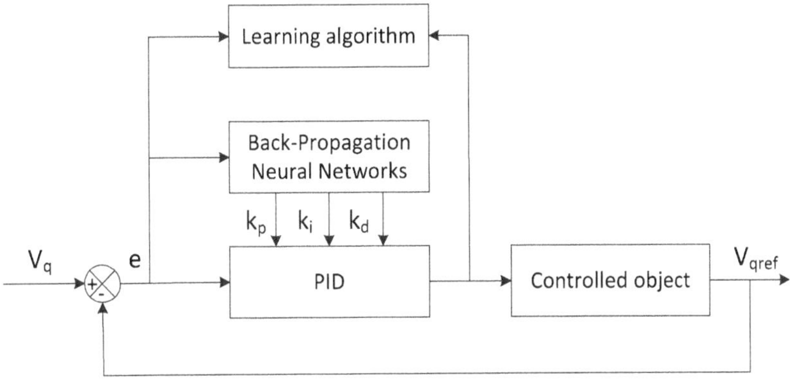

Figure 9. Block diagram of the PID control improved by the BP neural network algorithm.

Let i, j, and l be the number of neurons in the input layer, hidden layer, and output layer, respectively. The input layer of the BP neural network is the q-axis component Vq of the real-time AC voltage, the q-axis reference voltage Vqref, and the error e of both, and the output is the proportional coefficient kp, integral coefficient ki, and differential coefficient kd of the PID controller. The structure of the three-layer BP neural network is shown in Figure 10.

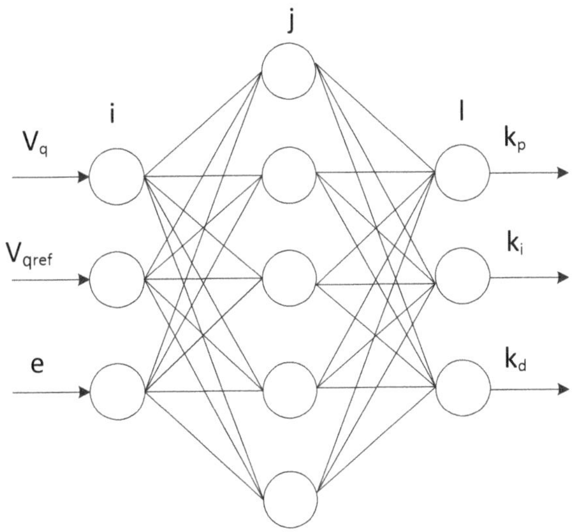

Figure 10. Structure of the BP neural network.

Formula (3) can be obtained by remembering that the JTH neuron of the hidden layer accepts the input layer as a_j.

$$a_j = \sum_{i=1}^{d} V_{ij} x_i \tag{3}$$

Formula (4) can be obtained by remembering that the KTH neuron in the output layer accepts the input from the hidden layer as β_k.

$$\beta_k = \sum_{j=1}^{a} W_{jk} b_j \tag{4}$$

The excitation function selected by the BP neural network is a sigmoid function, and the basic expression of the function is shown as Formula (5).

$$sigmoid(x) = \frac{1}{1+e^{-x}} \tag{5}$$

Then, the output of the neural network, namely the three parameters of the PID controller, can be expressed as Formula (6).

$$Y_k^l = f(\beta_k - \theta_k) \tag{6}$$

The backpropagation of the BP network is based on a gradient descent method to find the optimal solution. Through repeated backpropagation, the weight gradually tends toward the optimal solution by constantly adjusting the parameters, and the weight update follows Formula (7). The learning rate is η, and Formula (8) can be obtained.

$$w + \triangle w \rightarrow w \tag{7}$$

$$\triangle w_{jk} = -\eta \frac{\partial E_l}{\partial w_{jk}} \tag{8}$$

Here, from the properties of the activation function, that is Formula (9), by combining Formulas (9) and (10), Formula (11) can be obtained.

$$f(x)' = f(x)[1 - f(x)] \tag{9}$$

$$E_l = \frac{1}{2}\sum_{k=1}^{m}\left(Y_k^l - y_k^l\right)^2 \tag{10}$$

$$\frac{\partial E_l}{\partial w_{jk}} = \frac{\partial E_l}{\partial Y_k^l}\frac{\partial Y_k^l}{\partial \beta_k}\frac{\partial \beta_k}{\partial w_{jk}} \tag{11}$$

From Formula (11) and set b_j, Formula (12) can be obtained.

$$\frac{\partial E_l}{\partial Y_k^l}\frac{\partial Y_k^l}{\partial \beta_k} = g_k \tag{12}$$

$$\frac{\partial \beta_k}{\partial w_{jk}} = b_j \tag{13}$$

The expression for updating the weight of the output layer can be derived from Formula (13).

$$\triangle w_{jk} = -\eta \frac{\partial E_l}{\partial w_{jk}} = \eta g_k b_j = \eta Y_k^l(1 - Y_k^l)(y_k^l - Y_k^l)b_j \tag{14}$$

Once the gradient expression has been obtained, the optimal solution can be found by constantly updating the gradient. For the above formula, where η is the learning rate, the learning rate of the BP neural network algorithm was $\eta = 0.25$, and Y_k^l was the output value of the KTH neuron in the output layer of the neural network. Y_k^l was the result of the training output, i.e., the correct output given by the training set; b_j was the output of the JTH neuron in the hidden layer. The values of Vq and Vqref under stable operation of the system were recorded and sampled. A total of 100 values were selected as the training set, and the other 30 samples were selected as the test set. The weights were updated continuously through the above process until the error was sufficiently small and the learning was stopped. To measure the performance of the model and adjust the parameters of the model, improve the performance was improved by minimizing the loss function. The smaller the mean square error, the closer the prediction of the model to the true value, and the better the performance of the model.

The mean square error loss function of the actual output and the expected output is shown in Formula (10). After 100 iterations, the loss function of the training set and the test set tended to be stable. At this time, the mean square error loss function E_l of the training set was 0.0706 and the error was small.

3.2.2. Pre-Synchronization Control Method

The Clarke and Park transform was carried out to obtain the d-axis component Vgd and q-axis component Vgq, the deviation signal of the q-axis component was obtained through BP-improved PID adjustment, and the phase angle was finally integrated. Two synchronization signals controlled the start and end times of pre-synchronization to prevent integral saturation and disconnect the PID control. Subsequently, using this phase angle, and the $Vabc$ and $ILabc$ of the microgrid through Pike transformation, Vd, Vq, ILd, and ILq were calculated, achieving synchronization of the microgrid's frequency with the public grid. The relationship is expressed by the following equations.

$$f = \frac{Vq\left(Kp + \frac{Ki}{s}\right) + \omega 0}{2\pi} \tag{15}$$

$$\theta = mod\left[\left(\int \omega dt\right), 2\pi\right] \tag{16}$$

As shown in Figure 11, the pre-adjustment of the microgrid voltage frequency was conducted before the grid connection. When connected to the grid, the voltage frequency of the grid side was f_{01}, corresponding to the rated power P_0. Assuming that the microgrid operated in the islanded mode at point A, there existed a certain deviation between the load power P_{Load} at point A and the rated power P_0. Additionally, the frequency f_A at point A deviated from the grid voltage frequency f_{01}. Before transitioning from the islanded mode to the grid-connected mode, it was necessary to ensure the frequency adjustment while keeping the load power constant. The optimal frequency compensation value was obtained through the BP algorithm, and the operation point at point A, represented by the droop curve f_{01}, shifted upwards to curve f_{02}. At this point, the system's operating point moved from A to B. The working frequency f_B at point B coincided with the grid voltage frequency f_{01} during the grid connection, meeting the pre-synchronization conditions while keeping the power constant.

As shown in Figure 12, the pre-adjustment of the microgrid voltage before the grid connection is depicted. The initial droop curve is represented as V_{01}. When the microgrid operated in the islanded mode at point A, the system's rated voltage was V_{01}, and the reactive power Q_{Load} operated with a deviation from the grid's rated value. To achieve synchronization of the voltage with the grid while maintaining constant reactive power, it was necessary to shift the initial droop curve upwards. The optimal voltage compensation value was obtained through the BP algorithm. This adjustment took place in minimal time, moving the operating point from A to B. At point B, the working voltage V_B was exactly equal to the system's rated voltage V_{01}, achieving pre-synchronization of the voltage amplitude. After the completion of this synchronization process, the grid connection operations could be initiated.

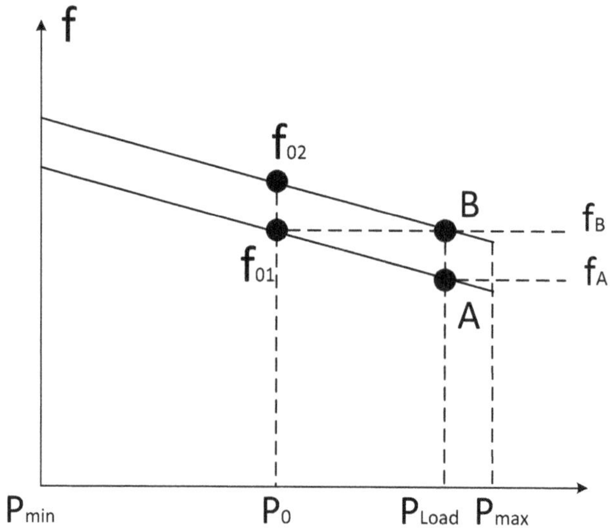

Figure 11. Secondary frequency control.

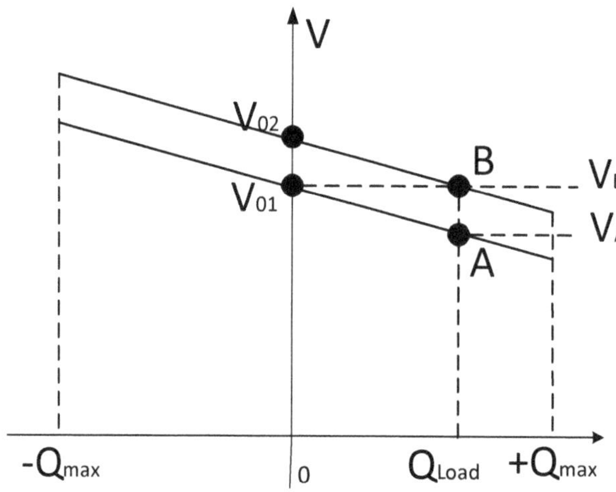

Figure 12. Secondary voltage control.

4. Simulation Results and Analysis

To verify the proposed wind–solar–storage hybrid AC/DC microgrid and its smooth grid connection and disconnection control strategy, an I-type microgrid was constructed, which included a DC microgrid, an AC microgrid, and bidirectional AC/DC converters, as shown in Figure 1. Experimental validation was conducted using the primary parameters listed in Table 1.

The microgrid was controlled to change from the grid-connected mode to the island mode in the first second, and from the island mode to the grid-connected mode in the second. This state transformation was realized by the opening and closing of the PCC points. Specifically, the PCC point switches were set to open and close automatically in the first and second seconds to simulate the unplanned off-grid and grid-connected states. According to the switching of the PCC states described above, the operational modes of the hybrid AC–DC microgrid were as follows. The microgrid was grid-connected within the first second, started islanding from the 1-s mark until the 2-s mark, resumed grid-connected operation starting at the 2-s mark, and continued until the 3-s mark when the operation concludes. The model was built and simulated in Matlab/Simulink and the fluctuations in bus voltage and current of the hybrid AC–DC microgrid were observed during the 0–3 s period.

Table 1. Parameters of the hybrid wind–solar–energy storage AC/DC microgrid system.

Argument	Parameter Value	Title 3	Parameter Value
Rated voltage of DC bus/V	220	ILC switching frequency/kHz	10
Rated voltage of AC bus/V	380	Alternating current load/kW	20~40
Rated frequency of AC bus/Hz	50	DC load/kW	20~40
Rated power of photovoltaic generation/kW	60	Two-way DC/DC converter inductance/mH	1
Rated wind power/kW	30	Two-way DC/DC converter capacitance/μF	1000
ILC filter inductor/mH	0.8	Boost inductance/mH	1
ILC filter capacitor/μF	82	ILC capacity/kVA	100

First, in the wind–solar–storage hybrid AC/DC microgrid, the wind power generation unit used traditional wind turbines and employed conventional voltage, current, and frequency control loops. The simulation results are shown in Figure 13.

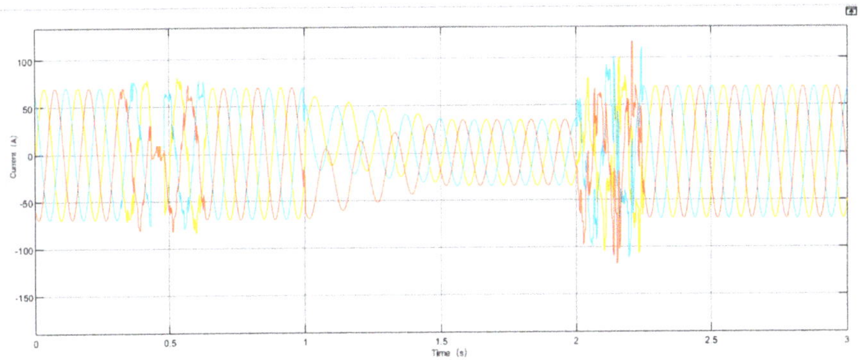

(**a**)

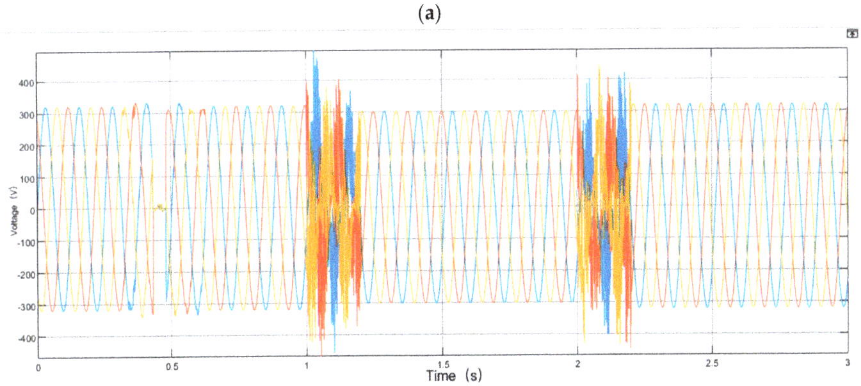

(**b**)

Figure 13. *Cont.*

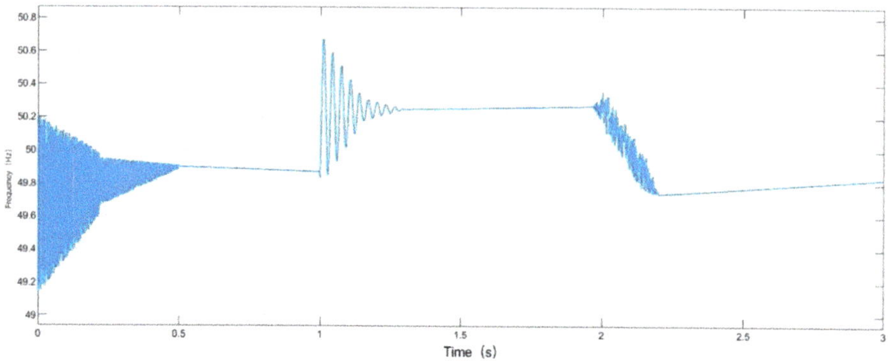

(c)

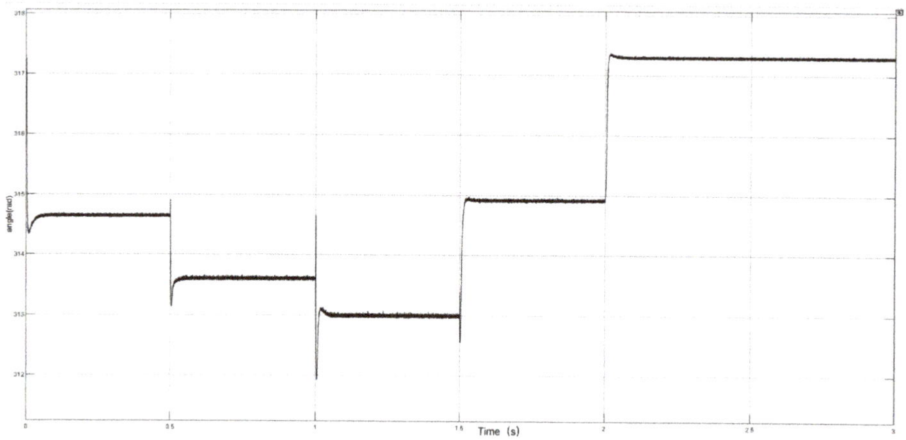

(d)

Figure 13. Simulation waveforms of the transition between the islanded mode and grid-connected mode in the hybrid AC/DC microgrid system. (**a**) Current waveform; (**b**) voltage waveform; (**c**) frequency waveform; (**d**) grid angle waveform.

As shown in Figure 13, the steady-state stability of the system was poor. Affected by factors such as wind speed and illumination, voltage and current fluctuations occurred even if the system was in stable operation, and the frequency was not constant. In terms of transient stability, if no additional control strategy was added, the AC/DC hybrid microgrid had huge fluctuations in the voltage, current, and frequency during the on-grid and off-grid switching moments. Specifically, at 1 s, that is, at the moment of the switch from a grid-connected to an off-grid state, due to the sudden loss of voltage support of the large power grid, the microgrid inevitably produced a voltage drop. If not properly controlled, the situation shown in Figure 13 will occur, where the current continues to drop, the current is much smaller than that when the power grid is supported, and the waveform is not a regular three-phase sine wave. The voltage experienced irregular shock fluctuations and could not be recovered in a short time. The frequency fluctuated significantly beyond the range allowed for stability. At 2 s, that is, the moment when the off-grid state was switched to the grid-connected state, the microgrid operated independently and normally in the island state after its own adjustment, but there was a certain gap between the voltage and current between the microgrid and the large power grid. At this time, if there was no pre-adjustment and the difference in the voltage and current parameters between the microgrid and the large power grid was reduced in advance, the situation shown in Figure 13 will occur. The current and voltage will have serious impact and distortion, and the frequency will also drop sharply until the microgrid is stabilized by the corresponding control after it is connected to the large grid. At the same time, the phase Angle of the grid fluctuated significantly between 312 rad and 317.2 rad, which also indicated the unstable operating state of the system. The above voltage, current, and frequency fluctuations generated at the switching moment were not only detrimental to the stable operation of the microgrid itself, but also excessive impulse voltage and current affected the safe operation of the large power grid. Therefore, appropriate control methods should be taken at the switching moment to make the microgrid quickly enter a stable operation state during state switching, so as to improve the stability of the system and the large power grid.

Figure 14 shows the simulation waveforms of the improved wind–solar battery hybrid AC/DC microgrid and its control strategy during the islanded and grid-connected mode transitions. The use of the DFIG enhanced the system's steady-state stability. Additionally, the integration of the signal compensation-based current–voltage–frequency adaptive control method and the improved pre-synchronization control method based on BP neural networks enabled smooth mode transitions between the islanded and grid-connected states. The fluctuation of the bus voltage and current in the AC/DC hybrid microgrid within 0–3 s was observed again.

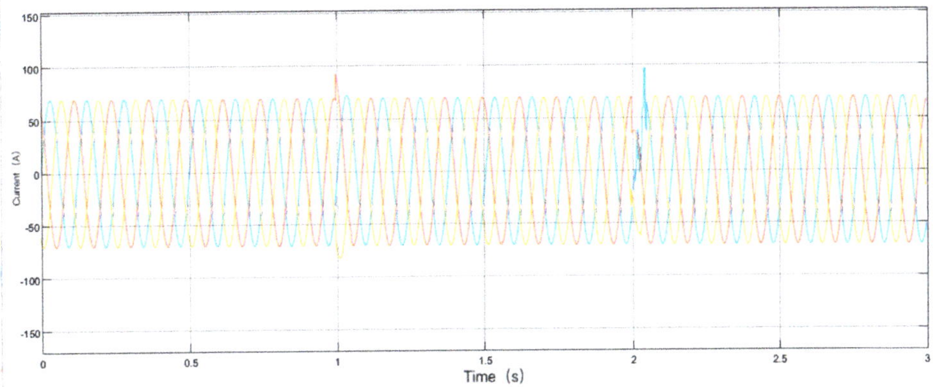

(**a**)

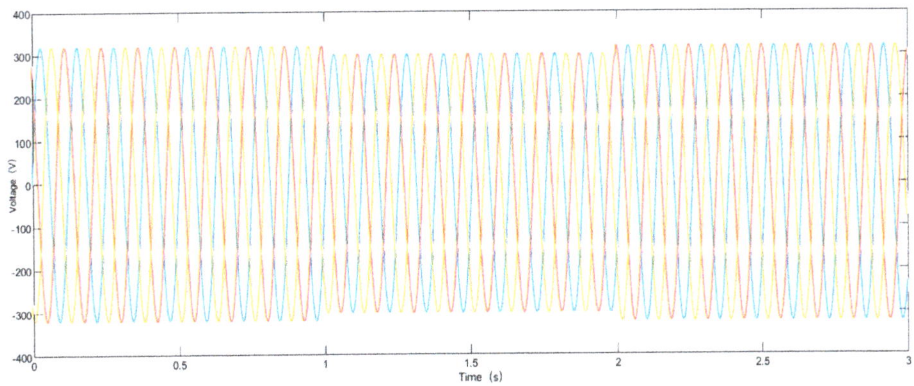

(**b**)

Figure 14. *Cont.*

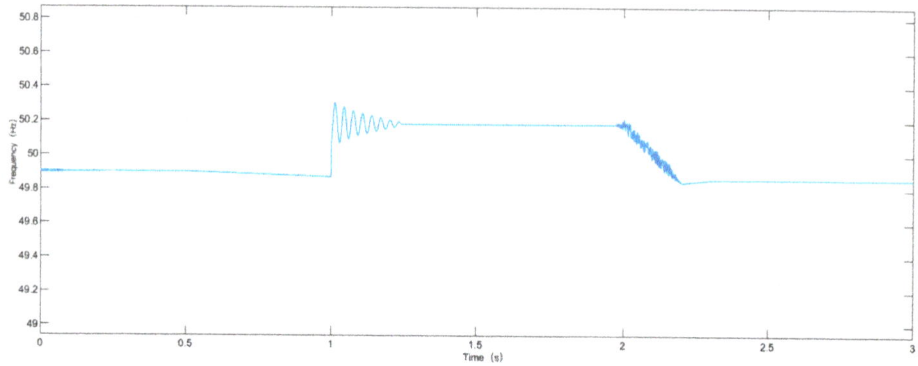

(**c**)

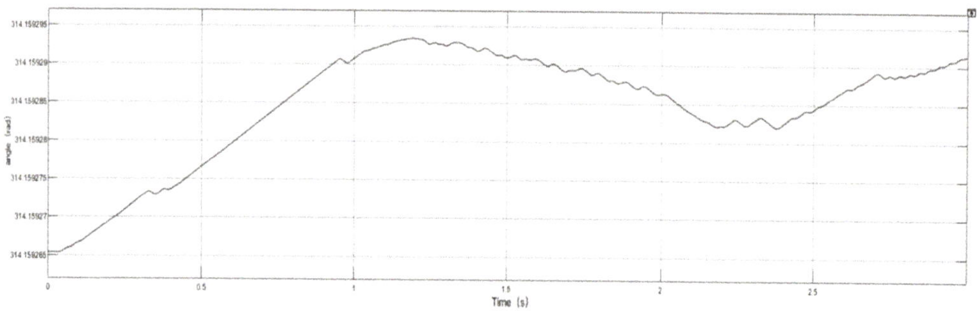

(**d**)

Figure 14. Simulation waveforms of the improved transition between the islanded mode and the grid-connected mode in the hybrid AC/DC microgrid system. (**a**) Current waveform; (**b**) voltage waveform; (**c**) frequency waveform; (**d**) grid angle waveform.

When the microgrid system was connected to the grid at 0–1 s, the output voltage amplitude of the inverter was approx. 320 V, the output current of the inverter was approx. 70 A, the system frequency was 50 Hz, and the system ran stably. At this stage, the photovoltaic power generation system and DFIG were the in MPPT mode, the bidirectional AC/DC converter worked in the V/F control mode, and the energy storage system was charged at a constant power of 20 kW and entered a floating charge state when the battery SOC reached 0.8 to prevent the battery from overcharging and ensure the battery was in the best working conditions.

At 1 s, the microgrid detected an islanded signal and prepared to enter the islanded mode. During this transition, the battery operated in a discharge mode to maintain the microgrid bus voltage, and it reached the switching condition as its output power approached zero, allowing a stable transition to the islanded mode. During this phase, the inverter output current underwent a brief fluctuation but ultimately stabilized back to its initial value of 70 A. The inverter output voltage remained stable at 300 V without voltage spikes. The system output frequency was 49.9 Hz.

At 2 s, the microgrid detected the parallel signal and the microgrid was ready to enter the grid-connected mode. Due to the difference in the voltage, current, and frequency between the microgrid system and the large grid, certain fluctuations occurred when the microgrid system was suddenly connected to the grid. At this time, the pre-synchronization control started. The microgrid started to connect to the grid, stopped the pre-synchronous control, and charged the battery pack until the SOC of the battery pack reached 0.8. The floating charge was carried out, and the power deficit in the microgrid was provided by the large power grid. When switching from off-grid to grid-connected, the control mode of the photovoltaic power generation system and wind power generation system switched back to the MPPT mode, the bidirectional AC/DC converter switched back to the V/F control mode, and the stable grid-connected state was finally completed. At this stage, the output current of the inverter was stabilized at 70 A after a short drop of 0.1 s, and there was no impulse current. The inverter output voltage was stable at 320 V with no fluctuation. The system output frequency was 50 HZ. At the same time, it can be seen from Figure 14d that during the operation of the microgrid and at the moment of parallel and off-grid switching, the change of the grid angle of the microgrid was less than 0.00003, which indicated that the system was relatively stable during operation.

In summary, the use of the smooth switching control strategy effectively avoided the occurrence of voltage and current spikes during the transitions between the islanded and grid-connected modes. The range of voltage and frequency fluctuations caused by the transitions was within the specified limits ($\Delta f < 0.1f_N$, $\Delta U < 7\%U_N$), meeting the requirements for the transitions.

5. Conclusions

This paper presented an enhanced model of a wind–solar storage mixed AC/DC microgrid based on a DFIG and its smooth transition control method for grid connection and disconnection.

1. Wind–solar storage mixed AC/DC microgrid based on a DFIG. By using the partial power transfer function of the DFIG, when the power grid failed or sudden load fluctuations occurred, the speed and magnetic field of the DFIG rotor could be adjusted to provide additional reactive power, reduce the risk of power failure, and stabilize the microgrid system.
2. Smooth transition control method for grid connection and disconnection in wind–solar storage mixed AC/DC microgrids. In view of the large fluctuations of the system voltage, current, and frequency caused by the change in the control strategy and PCC point voltage during the state switching process of the microgrid, a current–voltage–frequency adaptive integrated control method based on signal compensation and an improved pre-synchronization control method based on a BP neural network were proposed to effectively improve the transient stability during the state switching process of the microgrid.

The results of this work can be generalized to the grid-connected and off-grid transition control strategies suitable for more complex AC–DC hybrid microgrids, which are being explored by the authors for further research. However, there are still some problems that need further research.

Author Contributions: Conceptualization, Y.W. and Q.L.; methodology, Q.L.; software, Q.L.; validation, Q.L., M.Y. and Z.C.; formal analysis, Q.L.; investigation, Q.L.; resources, Q.L.; data curation, Q.L.; writing—original draft preparation, Q.L.; writing—review and editing, Q.L.; visualization, Q.L.; supervision, X.D. All authors have read and agreed to the published version of the manuscript.

Funding: This research received no external funding.

Data Availability Statement: Data are contained within the article.

Conflicts of Interest: The authors declare no conflict of interest.

References

1. Yang, X.; Su, J.; Lv, Z.; Liu, H.; Li, R. Review on microgrid technology. *Proc. CSEE* **2014**, *1*, 57–70.
2. Zhou, J.H.; Tian, L.G.; Pan, H. Study of planning and design of hybrid AC/DC micro-grid. *J. Hefei Univ. Technol. Nat. Sci.* **2015**, *38*, 166–170.
3. Chen, J.; Chen, X.; Feng, Z.Y.; Gong, C.Y.; Yan, Y.G. A control strategy of seamless transfer between grid-connected and islanding operation for microgrid. *Proc. CSEE* **2014**, *34*, 3089–3097.

4. Chen, J.; Niu, B.W.; Zhang, J.W.; Chen, X.; Gong, C.Y. Hybrid control strategy for the seamless transfer of microgrids. *Proc. CSEE* **2015**, *35*, 4379–4387.
5. Chen, X.; Ji, Q.H.; Liu, F. Smooth transferring control method of microgrids based on master-slave configuration. *Trans. China Electrotech. Soc.* **2014**, *29*, 163–170.
6. Wang, C.S.; Li, X.L.; Guo, L.; Li, Y.W. A seamless operation mode transition control strategy for a microgrid based on master-slave control. *Sci. China Technol. Sci.* **2012**, *55*, 1644–1654. [CrossRef]
7. Qiu, L.; Xu, L.; Zheng, Z.D.; Li, Y.D.; Zheng, Z.X. Control method of microgrid seamless switching. *Trans. China Electrotech. Soc.* **2014**, *29*, 171–176.
8. Zheng, T.W.; Chen, L.J.; Mei, S.W. Smooth transition control in a master-slave micro-grid considering timing-sequence-coordination. *Electr. Energy Manag. Technol.* **2014**, *10*, 44–49.
9. Bi, D.Q.; Zhou, W.; Dai, Y.X.; Li, X.G. Control strategies of seamless switching for energy storage converter in hybrid AC/DC microgrid. *Autom. Electr. Power Syst.* **2016**, *40*, 84–89.
10. Zeng, L.; Jinjun, L. A three-phase inverter control method which can realize smooth switching of distributed power generation system is presented. *Trans. China Electrotech. Soc.* **2011**, *26*, 52–61.
11. Shudong, W.; Wei, D.; Huanyu, W. Research on parallel/Off-grid switching Technology of Microgrid based on improved nonlinear Sag Control. *Electr. Meas. Instrum.* **2018**, *55*, 112–118.
12. Driesen, J.; Katiraei, F. Design for Distributed Energy Resources. *IEEE Power Energy Mag.* **2008**, *6*, 30–40.
13. Wei, T. Research on Operation Control Strategy of Hybrid Microgrid. Ph.D. Thesis, China University of Mining and Technology, Xuzhou, China, 2020.
14. Han, G. Research on Control Strategy of ACDC Bidirectional Power Converter in ACDC Hybrid Microgrid. Ph.D. Thesis, Taiyuan University of Technology, Taiyuan, China, 2019.
15. Wenchao, X.; Yongqiang, Z.; Ruihua, X. Power Hierarchical Control of AC-DC Hybrid Microgrid with Multiple interconnected converters Running side-by-side. *Shaanxi Electr. Power* **2016**, *44*, 9–13+56.
16. Yuan, F.; Yi, W.; Xiangyu, Z. Analysis and integrated control of inertia and primary frequency regulation for variable speed wind turbines. *Proc. CSEE* **2014**, *34*, 4706–4716.

Disclaimer/Publisher's Note: The statements, opinions and data contained in all publications are solely those of the individual author(s) and contributor(s) and not of MDPI and/or the editor(s). MDPI and/or the editor(s) disclaim responsibility for any injury to people or property resulting from any ideas, methods, instructions or products referred to in the content.

energies

MDPI

Article

Comparative Analysis of PWM Techniques for Interleaved Full Bridge Converter in an AC Battery Application

Tuan Anh Do [1,2], Quang Dich Nguyen [2], Phuong Vu [1,*], Minh Duc Ngo [3,*] and Seon-Ju Ahn [4,*]

1 School of Electrical and Electronic Engineering, Hanoi University of Science and Technology, Hanoi 100000, Vietnam; anh.dt220032d@sis.hust.edu.vn
2 Institute for Control Engineering and Automation, Hanoi University of Science and Technology, Hanoi 100000, Vietnam; dich.nguyenquang@hust.edu.vn
3 Department of Automation, Thai Nguyen University of Technology, Thai Nguyen 251750, Vietnam
4 Department of Electrical Engineering, Chonnam National University, Gwangju 61186, Republic of Korea
* Correspondence: phuong.vuhoang@hust.edu.vn (P.V.); ngoduc198-tdh@tnut.edu.vn (M.D.N.); sjahn@chonnam.ac.kr (S.-J.A.); Tel.: +098-925-88-54 (P.V.); +82-625-301-738 (S.-J.A.)

Abstract: The AC battery utilizing second-life time batteries has gained great interest currently with the advantages of both power solutions and economic benefits. In this system, the power converters play a crucial role in the stable and effective operation of the system. This paper focused on the AC/DC stage with the chosen topology being the interleaved full bridge (IFB) converter due to its flexibility and the ability to increase the power rate of the system. For the sake of high-performance operation, various pulse width modulation (PWM) methods for this converter are analyzed. First, based on the theory of the traditional PWM methods for a full bridge inverter in combination with the interleaved technique, this paper proposed three interleaved PWM methods for the IFB converter. Secondly, the proposed methods are theoretically compared in terms of the output current, common-mode voltage, and power losses. Finally, the evaluation is carried out by both the simulation and the experimental prototype, in which the results are in good agreement with the theoretical analysis.

Keywords: AC battery; power correction factor stage; interleaved full bridge converter; pulse width modulation

1. Introduction

Recently, along with the significant growth in the number of electric vehicles (EVs), EV batteries have become an interest of application. The capacity of these batteries after being discarded from EVs is approximately 70–80% of the initial value [1,2], corresponding to the impedance of the end-of-life lithium-ion battery, which normally is obtained from the battery manufacturers [3]. After that, they can be reused as second-life batteries (SLBs). These SLBs can store and deliver power in stationary applications [4] or energy storage systems [5]. Especially, the concept of the home battery storage system [6–8] has been concerned with consisting of multiple resources: renewable energy, the low voltage (LV) grid, domestic loads, and EVs and the SLBs from them. Among the two main structures for the home battery, the AC-coupled one is selected in this research because of its ability to retrofit the installed home PV system, its isolated battery system, and its flexibility with extended modules. To that extent, the AC battery, integrating both the battery and the power converters, has been introduced by many well-known companies.

To exchange power among all the sources in this resident application, the two-stage topology in Figure 1 is chosen as a suitable solution. This paper focuses on the power factor correction stage with the requirements of a multi-functional converter. First, in the US and some countries, while only the 1-phase grid power is allowed for the home, the 3-phase grid cord is acceptable for homes in many other countries for the higher power rate application. Therefore, the first function of the PFC stage is the ability to connect with both 1-phase and 3-phase grids. At the power rate of 6.6 kW for charging EVs, the TI prototype using a Totem-Pole converter is considered the best solution [9]. However, when the power rate increases, such as the 11 kW at level 2 of the charging standard, the high current stress and high current limit the advantage of the traditional Totem-Pole structure. A solution for this issue is using the interleaved converter, with the suggested one being the bridgeless interleaved boost [10]. The advantage of this converter is the significantly lower current ripple, which decreases the size of the electromagnetic interference (EMI) filter and improves the life of the systems. However, the disadvantage of this structure is that the upper switches are all four diodes, which means the power only can transfer in one direction and the system cannot supply the household loads. To ensure the second function of bidirectional power transfer for the home battery storage system, the IFB converter is chosen with the structure shown in Figure 2.

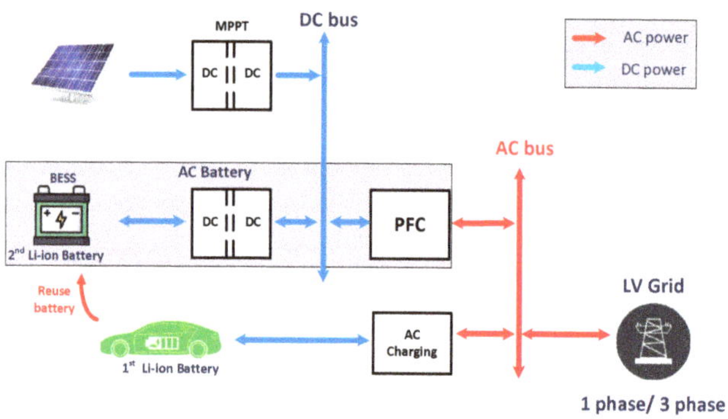

Figure 1. AC Battery application.

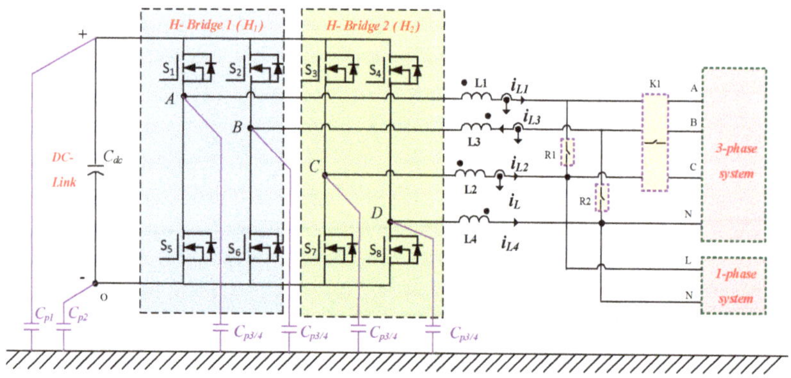

Figure 2. The overall structure of the IFB converter in distinct modes in an AC Battery.

This converter achieves the reduction in current stress and current ripple, hence, lowering the EMI circuit. Furthermore, this converter has the flexibility to transform into the three-phase four-leg converter so that the PFC can operate with the three-phase grid power, as indicated in the first function. The modulation for the IFB converter is the main focus of this research. The basic knowledge of the modulation rule for a single H-Bridge and the comparison among various PWM methods is presented in [11]. The unipolar PWM gives a lower THD current but the leakage current is high, while the bipolar PWM method achieves a low leakage current but the current ripple is high [12–15]. The discontinuous PWM method helps reduce power losses [16–18]. Based on that, several PWM techniques for the IFB converter were studied in [19]. Unipolar double-frequency PWM modulation has a high-frequency circulation current, and its influence on split-filter elements can be minimized. Unipolar frequency–PWM modulation has a high-frequency circulating current, but its effect on split inductors cannot be minimized. In particular, bipolar–PWM modulation has been proven to have no high-frequency circulating currents in any operating mode of the converters [19]. However, this research mainly focuses on the circulation of current issues and their solutions.

Taking advantage of the interleaved technique with the fixed phase difference of 180° [10], three PWM methods, named the interleaved bipolar PWM (IB-PWM), the interleaved unipolar PWM (IU-PWM), and the interleaved discontinuous PWM (ID-PWM), are proposed for the IFB converter based on the corresponding traditional PWM methods. Furthermore, a comprehensive comparison among different PWM methods for the IFB converter is presented in this paper. In detail, three main standards for comparison consist of the THD current, the common-mode voltage, and the power losses.

2. PWM Method Principle

2.1. Interleaved Bipolar PWM

- Pulse pattern analysis

In the IB-PWM method, the modulation indexes m_1, and m_2 are applied for H-Bridge 1 (H_1) and H-Bridge 2 (H_2), as shown in Figure 3, and they are assumed to be equal in the ideal case. Due to the rule of the traditional Bipolar PWM, in terms of the H_1, a sawtooth carrier-based signal is compared to the modulation index m_1 to create the control signal for switches S_1 and S_6, while inverted signals are used for S_2 and S_5. With regard to the H_2, the phase of the sawtooth signal is shifted by 180° from that of H_1, so-called the "Interleaved Bipolar PWM", and the control signals for switches S_3, S_8, S_4, and S_7 are created similarly. The pulse diagram for this modulation method is presented in Figure 4.

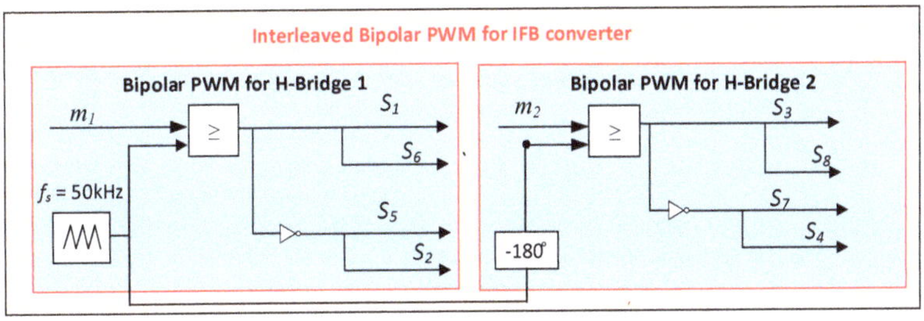

Figure 3. IB-PWM technique for the IFB converter.

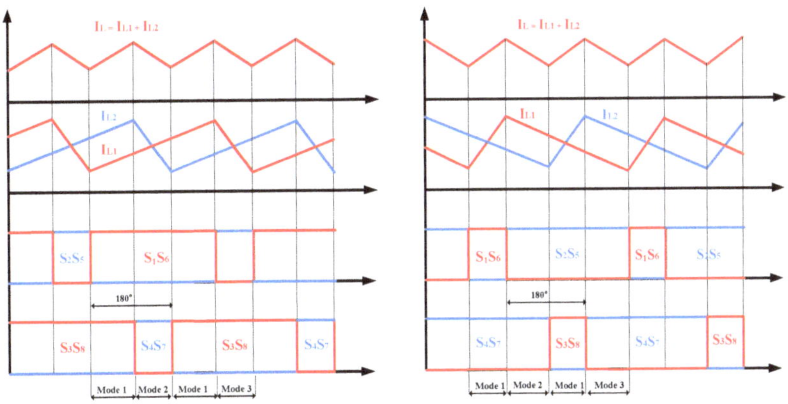

(**a**) (**b**)

Figure 4. Pulse pattern and current ripple of IB-PWM method (**a**) $D > 0.5$, (**b**) $D < 0.5$.

- Common-mode voltage analysis

Due to the 180° out of phase between the two carriers, the control signals for S_1 and S_3 are interleaved. In addition, assuming that the modulation indexes of two bridges are the same, the turn-on interval in a switching period of S_1 is equal to that of S_3. Therefore, the output voltages of two inverters are shifted by 180° in phase.

Defining that s_1, s_2, s_3, s_4 are the switching states of the switches S_1, S_2, S_3, S_4, correspondingly, the voltage $v_{AN}, v_{BN}, v_{CN}, v_{DN}$ can be written in terms of switching states, as follows:

$$\begin{cases} v_{AN} = s_1 V_{AN} + (1 - s_1) V_{ON} = s_1 V_{dc} + V_{ON} \\ v_{BN} = s_2 V_{dc} + V_{ON} \\ v_{CN} = s_3 V_{dc} + V_{ON} \\ v_{DN} = s_4 V_{dc} + V_{ON} \end{cases} \tag{1}$$

By applying Kirchhoff's voltage law at the output of the inverter:

$$\begin{cases} v_{AN} = L_1 \frac{di_{L1}}{2} + v_s \\ v_{BN} = -L_3 \frac{di_{L3}}{2} \\ v_{CN} = L_2 \frac{di_{L2}}{2} + v_s \\ v_{DN} = -L_4 \frac{di_{L4}}{2} \end{cases} \tag{2}$$

Assuming that the split-inductor currents are balanced: $i_{L1} = i_{L3}$, $i_{L2} = i_{L4}$, the common-mode voltage can be calculated from Equations (1) and (2) as follows:

$$v_{ON} = \frac{-(s_1 + s_2 + s_3 + s_4) V_{dc}}{4} + \frac{v_s}{2} \tag{3}$$

Each phase voltage takes the value of $+V_{dc}$ when the upper switch is on and 0 V when this switch is off. According to the pulse diagram in Figure 4, in a switching period, there are always two turned-on upper switches and two turned-off upper switches, which means the common voltage can be defined as

$$v_{CM\_IBPWM} = \frac{v_s}{2} - \frac{V_{dc}}{2} \tag{4}$$

Because $-\frac{V_{dc}}{2}$ is the constant component, the leakage current with the IB-PWM method only depends on the sinusoidal output voltage at the fundamental frequency of 50 Hz.

- Ripple of the split-inductor current and the output current

According to the traditional bipolar PWM, each bridge has two operation modes. In mode 1, taking H_1 for instance, as depicted in Figure 4, switches S_1 and S_6 are on, the output voltage of H_1 is $+V_{dc}$, and the inductor currents i_{L1} and i_{L3} increase, while in mode 2, switches S_2 and S_5 are on, the output voltage of H_1 is $-V_{dc}$, and the inductor currents i_{L1} and i_{L3} decrease. Because of the interleaved feature of the IB-PWM method, the two bridges of the IFB converter operate independently; therefore, the IFB converter with the IB-PWM method has four operation modes.

Table 1 synthesizes the comprehensive operation modes, along with the switching states of all switches of the IFB converter, applying various PWM techniques: the IB-PWM, the IU-PWM, and the ID-PWM. In this section, the equivalent circuit of each operation mode and the current ripple with the IB-PWM in each case are focused. The time intervals of mode 1, 2, 3, 4, 5, 6, 7, and 8 in a switching period T_s are defined as D_1T_s, D_2T_s, D_3T_s, D_4T_s, D_5T_s, D_6T_s, D_7T_s, and D_8T_s, respectively.

Table 1. Operation modes of IFB converter with different PWM methods.

Mode	H-Bridge 1	H-Bridge 2	Method
1	S_1 on, S_2 off, S_5 off, S_6 on	S_3 on, S_4 off, S_7 off, S_8 on	IB-PWM [1,2], IU-PWM [1]
2	S_1 on, S_2 off, S_5 off, S_6 on	S_3 off, S_4 on, S_7 on, S_8 off	IB-PWM [1,2]
3	S_1 off, S_2 on, S_5 on, S_6 off	S_3 on, S_4 off, S_7 off, S_8 on	IB-PWM [1,2]
4	S_1 off, S_2 on, S_5 on, S_6 off	S_3 off, S_4 on, S_7 on, S_8 off	IB-PWM [1,2], IU-PWM [2]
5	S_1 on, S_2 on, S_5 off, S_6 off	S_3 off, S_4 off, S_7 on, S_8 on	IU-PWM [1,2],ID-PWM [2]
6	S_1 off, S_2 off, S_5 on, S_6 on	S_3 on, S_4 on, S_7 off, S_8 off	IU-PWM [1], ID-PWM [2]
7	S_1 on, S_2 on, S_5 off, S_6 off	S_3 on, S_4 on, S_7 off, S_8 off	ID-PWM [1]
8	S_1 off, S_2 off, S_5 on, S_6 on	S_3 off, S_4 off, S_7 on, S_8 on	ID-PWM [2]

[1] D > 0.5, [2] D < 0.5.

To determine the current ripple, the modulation indexes for the two bridges are equal, and the effect of deadtime is neglected. In this case, the turn-on interval in one switching period of two switches S_1 and S_3 are the same, which is defined as DT_s.

- **With $D > 0.5$ ($Vs > 0$)**: The IFB converter works in modes 1, 2, and 3.

In mode 1, both the split-inductor current i_{L1}, i_{L2}, and the output current i_L increase. For the sake of intuition, the current ripple of i_L is calculated in this case.

Because two bridges operate separately with the IB-PWM, based on the equivalent circuit in Figure 5a–c, all the split-inductor currents increase in the interval of DT_s. Hence, the ripple of each inductor current is given as

$$\begin{cases} \Delta i_{L1\_IB} = \Delta i_{L3\_IB} = \frac{V_{dc}-v_s}{L_1+L_3}DT = \frac{V_{dc}-v_s}{2L}DT_s \\ \Delta i_{L2\_IB} = \Delta i_{L4\_IB} = \frac{V_{dc}-v_s}{L_2+L_4}DT_s = \frac{V_{dc}-v_s}{2L}DT_s \end{cases} \quad (5)$$

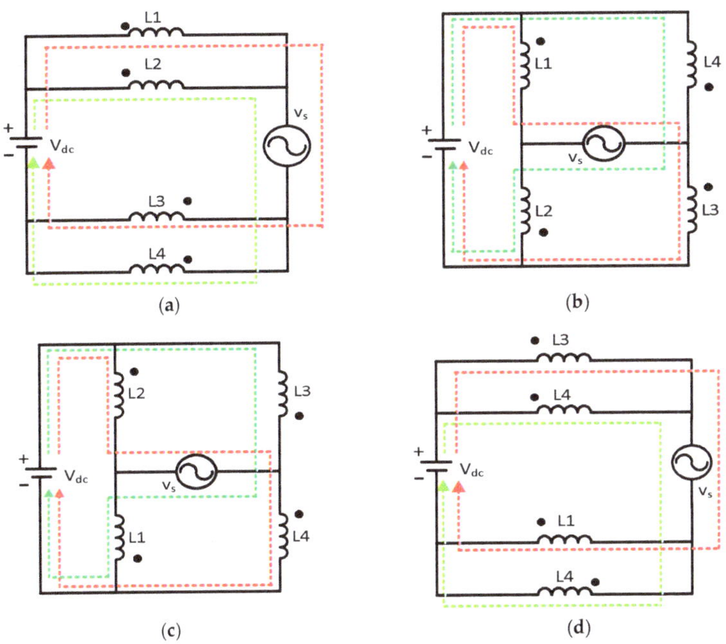

Figure 5. Equivalent circuit of the IFB converter applied to the IB-PWM method: (**a**) mode 1, (**b**) mode 2, (**c**) mode 3, (**d**) mode 4.

The time interval for the increase in the output current in a switching period is determined in the Figure 4a: $D_sT_s = (D-0.5)T_s$. Therefore, the output current ripple when $D > 0.5$ is determined as

$$\begin{aligned} \Delta i_{L\_IB} &= (\tfrac{\Delta i_{L1\_IB}}{DT_s} + \tfrac{\Delta i_{L2\_IB}}{DT_s})D_sT_s = \tfrac{V_{DC}-v_s}{2L} \times 2 \times (D-0.5)T_s \\ &= \tfrac{V_{DC}-v_s}{2L} \times (2D-1)T_s \end{aligned} \quad (6)$$

Inherently, $\Delta i_{L\_IB} < \Delta i_{L1\_IB}$ when $D < 1$, which means the interleaved PWM method has the ability to reduce the output current ripple. The decrease in the current ripple significantly reduces the THD of the output current, which is the advantage of the interleaved PWM method in general and the IB-PWM in particular.

- **With $D < 0.5$ ($Vs < 0$):** the IFB converter operates in modes 2, 3, and 4.

In mode 4, both the split-inductor current i_{L1}, i_{L2}, and the output current i_L decrease. For the intuitive approach, the ripple of i_L is calculated in this case.

Because two bridges operate separately with the IB-PWM, based on the equivalent circuit in Figure 5b–d, the split-current increases in the interval of (1 - D)T_s. Hence, the ripple of each inductor current is given as

$$\begin{cases} \Delta i_{L1\_IB} = \Delta i_{L3\_IB} = \frac{-V_{dc}-v_s}{L_1+L_3}(1-D)T_s = \frac{-V_{dc}-v_s}{2L}(1-D)T_s \\ \Delta i_{L2\_IB} = \Delta i_{L4\_IB} = \frac{-V_{dc}-v_s}{L_2+L_4}(1-D)T = \frac{-V_{dc}-v_s}{2L}(1-D)T_s \end{cases} \quad (7)$$

The time interval for the decrease in the output current in a switching period is determined from the Figure 4b: $D_s = 0.5 - D$. Therefore, the output current ripple when $D < 0.5$ is calculated as:

$$\begin{aligned} \Delta i_{L\_IB} &= (\tfrac{\Delta i_{L1\_IB}}{DT_s} + \tfrac{\Delta i_{L3\_IB}}{DT_s})D_sT_s = \tfrac{-V_{DC}-v_s}{2L} \times 2 \times (0.5-D)T_s \\ &= \tfrac{-V_{DC}-v_s}{2L} \times (1-2D)T_s \end{aligned} \quad (8)$$

Here, we have $\Delta i_{L\_IB} < \Delta i_{L1\_IB}$ since $D > 0$, which means the ripple of the output current is smaller than that of the split-inductor current.

2.2. Interleaved Unipolar PWM

- Pulse pattern analysis

The modulation rule for each H-Bridge is based on the traditional Unipolar PWM as shown in Figure 6. Taking the H-Bridge 1, for example, the modulation indexes m_1 and -m_1 are compared to the carrier sw_1 to generate the control signals for S_1 and S_2, respectively. By inverting the logic of control signals S_1 and S_2, the signals for S_5 and S_6 are achieved, respectively. In terms of H-Bridge 2, the control signals for S_3, S_4, S_7, and S_8 are generated by a similar rule with the modulation index m_2 (m_2 is chosen to be equal to m_1 in the ideal case), and the carrier sw_2 shifted 180° in phase with the sw_1 of H_1.

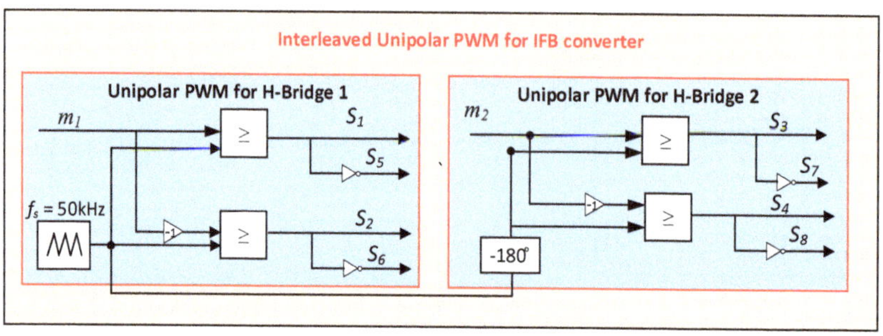

Figure 6. IU-PWM technique for the IFB converter.

The pulse pattern of the IB-PWM method is shown in Figure 7. An important feature that should be noted is that due to the symmetric feature, for example, when $D > 0.5$ (D is defined as in Section 2.1), the turn-on interval of both switches S_1 and S_2 overlaps the turn-off interval of both switches S_3 and S_4. As a result, at a moment in a switching period, there are two turn-on upper switches and two turn-off upper switches. Therefore, similar to the IB-PWM method, the common voltage of the IFB converter with the IU-PWM method can be defined from Equation (3):

$$v_{CM\_IUPWM} = \frac{v_s}{2} - \frac{V_{DC}}{2} \tag{9}$$

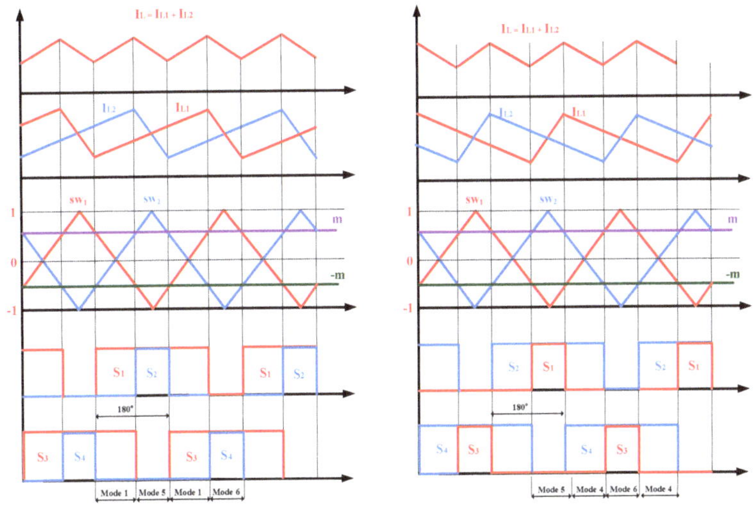

(a) (b)

Figure 7. Pulse pattern and the current ripple with the IU-PWM method: (**a**) $D > 0.5$, (**b**) $D < 0.5$.

As a result, the common-mode voltage with the IU-PWM is high-frequency pulse-free and only depends on the sinusoidal output voltage at the fundamental frequency of 50 Hz.

- Ripple of the split-inductor current and the output current
    - **With $D > 0.5$ ($Vs > 0$):** The IFB converter operates in modes 1, 5, and 6.

Different from the IB-PWM method, there is a current loop flowing through one branch of the H_1 and another branch of the H_2, as shown in mode 5 and mode 6 in Figure 8.

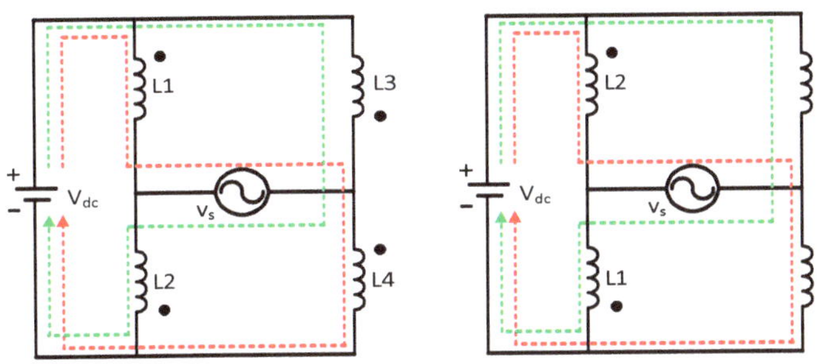

(a) (b)

Figure 8. Equivalent circuit of the IFB converter with the IU-PWM method: (**a**) mode 5, (**b**) mode 6.

Hence, the current ripple with the IU-PWM should be calculated in each mode instead of the whole DT_S interval with the IB-PWM. The split-inductor current ripple in mode 1 is given by

$$\begin{cases} \Delta i_{L1\_IU} = \Delta i_{L3\_IU} = \frac{V_{dc}-v_s}{L_1+L_3} D_1 T_s = \frac{V_{dc}-v_s}{2L}(D-0.5)T_s \\ \Delta i_{L2\_IU} = \Delta i_{L4\_IU} = \frac{V_{dc}-v_s}{L_2+L_4} D_1 T_s = \frac{V_{dc}-v_s}{2L}(D-0.5)T_s \end{cases} \quad (10)$$

The split-inductor current ripple in mode 5 is given by

$$\begin{cases} \Delta i_{L1\_IU} = \Delta i_{L4\_IU} = \frac{V_{dc}-v_s}{L_1+L_4} D_5 T_s = \frac{V_{dc}-v_s}{2L}(1-D)T_s \\ \Delta i_{L2\_IU} = \Delta i_{L3\_IU} = \frac{-V_{dc}-v_s}{L_2+L_3} D_5 T_s = \frac{-V_{dc}-v_s}{2L}(1-D)T_s \end{cases} \quad (11)$$

The split-inductor current ripple in mode 6 is given by

$$\begin{cases} \Delta i_{L1\_IU} = \Delta i_{L4\_IU} = \frac{-V_{dc}-v_s}{L_1+L_4} D_6 T_s = \frac{-V_{dc}-v_s}{2L}(D-0.5)T_s \\ \Delta i_{L2\_IU} = \Delta i_{L3\_IU} = \frac{V_{dc}-v_s}{L_2+L_3} D_6 T_s = \frac{V_{dc}-v_s}{2L}(D-0.5)T_s \end{cases} \quad (12)$$

Because all the currents flowing through the four split-inductors are balanced, the current ripple Δi_{L1_IU} of split-inductor L_1 is chosen. The split-inductor current ripple is determined from Equations (10) and (11):

$$\Delta i_{L1\_IU} = \frac{V_{dc}-v_s}{2L}(2D_1+D_5)T_s = \frac{V_{dc}-v_s}{2L}(2(D-0.5)+1-D)T_s = \frac{V_{dc}-v_s}{2L}DT_s \quad (13)$$

In mode 1, both the split-inductor current i_{L1}, i_{L2} and the output current i_L increase. For the intuitive approach, the output current ripple is determined in this mode:

$$\Delta i_{L\_IU} = \left(\frac{\Delta i_{L1\_IU}}{DT_s} + \frac{\Delta i_{L2\_IU}}{DT_s}\right)D_s T_s = \frac{V_{dc}-v_s}{2L} \times 2 \times (D-0.5)T_s = \frac{V_{dc}-v_s}{2L} \times (2D-1)T_s \quad (14)$$

Compared to Equations (4) and (6), $\Delta i_{L\_IB} = \Delta i_{L\_IU}$ and $\Delta i_{L1\_IB} = \Delta i_{L1\_IU}$. It can be noted that even though the pulse patterns of the IB-PWM and IU-PWM techniques are different when $D > 0.5$, the inductor current ripple and the output current ripple of the two modulation methods are the same.

- **With $D < 0.5$ ($Vs < 0$):** The IFB converter operates in modes 4, 5, and 6.

The current ripple through split-inductor L_1 is determined from Equations (12) and (13):

$$\Delta i_{L1\_IU} = \frac{-V_{dc}-v_s}{2L}(2D_4+D_6)T_s = \frac{-V_{dc}-v_s}{2L}(1-2D+D)T_s = \frac{-V_{dc}-v_s}{2L}(1-D)T_s \quad (15)$$

The output current ripple is determined as

$$\Delta i_s = \frac{V_{dc}-v_s}{2L} \times 2 \times (0.5-D)T_s = \frac{V_{dc}-v_s}{2L} \times (1-2D)T_s \quad (16)$$

2.3. Interleaved Discontinuous PWM

- Pulse pattern analysis

The discontinuous PWM method consists of two types: the DPWM1P method and the DPWM2P method [16], as shown in Figure 9, which is applied for a single H-Bridge converter. Because of the similar features of these types, this paper chooses the DPWM1P in combination with the interleaved technique, the so-called "Interleaved Discontinuous PWM", for the IFB converter to compare the other interleaved PWM methods. In the ID-PWM method, a clamp signal V_c, which is synchronous with the AC voltage Vs, is created. The double modulation index is added with the clamp signal, which is saturated from -1 to 1 to obtain the discontinuous signal V_{DIS}. This signal is compared to the carrier, taking the H_1, for instance, to generate control signals for switches S_1 and S_3, while those for switches S_5 and S_6 are inverted. Similarly, the control signal for the switches S_3, S_4, S_7, and S_8 is generated with the carrier being shifted 180° in phase from that of H_1.

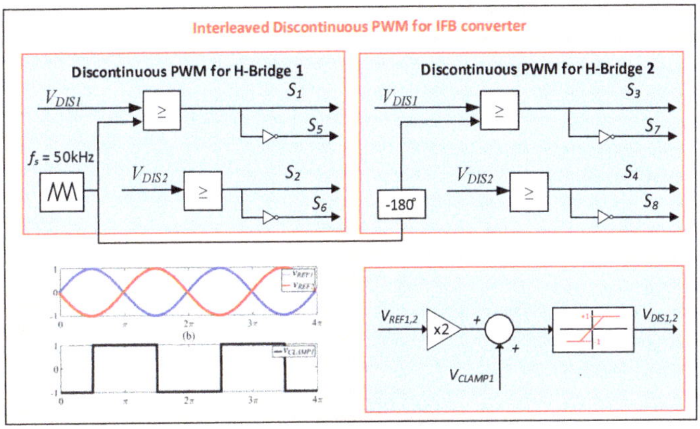

Figure 9. ID-PWM technique for the IFB converter.

- Common-mode voltage analysis

The pulse pattern of the IU-PWM is shown in Figure 10. As can be seen, in a switching period, there are two or no turn-on upper switches when $D < 0.5$, while there are two or four turned-on upper switches when $D > 0.5$. Therefore, unlike the IB-PWM and IU-PWM, the common mode voltage of the IFB converter can take one of these values: 0, $V_{dc}/2$, and V_{dc}. This high-frequency voltage component increases the leakage current of the converter [20]:

$$i_{lk} = C_p \frac{dv_{CMV}}{dt} \quad (17)$$

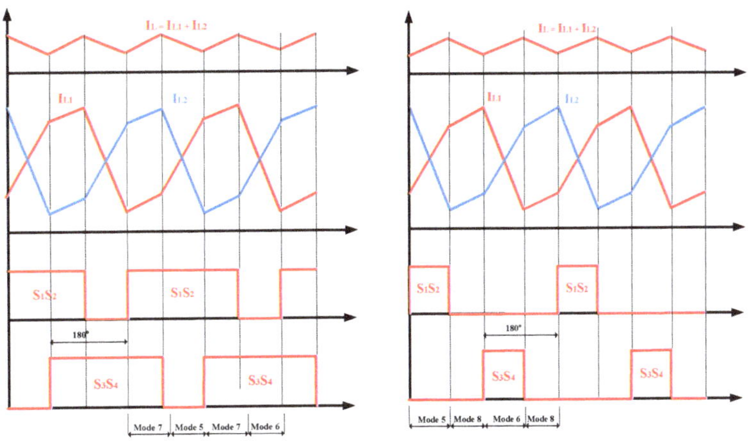

(a) (b)

Figure 10. Pulse pattern and current ripples with the IU-PWM method; (**a**) $D > 0.5$, (**b**) $D < 0.5$.

In which the parasitic capacitor C_p consists of several parasitic capacitors C_{p1}, C_{p2}, and C_{p3}, as shown in Figure 2.

- Ripple of the split-inductor current and output current

The current ripple with the IU-PWM is analyzed in two intervals: non-switching interval (accounts for half of the *Vs* cycle) and switching interval (accounts for the other half of the *Vs* cycle).

a. Non-switching interval: Due to the discontinuous PWM rule, all the upper switches S_1, S_2, S_3, and S_4 are on or off in this interval, hence, the currents through the four split-inductors are in phase and the current ripples seem to be zero.

b. Switching interval: In some modes of this interval, the currents only flow through the inductor and Vs. Therefore, the sign of Vs decides whether the current increases or decreases. In this section, the current ripples are analyzed when $Vs < 0$, while the analysis when $Vs > 0$ is similar. Unlike the previous interleaved PWM method, the inductor currents with the ID-PWM increase or decrease with the different rates in the different modes.

- **With $D > 0.5$:** The IFB converter operates in modes 5, 6, and 7.

Similar to the IU-PWM, there is a current loop flowing through one branch of the H_1 and another branch of the H_2, as shown in mode 5 and mode 6 in Figure 10.

The split-inductor current ripple in mode 5 is given by

$$\begin{cases} \Delta i_{L1\_ID} = \Delta i_{L4\_ID} = \frac{V_{dc}-v_s}{L_1+L_4} D_5 T_s = \frac{V_{dc}-v_s}{2L}(1-D)T_s \\ \Delta i_{L2\_ID} = \Delta i_{L3\_ID} = \frac{-V_{dc}-v_s}{L_2+L_3} D_5 T_s = \frac{-V_{dc}-v_s}{2L}(1-D)T_s \end{cases} \quad (18)$$

The split-inductor current ripple in mode 6 is given by

$$\begin{cases} \Delta i_{L1\_ID} = \Delta i_{L4\_ID} = \frac{-V_{dc}-v_s}{L_1+L_4} D_6 T_s = \frac{-V_{dc}-v_s}{2L}(1-D)T_s \\ \Delta i_{L2\_ID} = \Delta i_{L3\_ID} = \frac{V_{dc}-v_s}{L_2+L_3} D_6 T_s = \frac{V_{dc}-v_s}{2L}(1-D)T_s \end{cases} \quad (19)$$

The split-inductor current ripple in mode 7 is given by

$$\begin{cases} \Delta i_{L1} = \Delta i_{L3} = \frac{-v_s}{L_1+L_3}(D-0.5)T_s = \frac{-v_s}{2L}(D-0.5)T_s \\ \Delta i_{L2} = \Delta i_{L4} = \frac{-v_s}{L_2+L_4}(D-0.5)T_s = \frac{-v_s}{2L}(D-0.5)T_s \end{cases} \quad (20)$$

In mode 8, both the split-inductor currents i_{L1}, i_{L2} and the output current i_L decrease as shown in Figure 11. For the intuitive approach, the output current ripple is determined in this mode.

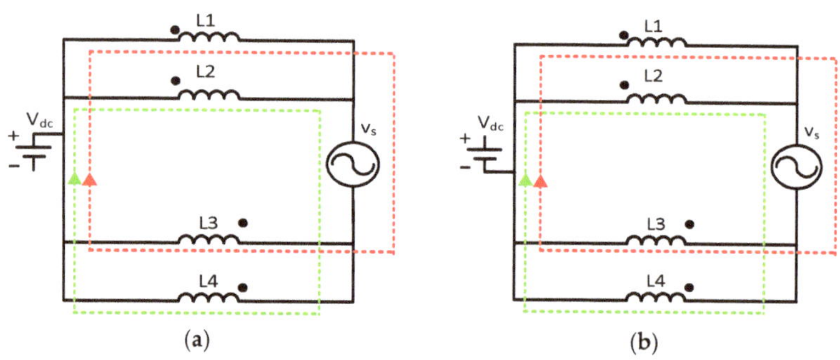

Figure 11. Equivalent circuit of the IFB converter with the ID-PWM method: (**a**) mode 7, (**b**) mode 8.

Because all the currents flowing through the four split-inductors are balanced, the current ripple of split-inductor L_1 is chosen. The split-inductor current ripple is determined from Equations (18) and (19):

$$\begin{aligned} \Delta i_{L1\_ID} &= \frac{V_{dc}-v_s}{2L} \times D_5 T_s + \frac{-v_s}{2L} \times 2D_7 T_s = \frac{V_{dc}-v_s}{2L} \times (1-D)T_s + \frac{-v_s}{2L} \times (2D-1)T_s \\ &= \frac{V_{dc}-v_s}{2L} \times DT_s + \frac{v_s}{2L} \times (2D-1)T_s \end{aligned} \quad (21)$$

The output current ripple is determined as

$$\Delta i_L = \Delta i_{L1} + \Delta i_{L2} = \frac{v_s}{2L} \times 2D_7T_s = \frac{-v_s}{2L} \times (2D-1)T_s \tag{22}$$

- **With $D < 0.5$:** The IFB converter operates in modes 5, 6, and 8.

The split-inductor current ripple in mode 8 is given by

$$\begin{cases} \Delta i_{L1} = \Delta i_{L3} = \frac{v_s}{L_1+L_3}(0.5-D)T_s = \frac{v_s}{2L}(0.5-D)T_s \\ \Delta i_{L2} = \Delta i_{L4} = \frac{v_s}{L_2+L_4}(0.5-D)T_s = \frac{v_s}{2L}(0.5-D)T_s \end{cases} \tag{23}$$

The split-inductor current ripple is determined from Equations (19) and (20):

$$\Delta i_{L1\_ID} = \frac{-V_{dc}-v_s}{2L} \times D_6T_s = \frac{-V_{dc}-v_s}{2L} \times DT_s \tag{24}$$

The output current ripple is determined as

$$\Delta i_L = \Delta i_{L1} + \Delta i_{L2} = \frac{v_s}{2L} \times 2D_8T_s = \frac{-v_s}{2L} \times (1-2D)T_s \tag{25}$$

2.4. *Theorectical Comparion among Three Interleaved PWM Methods for the IFB Converter*

To begin with, the stress voltage on the switches of the IFB converter is the same for all PWM methods. In terms of current, the current stress on each switch depends on the form of the corresponding split-inductor current, which was analyzed in each PWM method. The inductor current ripple Δi_{L1}, output current ripple Δi_L, and the high-frequency component of the common voltage v_{CM} of the IFB converter, applying three PWM techniques, are synthesized in Table 2. As can be seen in Table 2, first, the output current ripple is always smaller than the inductor current ripple for all methods, proving the effectiveness of the proposed interleaved modulation technique for the IFB converter particularly. In addition, even though the pulse patterns of the IB-PWM method and the IU-PWM method are different, the ripple of their inductor currents and output currents are the same.

Table 2. The performance of three proposed PWM methods.

		Pulse Width Modulation Method		
Criteria		IB-PWM	IU-PWM	ID-PWM
Δi_{L1}	$D > 0.5$	$\frac{V_{dc}-v_s}{2L}DT_s$	$\frac{V_{dc}-v_s}{2L}DT_s$	$\frac{V_{dc}-v_s}{2L} \times DT_s + \frac{v_s}{2L} \times (2D-1)T_s$
	$D < 0.5$	$\frac{-V_{dc}-v_s}{2L}(1-D)T_s$	$\frac{-V_{dc}-v_s}{2L}(1-D)T_s$	$\frac{-V_{dc}-v_s}{2L}DT_s$
Δi_L	$D > 0.5$	$\frac{V_{DC}-v_s}{2L} \times (2D-1)T$	$\frac{V_{dc}-v_s}{2L} \times (2D-1)T_s$	$\frac{-v_s}{2L} \times (2D-1)T_s$
	$D < 0.5$	$\frac{-V_{DC}-v_s}{2L} \times (1-2D)T_s$	$\frac{-V_{DC}-v_s}{2L} \times (1-2D)T_s$	$\frac{-v_s}{2L} \times (1-2D)T_s$
$\frac{(s_1+s_2+s_3+s_4)V_{dc}}{4}$		$\frac{V_{dc}}{2}$	$\frac{V_{dc}}{2}$	$0, \frac{V_{dc}}{2}, \frac{V_4}{2}$

Meanwhile, in this aspect, the ID-PWM method offers the smallest output current ripple, although its inductor current ripple could be smaller than that of other methods in the non-switching interval and be larger in the switching interval. As a result, the THD of the output current with the ID-PWM achieves the smallest value, while those of the IB-PWM and IU-PWM are equal. Second, while the common mode voltage with the IB-PWM and IU-PWM methods are free of high-frequency pulse, that of the ID-PWM additionally depends on the high-frequency pulse, which results in higher leakage current. Third, the reduction of the switching loss is the essential advantage of the ID-PWM compared to other methods. Obviously, the number of the commutations in one switching period is the same in the case of the IB-PWM and the IU-PWM method, while that of the ID-PWM is lower because its modulation index is clamped at the value of −1 or 1. Hence, the ID-PWM method has the smallest switching losses.

In the following sections, the power losses of three PWM methods are verified by simulation results, while the comparison of the current performance and common-mode voltage are verified by the experimental results.

3. Simulation Verification

3.1. Power Losses Evaluation

To determine the Mosfet power loss and thermal analysis, a thermal model of the IFB converter was built along with the thermal model using PLECS. The print screen of the simulated circuit is shown in Figure 12. PLECS allows researchers to provide the necessary parameters for the loss estimation of power semiconductors as the threshold voltage, thermal impedance, and drain-source on-state resistance. These are and can be used for the numerical estimation of losses using equations or implemented in the form of lookup tables for the turn-on loss, the turn-off loss, and conduction loss [16,20]. The simulation parameters are presented in Table 3. The power loss is compared among three PWM methods with 2 scenarios in the open-loop mode: variable switching frequency and variable power rate.

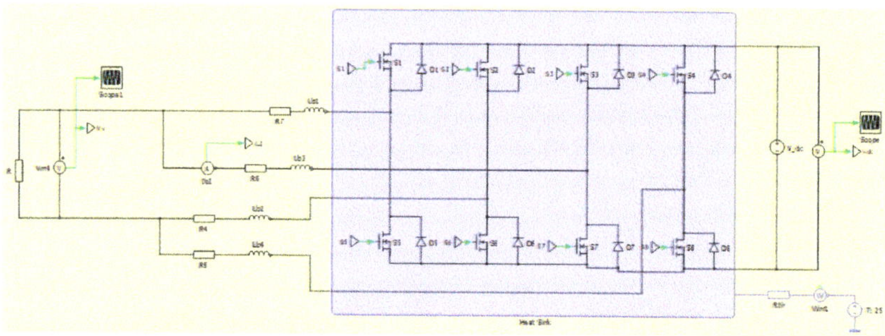

Figure 12. Simulation of the IFB converter by PLECS.

Table 3. Key specifications of the IFB converter.

Parameter	Unit	Value
Nominal power	kW	10
AC voltage	V	220
AC frequency	Hz	50
DC voltage	V	400
Switching frequency	kHz	30
L filter	μH	330
DC-link capacitor	μF	1000

- Scenario 1:

In this scenario, while the power rate is set up at 1.5 kW, which is similar to the experimental scenario, the switching frequency varies among 10 kHz, 30 kHz, 50 kHz, and 100 kHz.

As shown in Figure 13, the conduction losses of all three PWM methods for the IFB converter are similar with multiple chosen switching frequencies. Whereas, the switching losses increases significantly when the switching frequency increases. Moreover, at the low frequency of 10 kHz, the switching loss in three PWM methods is almost the same; however, that of the ID-PWM is significantly smaller than those of other interleaved methods. In detail, the reductions of around 23%, 30% and 27% in the switching loss of the IU-PWM are achieved in comparison with other methods at the switching frequency of 30 kHz, 50 kHz and 100 kHz, respectively.

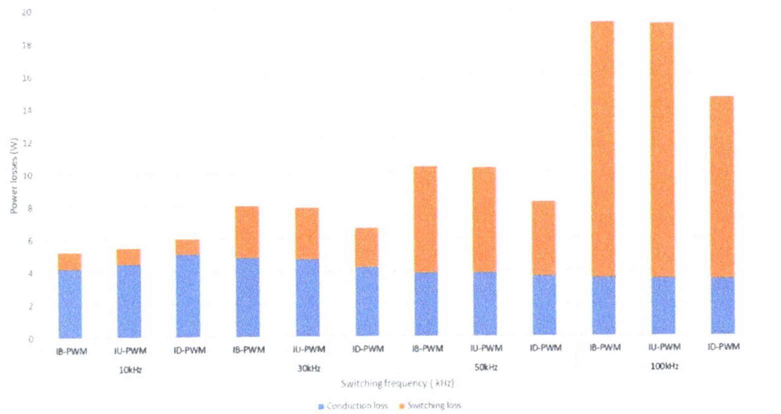

Figure 13. Power losses comparison with multiple switching frequency.

- Scenario 2:

In this scenario, while the switching frequency is set up at 30 kHz, which is similar to the experimental scenario, the power rate varies among 1.5 kW, 5 kW, 7.5 kW, and 10 kW. As shown in Figure 14, while the conduction losses of all three PWM methods for the IFB converter are still similar at any specific power rate, there is a difference among the switching losses. In detail, the reductions of around 25%, 24.8%, 25%, and 28% in the switching loss of the IU-PWM are achieved in comparison with those of other methods at the power rate of 1.5 kW, 5 kW, 7.5 kW, and 10 kW, respectively.

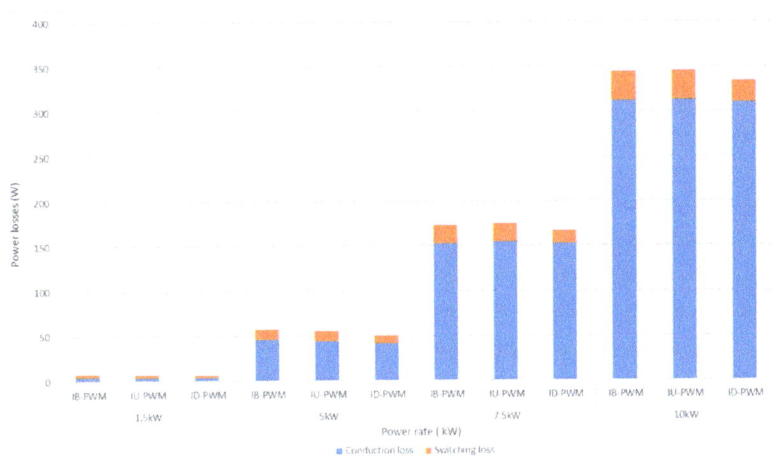

Figure 14. Power losses comparison with multiple power rate.

3.2. Leakage Current Comparisons

The parasitic capacitors normally have a small value; however, the accurate value is difficult to determine in a practical product. In this research, for simulation verification, the parameter of a parasitic RC branch is taken, as in [11] (R = 5 Ohms and C = 800 pF), which connects point O and point N, as in Figure 2. The high-frequency component in the common-mode voltage in the previous theoretical analysis leads to the different results among the leakage current of three methods, as shown in Figure 15.

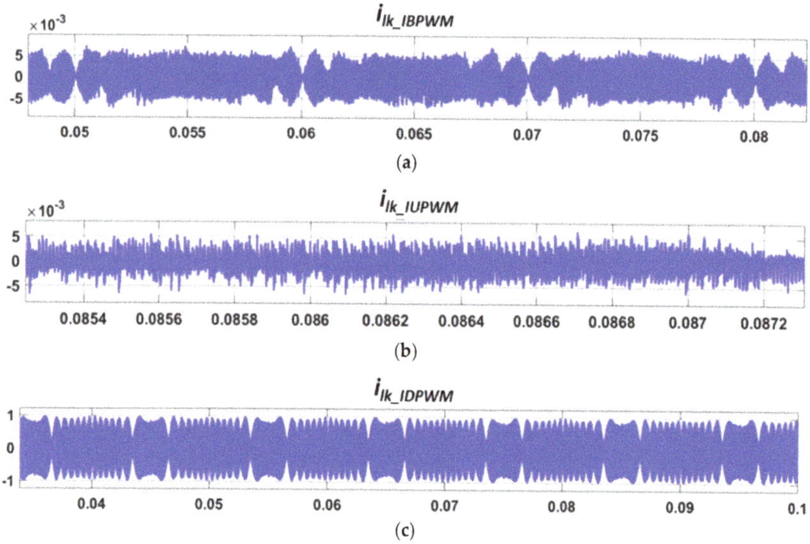

Figure 15. The leakage current of three PWM methods: (**a**) IB-PWM, (**b**) IU-PWM, (**c**) ID-PWM.

In detail, the root mean square (RMS) values of the leakage current with the IB-PWM and IU-PWM methods are relatively small at 2.29 mA and 2.39 mA, and the statistics for peak value are just 7.2 mA and 6.7 mA, respectively. Additionally, the RMS value and peak value of the leakage current in the case of the IU-PWM are much higher, at 470 mA and 980 mA. These values are out of standard IEC/EN 60335-1 [21] for leakage current in the domestic applications, which results in safety problems and electromagnetic emissions.

4. Experiment Verification

To clarify the comparison among the three proposed PWM methods, a 1.5 kW experimental prototype for the IFB converter was implemented. The experimental system shown in Figure 16 includes the IFB converter inverter, filtered inductor, a programmable electronic load ITECH IT8617, and a DC power supply model ITECH IT6018C-1500-40. The control algorithm is programmed in control kit Launchpad TMS320F28379D. To display and collect the data, a GW INSTEK GDS-2104A Digital Oscilloscopes is utilized with the current probe Micsig CP2100A and the isolated voltage probe Micsig DP10013. Except for a lower power rate of 1.5 kW, the two experimental scenarios in open-loop mode are carried out with similar parameters of the PLECS simulation in Table 3.

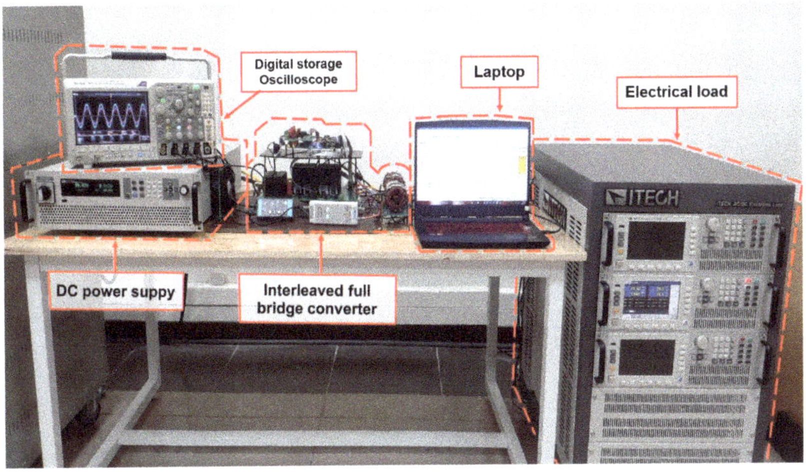

Figure 16. Experiment system for the IFB converter.

- Output current performance evaluation

The current responses are shown in Figures 17–19. Overall, the currents through the split inductor L_1 and L_2 are interleaved, which leads to the output current ripple being smaller than the inductor current ripple. In detail, the largest peak-to-peak current ripples of the split-inductor with the IB-PWM, IU-PWM, and ID-PWM are approximately 3 A, 3 A, and 5 A, respectively. These results show good agreement with the theoretical analysis in Section 2. In addition, the output current ripples with these three PWM methods are 2.5 A, 2.5 A and 1.8A, respectively, which proves the effectiveness of the interleaved technique for reducing the output current ripple. Moreover, the ID-PWM gives the smallest output current ripple as the aforementioned theory.

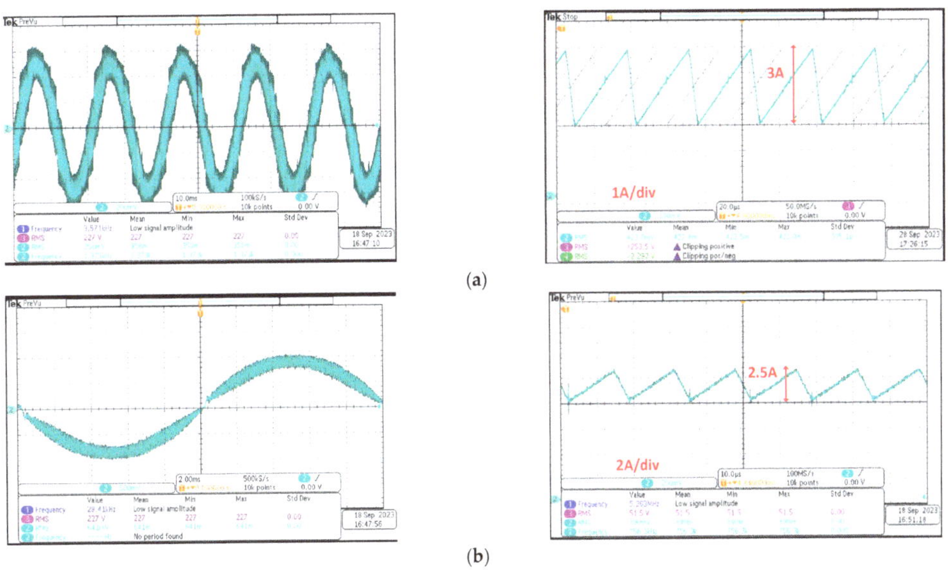

Figure 17. Current response with the IB-PWM method: (**a**) split-inductor current, (**b**) output current.

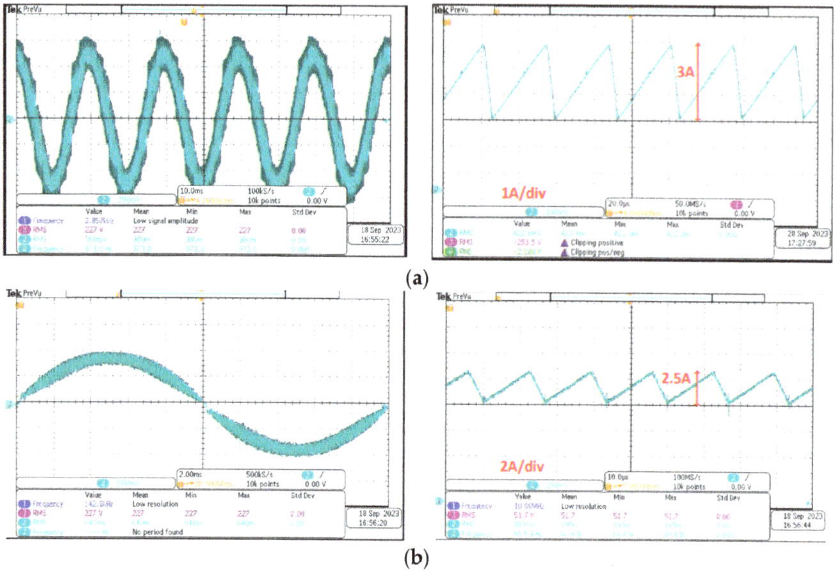

Figure 18. Current response with the IU-PWM method: (**a**) split-inductor current, (**b**) output current.

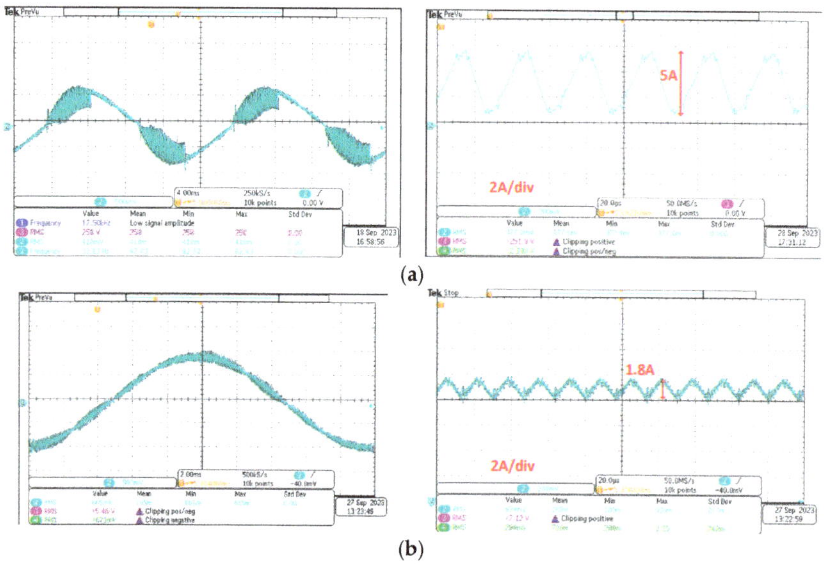

Figure 19. Current response with the ID-PWM: (**a**) split-inductor current, (**b**) output current.

To that extent, the THD of the output currents with three methods are compared in Figure 20. The part highlighted in red indicates the portion used for THD analysis. The harmonic spectrum is displayed as a bar graph relative to the fundamental frequency. While the THD current of the IB-PWM and the IU-PWM are equal at 9.76%, that of the IU-PWM is significantly smaller at 5.85%. To achieve a similar THD current as the ID-PWM, the switching frequency has to increase by 1.7 times or the inductance increases by 1.7 times. This produces higher switching loss, which decreases the total efficiency or increases the volume of the passive component, as well as reduces the power density. This also inherently illustrates the economic benefit of the IU-PWM for the IFB converter.

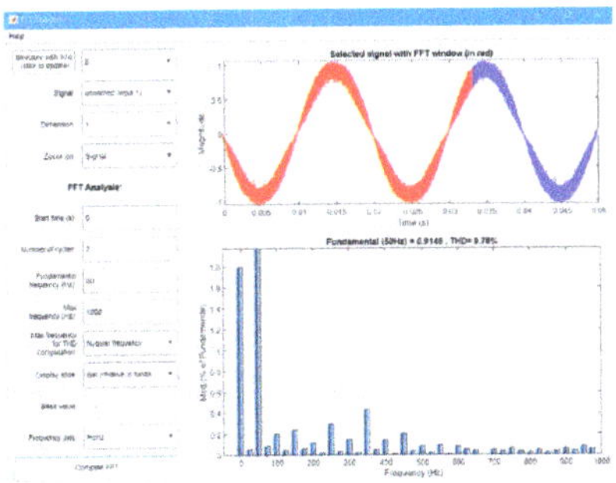

(**a**)

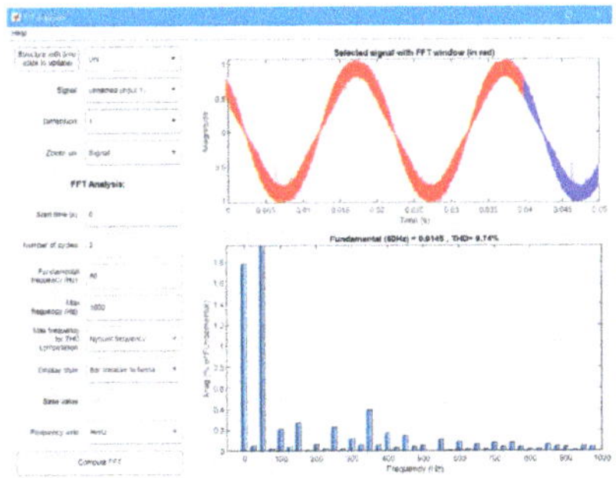

(**b**)

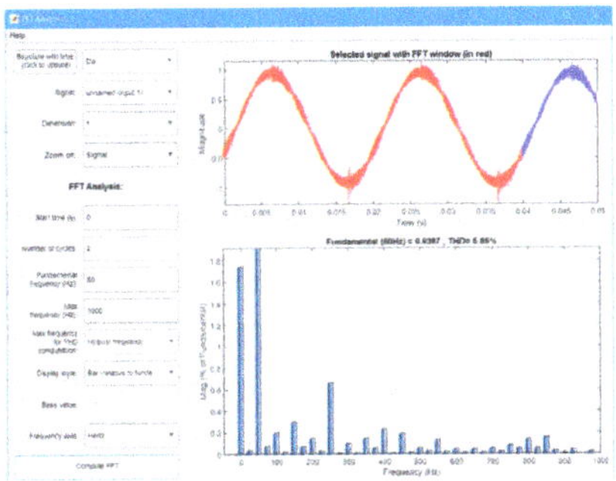

(**c**)

Figure 20. THD of the output current with three PWM methods: (**a**) IB-PWM, (**b**) IU-PWM, (**c**) ID-PWM.

Notably, the experimental and simulation results have proven the theoretical analysis. The advantage of lower power losses is achieved with the ID-PWM method for the IFB converter. Moreover, while the traditional discontinuous PWM does not improve the performance of the output current of the full bridge converter, the proposed ID-PWM offers a significant reduction in the THD current of the IFB converter compared to other interleaved PWM methods.

- Common-mode voltage discussion

In the experimental prototype, an open-frame IFB inverter is tested in which the parasitic capacitor is not taken into consideration due to the lack of the product chassis. Therefore, the leakage current responses are evaluated through the common-mode voltage. It can be seen that the common-mode voltage in Figure 21a for the IB-PWM method and in Figure 21b for the IU-PWM method are free of high-frequency pulses and they have the sinusoidal waveform for half of the grid voltage, with the offset being half of the DC voltage. However, high-frequency noise superimposes on a sinusoidal voltage in the measured negative poles of DC bus-to-ground voltage with IU-PWM in Figure 21c. In this case, high dv/dt could inject a spiky current with large di/dt through the parasitic capacitances and exceed the safety limitation for the leakage current in a household appliance in the IEC/EN 60335-1 standard [21]. To maintain the advantage of the ID-PWM method for the IFB converter, some hardware solutions could be considered.

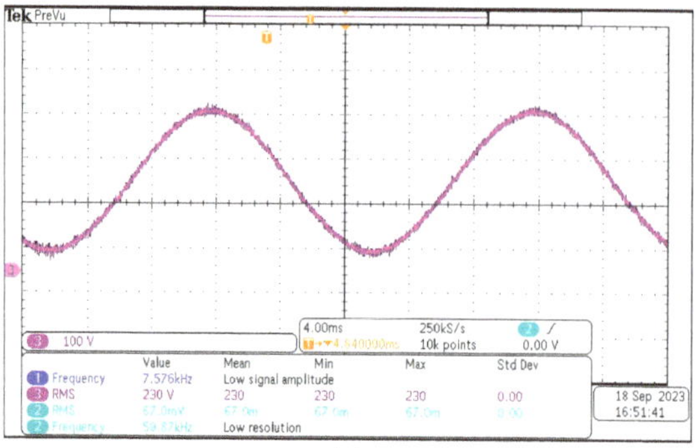

(a)

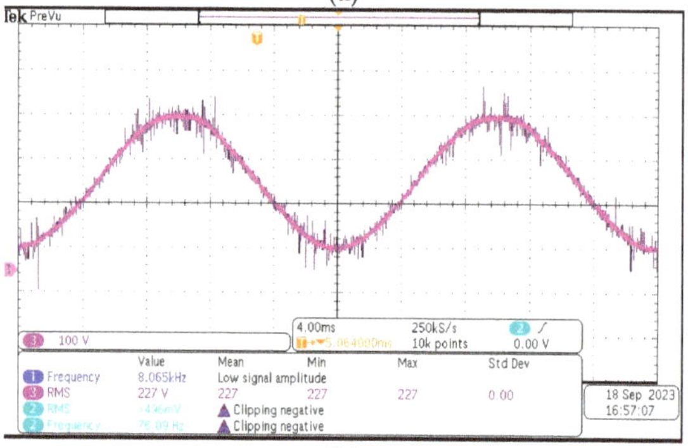

(b)

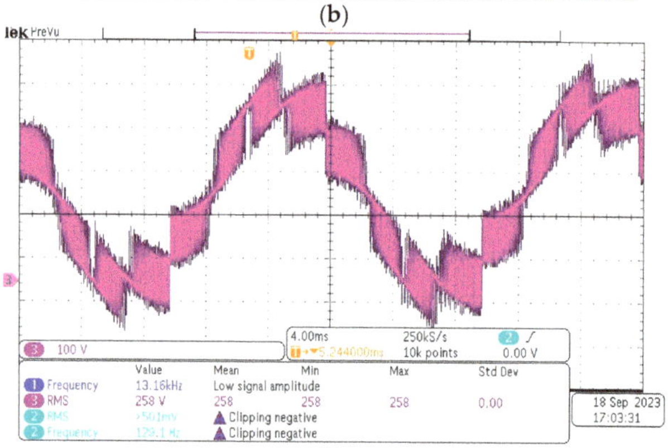

(c)

Figure 21. Common-mode voltage (inverted signal v_{NO}) with three PWM methods: (**a**) IB-PWM, (**b**) IU-PWM, (**c**) ID-PWM.

An EMI filter [22] is a traditional solution for the leakage current. In [23,24], the DC–bus midpoint is connected to the LCL filter capacitor midpoint in order to effectively reduce the DC–side leakage current. An integrated common-mode (CM) and different-mode filter with passive damping is proposed in [25] for grid-connected single-phase power converters for PV systems. Furthermore, an active CM filter is proposed in [26] for reducing the ground leakage current, but additional magnetic components and computational burden are required. Additionally, some modifications in structure by adding extra devices are applied in much of the research concerning full bridge converters. The H6 converter in [27] uses two additional switches, which would be turned off during the freewheeling period to disconnect the inverter from the DC source. The improvement of this structure can be mentioned in [28], with two additional diodes in the FB-DCBP inverter. Similarly, the AC side could also be bypassed with the HERIC inverter with two extra switching devices at the AC side [29].

5. Conclusions

This paper proposes three PWM methods, along with a comprehensive comparison of the interleaved full bridge converter in an AC Battery application. These PWM methods are IB-PWM, IU-PWM, and ID-PWM, namely, based on the traditional PWM method and the interleaved techniques method. The theoretical comparison among these PWM methods is also presented, and then, the simulation and experimental prototype are carried out to verify. The results prove that the IB-PWM and IU-PWM give a similar performance in output current ripple and THD, common-mode voltage, and power loss. The ID-PWM method not only achieves a reduction in the switching loss but also the lowest THD output current. However, different from the IB-PWM and IU-PWM, the common-mode voltage of the IFB converter using the ID-PWM consists of a high-frequency component, which results in a higher leakage current. This issue could be handled in a future work by hardware solutions, which allows this method to be the most suitable PWM method for the IFB converter.

Author Contributions: Conceptualization, P.V. and S.-J.A.; methodology, T.A.D.; software, T.A.D. and M.D.N.; validation, T.A.D., Q.D.N. and P.V.; formal analysis, T.A.D. and S.-J.A.; investigation, S.-J.A.; resources, Q.D.N.; data curation, M.D.N.; writing—original draft preparation, T.A.D.; writing—review and editing, S.-J.A. and P.V.; visualization, Q.D.N.; supervision, M.D.N.; project administration, S.-J.A. and P.V.; funding acquisition, S.-J.A. All authors have read and agreed to the published version of the manuscript.

Funding: This research was supported by "Regional Innovation Strategy (RIS)" through the National Research Foundation of Korea (NRF) funded by the Ministry of Education (MOE) (2021RIS-002); in part by the Korea Institute of Energy Technology Evaluation and Planning (KETEP) grant funded by the Korea government (MOTIE) (20225500000060).

Data Availability Statement: Data are contained within the article.

Conflicts of Interest: The authors declare no conflicts of interest.

References

1. Li, J.; He, S.; Yang, Q.; Wei, Z.; Li, Y.; He, H. A Comprehensive Review of Second Life Batteries Towards Sustainable Mechanisms: Potential, Challenges, and Future Prospects. *IEEE Trans. Transp. Electrif.* **2022**, *9*, 4824–4845. [CrossRef]
2. Zhao, Y.; Pohl, O.; Bhatt, A.I.; Collis, G.E.; Mahon, P.J.; Rüther, T.; Hollenkamp, A.F. A Review on Battery Market Trends, Second-Life Reuse, and Recycling. *Sustain. Chem.* **2021**, *2*, 167–205. [CrossRef]
3. Simatupang, D.; Benshatti, A.; Park, S.-Y. Embedded Electrochemical Impedance Spectroscopy into Battery Management System. In Proceedings of the IECON 2021—47th Annual Conference of the IEEE Industrial Electronics Society, Toronto, ON, Canada, 13–16 October 2021; pp. 1–6.
4. Hu, X.; Deng, X.; Wang, F.; Deng, Z.; Lin, X.; Teodorescu, R.; Pecht, M.G. A Review of Second-Life Lithium-Ion Batteries for Stationary Energy Storage Applications. *Proc. IEEE* **2022**, *110*, 735–753. [CrossRef]
5. Yang, S.; Ma, X.; Bao, W.; Liu, J. Multiple-Function Control Strategy for Hybrid SLB-NB Energy Storage System. In Proceedings of the 2022 7th Asia Conference on Power and Electrical Engineering (ACPEE), Hangzhou, China, 15–17 April 2022; pp. 318–323. [CrossRef]
6. Makhadmeh, S.N.; Al-Betar, M.A.; Alyasseri, Z.A.A.; Abasi, A.K.; Khader, A.T.; Damaševičius, R.; Mohammed, M.A.; Abdulkareem, K.H. Smart Home Battery for the Multi-Objective Power Scheduling Problem in a Smart Home Using Grey Wolf Optimizer. *Electronics* **2021**, *10*, 447. [CrossRef]
7. Zich, J.; Jandik, J. Active Battery Management System for Home Battery Energy Storage. In Proceedings of the 2020 21st International Scientific Conference on Electric Power Engineering (EPE), Prague, Czech Republic, 19–21 October 2020; pp. 1–4. [CrossRef]
8. Jiang, L.; Wang, X. Research on the Participation of Household Battery Energy Storage in the Electricity Peak Regulation Ancillary Service Market. *Processes* **2023**, *11*, 794. [CrossRef]

9. Inamdar, S.; Thosar, A.; Mante, S. Literature Review of 3.3kW On Board Charger Topologies. In Proceedings of the 2019 3rd International conference on Electronics, Communication and Aerospace Technology (ICECA), Coimbatore, India, 12–14 June 2019; pp. 276–281.
10. FMusavi, F.; Eberle, W.; Dunford, W.G. A high-performance single-phase AC-DC power factor corrected boost converter for plug in hybrid electric vehicle battery chargers. In Proceedings of the 2010 IEEE Energy Conversion Congress and Exposition, Atlanta, GA, USA, 12–16 September 2010; pp. 3588–3595. [CrossRef]
11. Xia, Y.; Ayyanar, R. Comprehensive comparison of THD and common mode leakage current of bipolar, unipolar and hybrid modulation schemes for single phase grid connected full bridge inverters. In Proceedings of the 2017 IEEE Applied Power Electronics Conference and Exposition (APEC), Tampa, FL, USA, 26–30 March 2017; pp. 743–750. [CrossRef]
12. Marandi, D.; Sowmya, T.N.; Babu, B.C. Comparative study between unipolar and bipolar switching scheme with LCL filter for single-phase grid connected inverter system. In Proceedings of the 2012 IEEE Students' Conference on Electrical, Electronics and Computer Science, Bhopal, India, 1–2 March 2012; pp. 1–4. [CrossRef]
13. Awais, M.; Yasin, A.R.; Riaz, M.; Saqib, B.; Zia, S.; Yasin, A. Robust Sliding Mode Control of a Unipolar Power Inverter. *Energies* **2021**, *14*, 5405. [CrossRef]
14. Algaddafi, A.; Elnaddab, K.; Ma'Mari, A.; Esgiar, A.N. Comparing the Performance of Bipolar and Unipolar Switching Frequency to Drive DC-AC Inverter. In Proceedings of the 2016 International Renewable and Sustainable Energy Conference, Marrakech, Morocco, 14–17 November 2016. [CrossRef]
15. Behera, P.K.; Das, S.; Pattnaik, M. Performance Comparison Between Bipolar and Unipolar Switching Scheme for a Single-Phase Inverter Based Stand-alone Photovoltaic System. In Proceedings of the 2019 IEEE 16th India Council International Conference, Rajkot, India, 13–15 December 2019; pp. 1–4. [CrossRef]
16. Ferrer-Arnau, L.; Berbel, N.; Capella, G.J.; Zaragoza, J. Study of modulation techniques applied to full bridge single-phase inverters based on wide-bandgap semiconductors. In Proceedings of the IECON 2019—45th Annual Conference of the IEEE Industrial Electronics Society, Lisbon, Portugal, 14–17 October 2019; pp. 2032–2037. [CrossRef]
17. Fernandez, M.; Robles, E.; Aretxabaleta, I.; Kortabarria, I.; Martín, J.L. Proposal of Hybrid Discontinuous PWM Technique for Five-Phase Inverters under Open-Phase Fault Operation. *Machines* **2023**, *11*, 404. [CrossRef]
18. Hossameldin, A.A.; Abdelsalam, A.K.; Ibrahim, A.A.; Williams, B.W. Enhanced Performance Modified Discontinuous PWM Technique for Three-Phase Z-Source Inverter. *Energies* **2020**, *13*, 578. [CrossRef]
19. Zhang, L.; Sun, K.; Xing, Y.; Zhao, J. Parallel Operation of Modular Single-Phase Transformerless Grid-Tied PV Inverters with Common DC Bus and AC Bus. *IEEE J. Emerg. Sel. Top. Power Electron.* **2015**, *3*, 858–869. [CrossRef]
20. Kiranmai, K.S.P.; Damodaran, R.V.; Hushki, M.; Shareef, H. An alternate hybrid PWM for uniform thermal sharing in single phase voltage-source inverter. *Sci. Rep.* **2023**, *13*, 3348. [CrossRef] [PubMed]
21. Available online: https://www.elipse.eu/wp-content/uploads/2020/10/White-Paper-Safety-Standards-Elipse-Oct-2020.pdf (accessed on 1 September 2023).
22. Yang, Y.; Blaabjerg, F.; Wang, H. Low voltage ride-through of single-phase transformerless photovoltaic inverters. In Proceedings of the 2013 IEEE Energy Conversion Congress and Exposition, Denver, CO, USA, 15–19 September 2013; pp. 4762–4769. [CrossRef]
23. Trzynadlowski, A.M. EMI Effects of power converters. In *Power Electronics Handbook*; Butterworth-Heinemann: Oxford, UK, 2007. [CrossRef]
24. Wu, W.; Sun, Y.; Lin, Z.; He, Y.; Huang, M.; Blaabjerg, F.; Chung, H.S.-H. A Modified LLCL Filter With the Reduced Conducted EMI Noise. *IEEE Trans. Power Electron.* **2014**, *29*, 3393–3402. [CrossRef]
25. Hedayati, M.H.; John, V. EMI and Ground Leakage Current Reduction in Single-Phase Grid-Connected Power Converter. *IET Power Electron.* **2017**, *10*, 938–944. [CrossRef]
26. Figueredo, R.S.; de Carvalho, K.C.M.; Matakas, L. Integrated common and differential mode filter applied to a single-phase transformerless PV microinverter with low leakage current. In Proceedings of the 2014 International Power Electronics Conference (IPEC-Hiroshima 2014—ECCE ASIA), Hiroshima, Japan, 18–21 May 2014; pp. 2618–2625.
27. Barater, D.; Buticchi, G.; Lorenzani, E.; Concari, C. Active common-mode filter for ground leakage current reduction in grid-connected PV converters operating with arbitrary power factor. *IEEE Trans. Ind. Electron.* **2014**, *61*, 3940–3950. [CrossRef]
28. Barater, D.; Buticchi, G.; Crinto, A.S.; Franceschini, G.; Lorenzani, E. A New Proposal for Ground Leakage Current Reduction in Transformerless Grid-Connected Converters for Photovoltaic Plants. In Proceedings of the 35th Annual Conference of IEEE Industrial Electronics (IECON), Porto, Portugal, 3–5 November 2009; pp. 4567–4572.
29. Gonzalez, R.; Lopez, J.; Sanchis, P.; Marroyo, L. Transformerless inverter for single-phase photovoltaic systems. *IEEE Trans. Power Electron.* **2007**, *22*, 693–697. [CrossRef]

Disclaimer/Publisher's Note: The statements, opinions and data contained in all publications are solely those of the individual author(s) and contributor(s) and not of MDPI and/or the editor(s). MDPI and/or the editor(s) disclaim responsibility for any injury to people or property resulting from any ideas, methods, instructions or products referred to in the content.

energies

Article

Deep Learning-Based Algorithm for Internal Fault Detection of Power Transformers during Inrush Current at Distribution Substations

Sopheap Key, Gyu-Won Son and Soon-Ryul Nam *

Department of Electrical Engineering, Myongji University, Yongin 17058, Republic of Korea; keysopheap123@gmail.com (S.K.); onyang01111@gmail.com (G.-W.S.)
* Correspondence: ptsouth@mju.ac.kr; Tel.: +82-31-330-6361

Abstract: The reliability and stability of differential protection in power transformers could be threatened by several types of inferences, including magnetizing inrush currents, current transformer saturation, and overexcitation from external faults. The robustness of deep learning applications employed for power system protection in recent years has offered solutions to deal with several disturbances. This paper presents a method for detecting internal faults in power transformers occurring simultaneously with inrush currents. It involves utilizing a data window (DW) and stacked denoising autoencoders. Unlike the conventional method, the proposed scheme requires no thresholds to discriminate internal faults and inrush currents. The performance of the algorithm was verified using fault data from a typical Korean 154 kV distribution substation. Inrush current variation and internal faults were simulated and generated in PSCAD/EMTDC, considering various parameters that affect an inrush current. The results indicate that the proposed scheme can detect the appearance of internal faults occurring simultaneously with an inrush current. Moreover, it shows promising results compared to the prevailing methods, ensuring the superiority of the proposed method. From sample *N*–3, the proposed DNN demonstrates accurate discrimination between internal faults and inrush currents, achieving accuracy, sensitivity, and precision values of 100%.

Keywords: inrush current; data window; deep learning; differential protection; internal fault

1. Introduction

A power transformer is an essential component used in power systems where voltage conversion is required. To ensure efficient operation in power systems, current differential protection is conventionally adopted as the primary protection, which is based on Kirchoff's current law. However, it is susceptible to unwanted abnormalities such as magnetizing inrush currents during transformer energization and a parallel connection of transformers under normal operations, as well as CT saturation due to overexcitation. These abnormalities might negatively result in the mis-operation of the current differential protection. An inrush current is a non-sinusoidal and high-magnitude current generated due to flux saturation in the transformer during energization. The magnitude of an inrush current is highly dependent on the switching angle, the amount of residual flux, and the sizes of the transformers. The fundamental principles and derivation of magnetizing an inrush current are presented in [1].

Since magnetizing inrush currents generally has a large ratio of the second-harmonic component compared to an internal fault and normal conditions, harmonic blocking and restraint have been designed to avoid false operations due to inrush currents [2] and have been widely employed in commercial relays [3]. Moreover, with the newly improved material of modern transformers, second-harmonic restraint/blocking faces the downside of lower second-harmonic components during transformer energization [4]. Therefore, the conventional scheme in transformer protection can be blocked for several cycles due to an indecisive threshold. In addition, the energization of a faulty transformer may reduce the sensitivity of harmonic restraint due to the high ratio of the second harmonic in healthy phases and leads to low reliability during the energization of a faulty transformer. Thus, novel functionalities must be proposed or integrated with the existing differential protection to enhance reliability and security in detecting internal faults during energization. Recently, the Korean Electric Power Company (KEPCO) reported numerous failures of the differential relay in the field when internal faults occurred during an inrush current, leading to a malfunction in the differential relay, as conventional harmonic blocking could not respond to them and continuously blocked the differential relay until the presence of the second harmonic fell under a set value. Therefore, a new scheme for power transformer protection is urgently required to secure the stability of power systems.

1.1. Literature Review and Related Works

Conventionally, the utilization of the second-harmonic principle is widely adopted in power transformer protection against inrush currents, as described in the above section. However, this method has been proven to be ineffective in several circumstances [5,6]. During internal faults, there is a large ratio of the second harmonic in a few cycles, which blocks the differential relay from operating, resulting in damage to power transformers. An extensive outage and a blackout were reported in [7] when the power transformer protection mis-operated under inrush conditions. Moreover, as the power system expands, the second-harmonic components increase on long transmission lines when the transformers are connected to shunt reactors or series capacitors [8]; as a result, differential protection is bypassed when this scenario occurs.

Several transformer-protection techniques have been actively proposed to identify the inrush condition, such as artificial neural networks, fuzzy logic, wavelet transform, and mathematical-based algorithms. A statistical approach based on Principle Component Analysis (PCA) was described in [9] to differentiate inrush currents, internal faults, and overexcitation conditions. It captures 2D feature space as a pattern recognition for each abnormal condition. Methods based on fuzzy and artificial neural networks were proposed in [10,11], and a correlation-based algorithm was developed for inrush current discrimination [12]. For a similar purpose, a method combining a support vector machine as the classifier and a wavelet transform for feature extraction was also proposed in [13]. A deep learning application was proposed in [14,15] to address current transformer saturation on transmission lines, and another deep learning-based approach was also proposed in [16] to remove the decaying DC offset in a power system.

As signal processing techniques based on wavelet transforms have proven to be efficient tools for the analysis, detection, and classification of non-stationary signals at various levels of time–frequency resolution in the literature, they could be applicable in real-time devices. For instance, a wavelet transform has been utilized to address existing issues in power systems such as fault detection, location, and classification [17,18], as well as in the differential protection of power transformers [19–23]. Although it has good performance without the need of harmonic information, there are some limitations for practical applications in power system protection, such as the strong influence of the mother wavelet and time delay. However, it does not provide an answer for internal fault detection during inrush conditions, which is a significant concern in transformer differential protection. An improved wavelet transformation, namely the Real-Time Boundary Stationary Wavelet Transform (RT-BSWT), was proposed in [24] to detect internal faults during inrush currents. Despite the improvement made, a high sampling rate is required, and it may be susceptible to noise. A process to identify an inrush current based on the enhanced GSA-BP approach was proposed in [25] to discriminate inrush currents from fault currents in transformers.

A low-computation method based on a fault component network was developed in [26] to enhance the accuracy of transformer protection, regardless of magnetizing inrush conditions. A method based on the current and voltage ratio was demonstrated in [27], where it deployed the absolute difference of the current and voltage to differentiate inrush currents from internal faults. A unidirectional index was utilized to detect the direction of magnetizing inrush currents in power transformers [28]. The detection of inrush currents based on the dead angle was introduced in [29]. If the waveform distortion is so severe that the wave width is less than 140°, it will cause a delay in protection or even a wrong judgment; therefore, the efficacy of this method presents a drawback. A new adaptive coordination approach between generator and transformer was proposed to enhance the abnormal operating conditions [30].

1.2. Key Contributions and Organization

Motivated by the above-mentioned problems with the conventional approach, this paper presents a protection scheme to discriminate internal faults and inrush currents by combining a data window with deep neural networks (DNNs). In recent years, new techniques based on intelligent methods have demonstrated a robust distinction between inrush currents and internal faults for power transformer protection, overcoming the drawbacks of traditional differential protection. To detect inrush currents and internal faults, the proposed scheme first utilizes the data window to obtain the distinctive feature signal that separates the region of internal faults and inrush currents. The proposed scheme can identify internal faults during inrush currents. It not only provides stability when these two abnormal conditions occur simultaneously but also improves the response time compared to conventional harmonic-blocking methods. Furthermore, the proposed scheme is applicable to inrush currents and internal faults of various magnitudes due to its normalization quantity during the preprocessing prior to deep-learning training. Then, a DNN is employed to discriminate internal faults from inrush currents. The key contributions of the proposed work can be highlighted as follows.

1. A wide range of applicability, regardless of inrush current magnitude, the residual flux in power transformers, internal fault magnitude, and fault angles;
2. An improved discrimination of internal faults, considering winding-ground faults during inrush currents;
3. A universal application for other power transformers with different characteristics;
4. A data window-based operation without the need for a threshold.

The rest of this paper is organized as follows. Section 2 highlights the literature review of the behavior of inrush currents using a data window and addresses issues related to the second-harmonic-blocking method. This section also includes information on data acquisition and dataset preparation for training, along with a detailed description of inrush current features. Section 3 presents the proposed deep neural network (DNN) method and its structure. The simulation setup, implemented in both Python and PSCAD/EMTDC, is detailed in Section 4. Section 5 addresses the results of the proposed method for inrush current and internal fault detection and provides a comparative analysis using the conventional approach. A discussion of the performance evaluation based on statistical percentages is demonstrated in Section 6. Lastly, Section 7 includes concluding remarks and information regarding potential future works.

2. Problem Statement

This section presents the principles and approaches utilized for internal fault and inrush-current detection based on a data window. To facilitate understanding in the subsequent sections, a list of relevant acronyms and their definitions are provided in Table 1.

Table 1. Relevant acronyms, units, and their definitions.

Acronym	Unit	Definition
DW		Data window
DNN		Deep neural network
AE		Autoencoder
SAE		Stacked autoencoder
HAR		Second-harmonic restraint
UNI		Unidirectional index
EKF		Extended Kalman filter
i_{diff}	A	Differential current
m		Size of data window
k		Last index of differential current
x_{norm}	A	Normalized differential current
θ_k		Parameter set model
y_i		Stochastic variable of the output class
S		Activation function
L_{AE}		Reconstruction loss for AE
L		Softmax loss function
σ		Weight decay

2.1. Overview of Magnetizing Inrush Current and Second Harmonic Ratio

An inrush current is the high current drawn by a transformer when it is initially energized. It is caused by an abrupt change in magnetic flux within the transformer core and is proportional to the current flowing through the primary winding. Figure 1 illustrates a differential current and the ratio of the second harmonic. As mentioned in the Introduction, a common approach to differential protection in power transformers involves second harmonic-based blocking to prevent unnecessary tripping. Due to the substantial ratio of the second harmonic during transformer energization, it can be effectively used to distinguish an inrush current from internal faults. However, as shown in Figure 1, a second harmonic may also be generated during internal faults due to a decaying DC component from faults. At the moment of transformer energization, the ratio of the second harmonic rapidly increases to approximately 60%. Consequently, the harmonic-blocking method blocks the operation of the differential relay in this scenario, leading to potential damage to the power transformer.

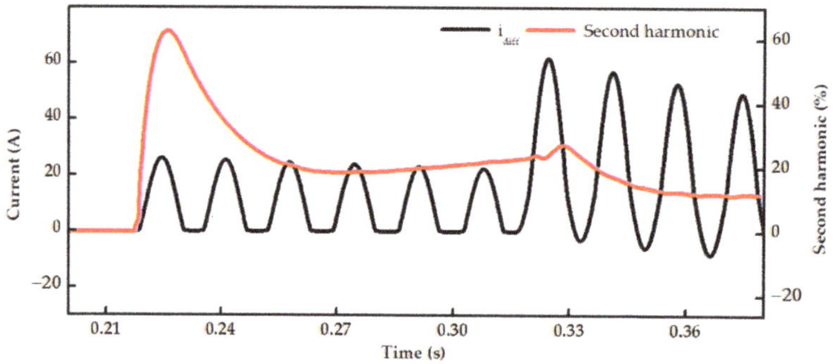

Figure 1. Waveform of differential current pertaining to inrush current and internal faults.

DW is a technique applied in power system protection for fault detection, direction estimation, time-series forecasting, and fault classification. It yields promising results at every instant when there is a significant fluctuation in the waveform. The results based on the data window from [14–16] are noticeable when dealing with abnormal conditions. Inspired by this concept, we develop a DW which was originally proposed in [31] to detect power swings on a transmission line. Considering the measured differential current

$i_{diff} = \{x_1, \ldots, x_k\}$, where k is the last index of the measured differential current, the equation derived from a set of DWs on the measured differential current is expressed in (1) as follows:

$$Set\ of\ DWs = \begin{bmatrix} x_1 & x_2 & \cdots & x_{m-1} & x_m \\ x_2 & x_3 & \cdots & x_m & x_{m+1} \\ \vdots & \vdots & \ddots & \vdots & \vdots \\ x_{k-m+1} & x_{k-m+2} & \cdots & x_{k-1} & x_k \end{bmatrix} \quad (1)$$

2.2. Data Window of Inrush Current and Internal Faults

The waveform in the DW, as described in (1), forms an abundance of distinctive waveform characteristics at each sample point. These characteristics enable DNN to capture the unique features distinguishing inrush currents from internal faults. Figure 2 illustrates a region of instantaneous differential currents under a DW with a length of one cycle. This figure clearly demonstrates the DW under conditions of an inrush current and an internal fault. Prior to a sudden spike in the current due to transformer energization, every value in the DW is zero in each sample. Upon closing the circuit breaker, there is a sudden change in the magnetic flux, leading to a significant increase in the differential current. Similarly, an internal fault also manifests a sudden change at the initial point in the differential current, posing a challenge for conventional methods to discriminate internal faults from inrush currents when both conditions coincide. As illustrated in Figure 2, every value in the DW before point A is zero, designating this region as the normal condition (state 0). Upon reaching point A, the value of the last index of the DW becomes positive, indicating the occurrence of a transient state (state 1). If an internal fault and an inrush current occur simultaneously at this point, it becomes challenging to determine each disturbance. Therefore, no action will be taken during this transition. At point B, an internal fault exhibits different characteristics from an inrush current. For internal faults, the value becomes zero for a fault inception angle of 0° or negative for a fault inception angle of 90°. When this behavior is detected, the algorithm promptly changes to state 3 (internal fault); otherwise, it identifies an inrush current (state 2). The sampling delay between points A and B is less than one cycle, specifically 58 samples, considering that one cycle corresponds to 64 samples. The reason for this delay is to achieve a clear discrimination between inrush currents and internal faults at a fault inception angle of 0°.

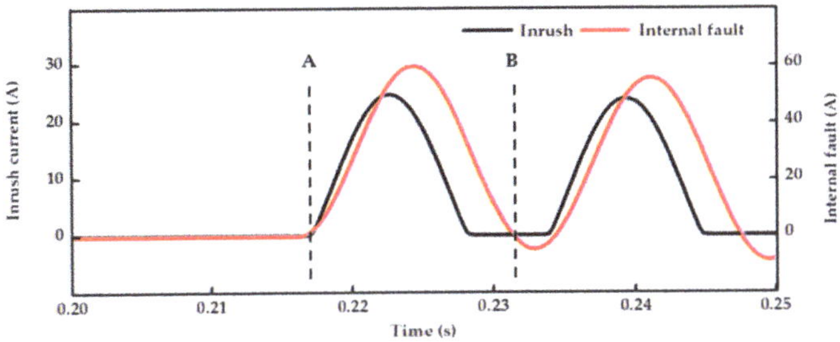

Figure 2. Illustration of a DW of a differential current under the conditions of an inrush current (upper) and an internal fault (below).

Therefore, it is apparent that this difference can be effectively used as an important feature to distinguish internal faults from inrush currents and to create learning labels for DNNs, which will be explained later in the following section.

2.3. Dataset Acquisition for Training and Testing Procedure

A thorough analysis is necessary to achieve high accuracy and generalization in the discrimination model for inrush currents and internal faults using a DNN-based method. The generation of sufficient datasets for training DNNs is crucial for accurately discriminating between the mentioned abnormalities. To obtain diverse datasets for inrush currents, extensive simulations are required for subsequent analysis. The training dataset considers influencing parameters in inrush conditions, such as the residual flux in the power transformer, the switching angle, and the polarity of the residual flux. The inrush current magnitude is at its maximum when the transformer switches on at 0°. Moreover, the polarity of the residual flux significantly impacts the magnitude of the inrush current. The influencing parameters for inrush currents and internal faults are listed in Table 2. The datasets for inrush currents accumulated 170 inrush conditions, corresponding to 228,140 datasets available for training and testing. The influencing parameters for the internal faults are given in Table 2. The datasets for the internal faults accumulated 90 cases of a–g faults, corresponding to 111,870 datasets available for training and testing. The inrush current and internal faults are randomly partitioned into training and testing datasets with an 80% to 20% ratio, respectively.

Table 2. Dataset of the inrush current and internal faults for the DNN procedure.

Case	Parameters	Value
Inrush	Switching angle (°)	0, 10, 20, 30, 40, 50, 60, 70, 80, 90
	Residual flux (%)	−80, −70, −60, −50, −40, −30, −20, −10, 0, 10, 20, 30, 40, 50, 60, 70, 80
Internal fault	Fault inception angle (°)	0, 10, 20, 30, 40, 50, 60, 70, 80, 90
	Winding location (%)	0, 10, 20, 30, 40, 50, 60, 70, 80

2.4. Dataset Preprocessing for Training

The preprocessing stage for training DNNs is the most crucial part, determining the outcome of the trained model. It serves as a platform for DNNs to quickly comprehend the problem statement and the approach to achieving the expected outcomes in the final stage. As the magnitude of an inrush current varies depending on the influencing parameters listed in Table 2, it is challenging to determine a specific threshold for the correct label for DNNs. Therefore, normalization is introduced to address the problem of numerical instability and uncertain thresholds caused by the varying magnitudes of an inrush current. The derived equation for normalizing the training input is given in (2) as follows.

$$x_{norm} = \frac{x_i}{x_{max}} \tag{2}$$

where x_{max} is the maximum value captured in the measured differential current. Normalizing the input dataset scales the training input within the range of [–1, 1]. Additionally, this process enhances the robustness and capability of the proposed DNN, making it applicable to datasets from different systems. Once a set of DWs are formed, as described in (1), and the label for each condition is defined, as described in Section 2.2, we convert the multi-class region into a binary form using one-hot encoding, as shown in Table 3.

Table 3. Binary form using one-hot encoding.

Class	Label	Binary Form
0	Normal condition	0 0 0 1
1	Transient	0 0 1 0
2	Inrush current	0 1 0 0
3	Internal fault	1 0 0 0

3. Deep Neural Network (DNN)-Based Discrimination

DNNs have undergone continuous evolution, demonstrating a strong capability to address challenging problems in recent years, particularly in cases where conventional methods struggle with nonlinear issues. This section introduces the concepts and strategies implemented to discriminate inrush currents and internal faults. To enhance the structure of DNNs, the proposed discrimination scheme adopts unsupervised pre-training using stacked autoencoders and supervised fine-tuning. The details of benchmark models are demonstrated well in [32,33].

3.1. Principle of Autoencoders

An autoencoder is the basic component of an SAE learning in an unsupervised way, typically containing an encoder and a decoder. In a simple autoencoder, the input $x \in R_n$ $(x_1, x_2, \ldots, x_n)$ is included in the training dataset. The input is then encoded to a low dimension and restored to its original dimension in the decoding part. The training uses backpropagation to minimize the reconstruction error of the input features. Once the training converges, the transformed features $(f_1, f_2, \ldots, f_n)$ are saved and used to train other autoencoders. The encoder employs a deterministic mapping function to map input x to the hidden layer f. The encoding process is given as follows in (3), where W_1 and b_1 are the weight and bias of the encoding parts.

$$f = S(W_1 x + b_1) \tag{3}$$

The decoder reconstructs the hidden layer representation (f) to obtain the output ($\hat{x}$), as shown in (4), where W_2 and b_2 are the weight and bias of the decoding parts. S denotes the activation function for training the AE, and ReLU is used for both the encoder and decoder.

$$\hat{x} = S(W_2 f + b_2) \tag{4}$$

The parameters of the AE are optimized to minimize the reconstruction error, as shown in (5).

$$L_{AE}\left(x, \hat{x}\right) = \frac{1}{N}\sum_{n=1}^{N}(x - \hat{x}_\theta(f_\theta(\hat{x}))) \tag{5}$$

3.2. Framework of Stacked Autoencoder

An SAE is a neural network consisting of multiple layers of AEs, where the features of each AE are stacked and fed as inputs to the successive AE. The first AE is trained in a bottleneck fashion with the initial weight and bias (w_1 and b_1). The input is compressed into a low-dimensional feature through the encoding function and then restored back to its original dimension in the decoding layer. After removing the decoding layer ($\hat{x}$) in the first AE, a new hidden layer (h_2) and an output ($\hat{h}_1$) are stacked onto the first AE. Using a similar process, many AEs are successively stacked together to form a deeper network structure. This process is commonly known as pre-training because it adopts a greedy-layer training method. Finally, an output layer is trained with the given label (binary form of the abnormality) to discriminate between inrush currents and internal faults. All optimal SAE weights and biases (w_i and b_i, where $i = 1, 2, \ldots, n$), which are obtained during the pre-training process, are fine-tuned using the backpropagation algorithm to achieve significant improvements in discrimination ability. The construction process of a three-layer SAE is depicted in Figure 3.

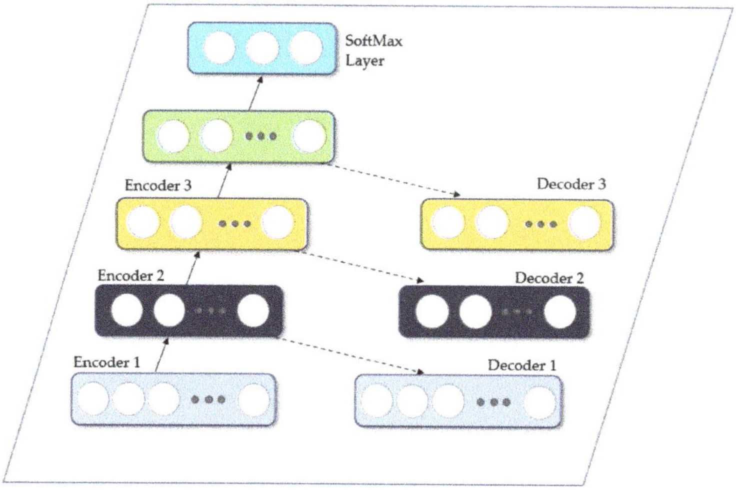

Figure 3. Construction process of *a three-layer* SAE used in simulations.

3.3. Fine-Tuning and SoftMax Classifier

The pre-trained model from the SAE can be further optimized by using parameters from all encoding layers during the pre-training phase with the backpropagation algorithm to minimize errors. Using the weights (W_i) and biases (b_i) from an SAE as initial values for fine-tuning enables the deeper network to generalize more effectively to other inrush variations produced by power transformers. We assign labels and extract features from the SAE to the classifier layer for the precise discrimination between normal conditions, inrush currents, and internal faults. Consequently, the classification outputs exhibit minimal errors, resulting in high accuracy.

A SoftMax classifier is employed in the classifier layer to discriminate among the four classes listed in Table 3. It estimates the posterior probabilities of each class in the range of [0,1], and the hypothesis is calculated as follows.

$$h_\theta(x^i) = \begin{bmatrix} p(y^i = 1|x^i,\theta) \\ p(y^i = 2|x^i,\theta) \\ \vdots \\ p(y^i = k|x^i,\theta) \end{bmatrix} = \frac{1}{\sum_{l=1}^{k} e^{\theta_l^T x^i}} \begin{bmatrix} e^{\theta_1^T x^i} \\ e^{\theta_2^T x^i} \\ \vdots \\ e^{\theta_k^T x^i} \end{bmatrix} \tag{6}$$

where y^i is the stochastic variable of the output class corresponding to input dataset x^i, and j represents the output class, encompassing four conditions: normal, transient, inrush, and internal fault. $\theta = \left(\theta_1^T, \theta_2^T, \ldots, \theta_k^T\right)^T$ denotes the parameter set of the model. Consequently, the output of the SoftMax classifier is given in a 4-dimensional vector containing four possible classes. The maximum probability of each class is determined as follows.

$$Classx^i = argmax_{j=1,\ldots,k} p(y^i = j|x^i, \theta) \tag{7}$$

Likewise, the SoftMax classifier converges to the global minimum by iteratively optimizing the cost function in (8) using categorical cross entropy.

$$L = -\frac{1}{N}\left[\sum_{i=1}^{m}\sum_{j=1}^{k} S\left\{y^i = j\right\} \log \frac{e^{\theta_j^T x^i}}{\sum_{l=1}^{k} e^{\theta_l^T x^i}}\right] + \frac{\sigma}{2}\sum_{i=1}^{k}\sum_{j=1}^{N} \theta_{ij}^2 \tag{8}$$

where y_i is the ith scalar value from the SoftMax output in (7), S represents the indicator function, σ is included in the cost function to penalize large values of the parameters, and L is strictly convex.

A flowchart of the proposed DNN is depicted in Figure 4.

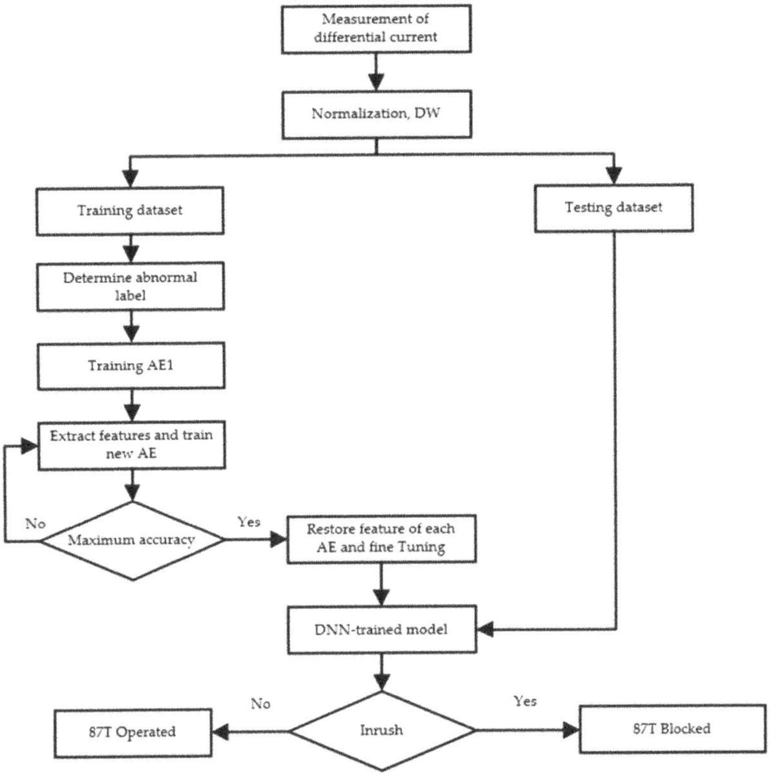

Figure 4. Flowchart of the proposed DNN to discriminate between inrush currents and internal faults.

4. Simulation Model

This section will highlight the simulation setup used to generate datasets for verifying the proposed DNN.

4.1. PSCAD/EMTDC Model

The effectiveness of the proposed technique was verified using a typical Korean 154 kV distribution substation. A simulation model of a 154/23 kV distribution system with 40 MVA power transformer and a Y-Y configuration was built in PSCAD/EMTDC, as illustrated in Figure 5. The sampling frequency was set to 3840 Hz or 64 samples per cycle in 60 Hz systems. The source was defined by the specific parameters listed in Table 4. In this study, only winding-ground faults were considered for evaluation, with variations in the fault inception angles and percentages of the winding faults. The winding faults were simulated by varying the fault location in the transformer, winding between 10% and 90%, in steps of 20%, from the winding terminal on the primary side of the transformer. The fault inception angle of the internal faults varied from 0° to 90°, in steps of 15°, with reference to the phase-A current. During the generation of magnetizing inrush currents, a residual flux was considered in the range of −80% to 80%, in steps of 10%, and different switching instances were considered between 0° and 90°.

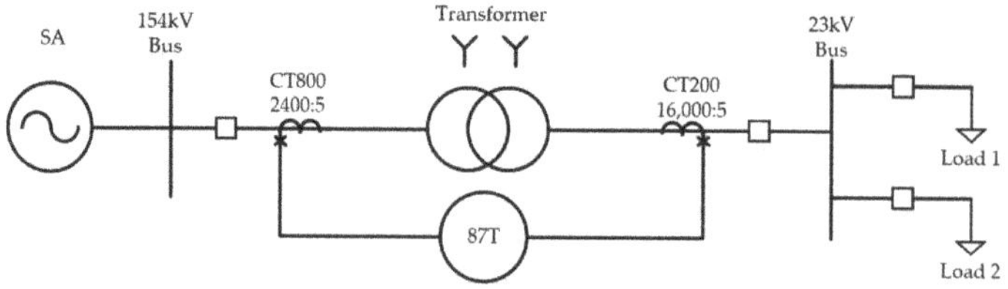

Figure 5. Single line of PSCAD modelling.

Table 4. Source and transformer parameters used in PSCAD modelling.

	Specification	Parameters	Value
Source	Positive and Negative	R_1, R_2 L_1, L_2 C_1, C_2	0.0419 0.8921 0.0128
	Zero	R_0 L_0 C_0	0.0293 2.6657 0.0042
Transformer 154/23 kV	Positive leakage reactance Air core reactance Magnetizing current	$\%Z$ $\%X$ $\%Im$	10.99 20 1

4.2. Deep Neural Network Model

Tensorflow is one of the most common deep learning platforms developed by Google. It offers a high-level API to optimize neural network models and the training procedure of the proposed DNN model. Therefore, the Tensorflow library is adopted in this paper to construct the network model and to train it to discriminate between inrush currents and internal faults.

In both the unsupervised and supervised learning modes, a categorical cross-entropy loss was employed to quantify the error between the network output and the reference output. The Adam optimizer was used to build the network for gradient backpropagation and parameter updates in every epoch. A decaying learning rate was applied to enhance convergence performance and to expedite the training process, preventing issues related to overfitting. It was initially set at 8×10^3 and then exponentially decreased with each iteration. The structure of the DNN and the training parameters for each AE are given in Table 5.

Table 5. Structure of the proposed DNN and training parameters.

	AE1	AE2	AE3	SoftMax Layer
Neuron	30	18	9	4
Batch size	128	64	64	64
Learning rate	0.001	0.0024	0.0019	0.0159

5. Simulation Results

In this section, the efficiency of the proposed DNN is verified and compared to the unidirectional index method in [28], the conventional harmonic-blocking scheme [34], and the Extended Kalman filter in [35]. Graphical illustration and evaluation metrics make it abundantly evident that the proposed method is effective against inrush currents and internal faults. In Figures 6–11, DNN, UNI, and HAR denote the proposed DNN method, the unidirectional index in [28], and the second-harmonic-blocking approach [34], respectively. The Extended Kalman filter in [35] is used for comparison when internal faults are present, because EKF only detects the instance of internal faults. It is generally known that protection relays in power system protections operate after one cycle. Therefore, the evaluation of the proposed DNN and alternative methods will be discussed based on the 58th (=N–6) and 61st (=N–3) samples from each abnormality.

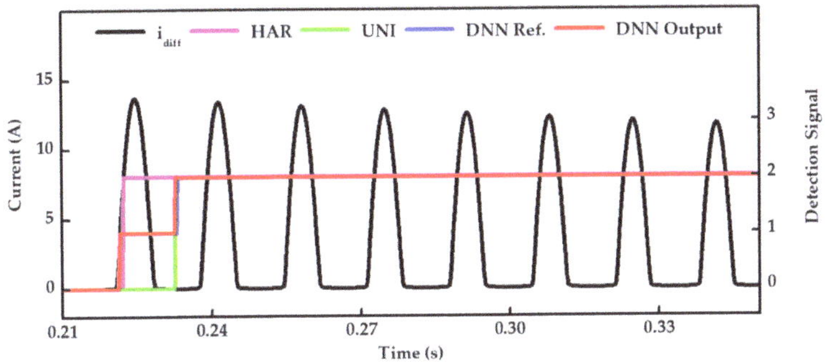

Figure 6. Results of inrush-current detection in a case with no residual flux and at a switching angle of 0°.

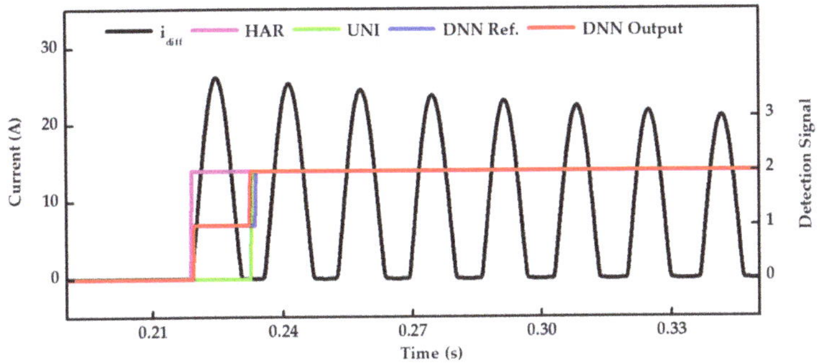

Figure 7. Results of inrush-current detection in a case with maximum residual flux and at a switching angle of 0°.

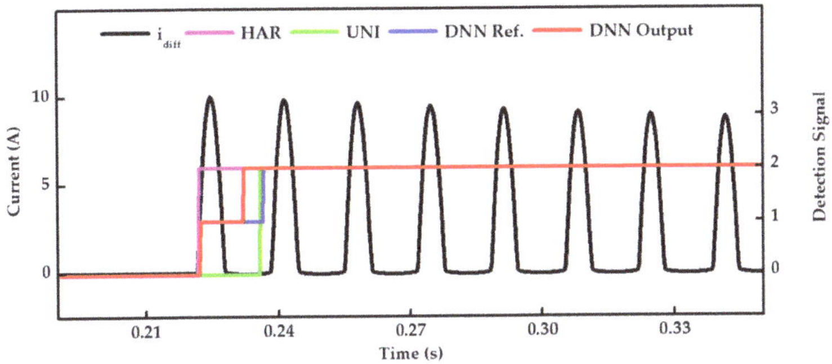

Figure 8. Results of inrush-current detection in a case with maximum residual flux and at a switching angle of 90°.

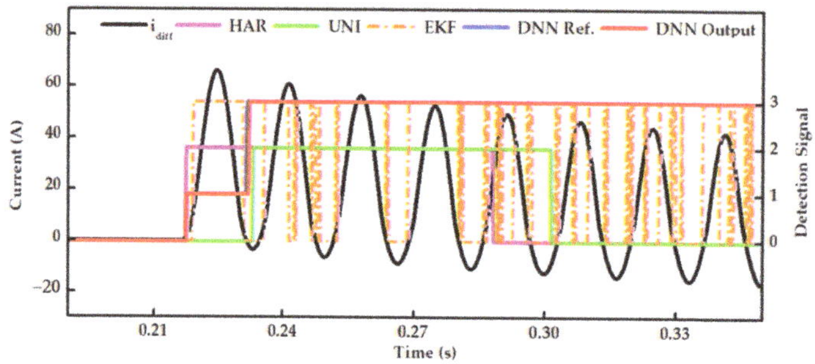

(a)

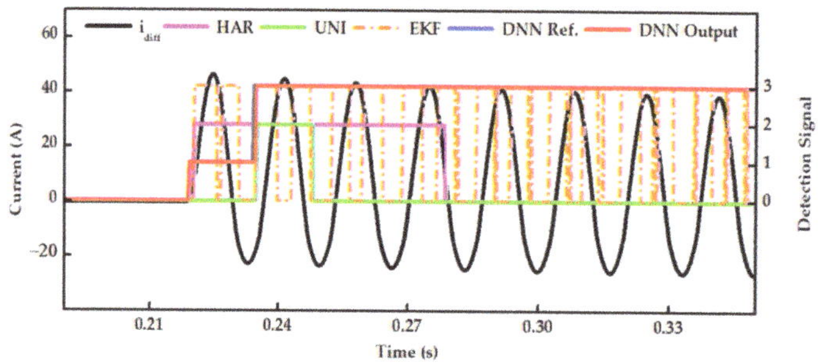

(b)

Figure 9. Results of internal-fault detection when a power transformer is energized in the presence of an internal fault: (**a**) fault inception angle of 0° and (**b**) fault inception angle of 90°.

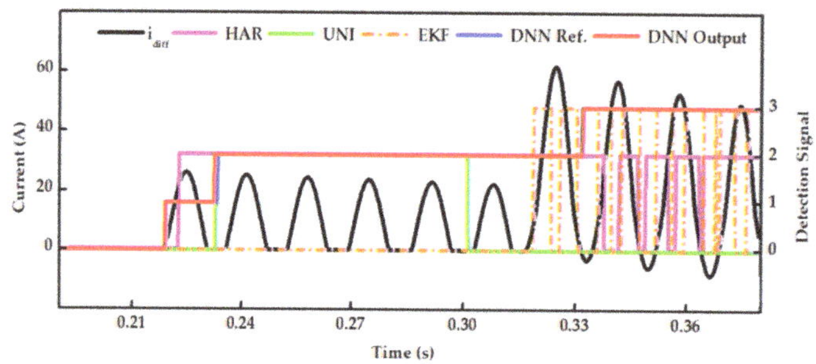

(a)

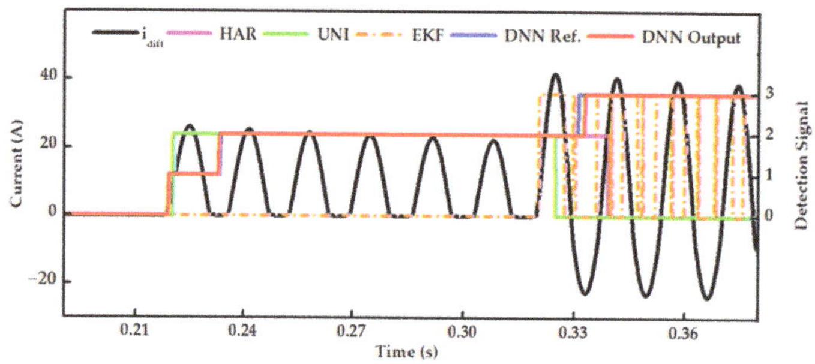

(b)

Figure 10. Results of the detection of phase-A-to-ground internal faults occurring during the energization of a power transformer: (**a**) fault inception angle of 0° and (**b**) fault inception angle of 90°.

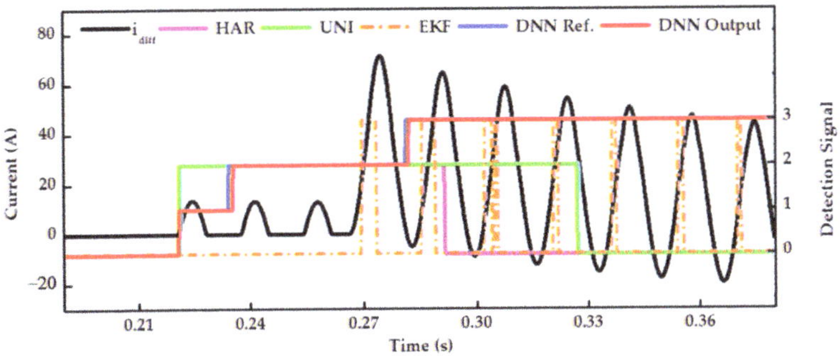

Figure 11. Results of the detection of a phase-B–C-to-ground internal faults occurring during the energization of a power transformer.

5.1. Case Study 1: Inrush Current at a Switching Angle of 0°

Magnetizing inrush currents are generated due to the remanent magnetism and no-load closing of a power transformer. The closing instance significantly influences the waveform characteristics of the inrush current, while its remanent magnetism mainly affects its amplitude.

Transformer energization cases without and with residual flux are studied in this section. Figure 6 shows the results when a transformer without residual flux was energized at a switching instance of 0°, corresponding to 0.2 s. As the ratio of the second harmonic sharply increased at the closing instance, the HAR was theoretically effective in quickly detecting the inrush current. The UNI detected the inrush current after a timing delay due to the data window, while the proposed DNN detected it at 0.231 s, with a slightly quicker response than the DNN reference. Based on Figure 6, it is evident that the proposed DNN presented a promising output in noticing the inrush current after sample N–6, which was comparable to the HAR and UNI.

The performance of the proposed DNN was also evaluated considering transformer energization with the maximum residual flux, which was approximately 80%. The amount of residual flux heavily influenced the magnitude of the inrush current; as a result, the magnitude of the inrush current nearly doubled in this case, as demonstrated in Figure 7. It can be seen that the HAR yielded the best output among the three approaches in this case. Considering a time delay, the UNI responded to the inrush current at 0.234 s, whereas the DNN demonstrated a quicker detection instance than the UNI. For instance, the DNN detected inrush currents faster (one sample) and more accurately than the UNI.

5.2. Case Study 2: Inrush Current at a Switching Angle of 90°

Switching a power transformer at 90° with no residual flux does not impact the operation of conventional differential relays and produces the least inrush currents. However, the maximum flux in the power transformer strongly influences the nonlinear nature of the magnetizing inrush current, as depicted in Figure 8. The magnitude of the inrush current in this case is similar to that depicted in Figure 6. Therefore, the detection of the inrush current was examined at the maximum switching angle and with residual flux. As displayed in Figure 8, the HAR showed the most promising outcome, as it reacted to the first instance of an inrush current due to the presence of the second harmonic ratio. Due to the data window used in the UNI and DNN, their detections showed a timing delay of less than 1 cycle. In particularly, the DNN yielded a more promising outcome than the UNI, as it was 8 samples quicker. That is, the DNN faultlessly detected the inrush current after the 61st (=N–3) sample from the switching instance.

5.3. Case Study 3: Energization of a Power Transformer in the Presence of an Internal Fault

Energizing a power transformer in the presence of an internal fault is a challenging task for conventional protections, as the ratio of the second harmonic may cause the differential relay to be blocked, potentially leading to severe damage to the power transformer. In this case, we consider a–g faults for internal faults. Figure 9 shows the results of internal-fault detection when a power transformer was energized in the presence of an internal fault. The evaluation was conducted in two different scenarios at fault inception angles of 0° and 90°.

As shown in Figure 9a, the conventional HAR method detected the inrush current rather than the internal fault due to the presence of the second harmonic in the decaying DC component generated during the internal fault. Consequently, it prevented the internal fault from being detected, resulting in the blocking of the differential relay operation. In contrast, the UNI detected the differential current as an inrush current instead of an internal fault. The EKF could not discriminate the internal fault from the inrush current. Moreover, the inaccuracy increased as the EKF estimated differential currents with noise. Unlike the conventional HAR and UNI methods, the proposed DNN demonstrated an impressive success rate in discriminating the internal fault from the inrush current after the 58th sample from the abnormality. In this manner, the DNN exhibited high sensitivity to internal faults, even though the HAR and UNI failed to detect them. As shown in Figure 9a, for the fault inception angle of 90°, the HAR failed to detect the internal fault for several cycles, highlighting a drawback of using HARs in modern transformers. In contrast, the proposed DNN successfully detected the internal fault, starting from just one sample later than the DNN reference. Similarly, as illustrated in Figure 9b, the DNN exhibited a promising output in discriminating between inrush currents and internal faults at a fault inception angle of 0°.

5.4. Case Study 4: Phase-A-to-Ground Internal Faults Occurring during the Energization of a Power Transformer

The proposed DNN was validated during an internal fault occurring a few cycles after the switching of a power transformer. The harmonic-blocking scheme blocked the operation of the differential relay due to the large ratio of the second harmonic at the onset of an internal fault. This could lead to damage to the power transformer and should be avoided.

A power transformer was switched on for energization at 0.22 s, and the internal fault occurred at 0.32 s, as demonstrated in Figure 10. With the interference of the internal fault, the HAR showed unsatisfactory results as soon as the internal fault occurred. The HAR blocked the differential relay from operating for around two cycles, which could negatively affect the power transformer. The UNI showed the worst results among the three methods, as it did not respond to the internal fault in this case. UNI is only applicable when there is a direction of the waveform on the positive or negative side, as its bidirectional index makes it vulnerable to internal faults. The proposed DNN could detect the internal fault with a time delay of less than one cycle from the fault inception. The evaluation was performed on internal faults at fault inception angles of 0° and 90°, as illustrated in Figure 10a,b, respectively. The results show that the proposed DNN can detect internal faults after a time delay of less than one cycle, regardless of the fault inception angle.

The influence of external faults on the proposed DNN can be ignored since the differential current will be zero during an external fault. Therefore, the DNN bypasses external faults and allows relevant protection schemes outside the protection zone to operate based on disturbance criteria.

5.5. Case Study 5: Phase-B–C-to-Ground Internal Faults Occurring during the Energization of a Power Transformer

To demonstrate the capability of the proposed DNN across different fault types, phase-B–C-to-ground internal faults are considered in this case. Figure 11 presents a case of a phase-B–C-to-ground internal fault at a different time node considering a fault inception angle of 0°. The internal fault depicted in Figure 11 occurs three cycles after the inrush current takes place. Similar to Case Study 4, the UNI successfully detects the instance of the inrush current; however, the operation of the differential protection is continually blocked for almost one cycle after an internal fault occurs. On the other hand, the UNI proves to be effective in responding to the inrush current but fails to detect the internal fault for several cycles. The EKF exhibits low sensitivity to the internal fault because the estimated current from the EKF produces noise. Unlike these three methods, the proposed DNN demonstrates an accurate and reliable output in discriminating internal faults with a given time delay.

6. Discussion on the Performance Evaluation Metrics

To effectively evaluate the performance of the proposed DNN, three indicators were selected as evaluation metrics: accuracy, sensitivity, and precision. Traditionally, accuracy alone is insufficient to determine whether the proposed DNN yields a promising outcome. To visualize the stability of the proposed DNN method, a confusion matrix was used, summarizing the classification performance and providing a visual representation of the actual and predicted classes. The evaluation matrix was assessed using the following four performance indices: *TP* (true positive), *TN* (true negative), *FP* (false positive), and *FN* (false negative).

$$ACC = \frac{TP+TF}{(TP + TF + FP + FN)} \tag{9}$$

$$SEN = \frac{TP}{(TP + FN)} \tag{10}$$

$$PRE = \frac{TP}{(TP + FP)} \tag{11}$$

Conventionally, accuracy (*ACC*) shows the authenticity of a detection method, defining the correct detections over the total numbers of detections, including correct and false ones. Sensitivity (*SEN*) measures the proportion of inrush and internal faults that were correctly identified among the actual labels. It is a crucial metric in discrimination, because it influences the decision to allow the differential relay to operate when an internal fault occurs during inrush currents. A high percentage of *SEN*s is essential to determine the stability of the proposed DNN. Precision (*PRE*) is another important metric required to affirm the correctness of the proposed DNN. For instance, it demonstrates the capability of the proposed DNN to isolate internal faults from inrush currents when both abnormalities occur simultaneously. In other words, it demonstrates the ability of internal-fault detection without mistakenly identifying it as an inrush current. A comparative analysis was conducted, and the evaluation metrics are presented in Table 6. The effectiveness of these metrics was assessed at the 58th and 61st samples from the beginning of each abnormality.

In cases where a power transformer is energized in the presence of an internal fault, the aim is to avoid a situation where the DNN mistakenly detects it as an inrush current instead of an internal fault. Therefore, the DNN places emphasis on minimizing FNs; otherwise, incorrect detections could lead to damage to the power transformer. The DNN detects the internal fault at the 61st sample, which is three samples later than the DNN reference; therefore, the DNN experienced three *FN*s in this case. The performance of the proposed DNN and the other methods was evaluated at the 58th (=N–6) and 61st (=N–3) samples from the beginning of each abnormality. It is noted that detection with a time delay of 61 samples will be sufficient to protect the power transformer, as the protection decision will be made after 64 samples.

According to the percentages presented in Table 6, it is evident that all four methods correctly classified the normal condition from the other two abnormalities without any defects. For inrush conditions, the HAR was undoubtedly proven to be effective, achieving the highest metrics at the 58th and 61st samples. The UNI exhibited good performance in detecting inrush currents, with *ACC*, *SEN*, and *PRE* values of 99.852%, 93.814%, and

95.724%, respectively. The UNI is unable to achieve the highest metric at the 61st sample, as inrush currents were detected at the 62nd sample in some cases. On the other hand, the UNI performed poorly when experiencing internal faults, as it was more sensitive to inrush currents. The DNN displayed a promising evaluation index in detecting the inrush duration at the 58th sample, yielding the highest *ACC*, *SEN*, and *PRE* values of 99.526%, 100%, and 99.523%, respectively. At the 61st sample, the DNN could accurately classify between inrush currents and internal faults, achieving 100% for all three metrics.

Table 6. Evaluation metrics at the 58th (=*N*–6) and 61st (=*N*–3) samples from the beginning of each abnormality.

Method	Case	Accuracy (%)		Sensitivity (%)		Precision (%)	
		N–6	*N*–3	*N*–6	*N*–3	*N*–6	*N*–3
HAR	Normal	100	100	100	100	100	100
UNI		100	100	100	100	100	100
EKF		100	100	100	100	100	100
DNN		100	100	100	100	100	100
HAR	Inrush	100	100	100	100	100	100
UNI		99.852	99.932	93.814	95.613	95.724	97.741
EKF		-	-	-	-	-	-
DNN		99.526	100	100	100	99.523	100
HAR	Internal fault	-	-	-	-	-	-
UNI		-	-	-	-	-	-
EKF		90.513	92.364	69.192	72.951	71.231	72.367
DNN		99.931	100	100	100	98.842	100
HAR	Inrush and Internal fault	-	-	-	-	-	-
UNI		-	-	-	-	-	-
EKF		91.103	92.136	69.583	71.369	70.124	70.364
DNN		99.651	100	99.642	100	100	100

Furthermore, the DNN demonstrates excellent performance in detecting internal faults during inrush currents. The evaluation index produced by the DNN outperformed the other three methods at sample *N*–6, achieving *ACC*, *SEN*, and *PRE* values of 99.651%, 99.642%, and 100%, respectively. At sample *N*–3, the DNN achieved the best metrics (*ACC*, *SEN*, and *PRE*), all at 100%. In contrast, the EKF showed worse performance compared to the DNN in this study, as it mis-detected the internal faults due to the difference between the measured and estimated currents. Moreover, EKF is inapplicable to other systems and significantly relies on a threshold to detect internal faults, presenting a less favorable discrimination between inrush currents and internal faults. At sample *N*–3, it yielded *ACC*, *SEN*, and *PRE* values of 92.136%, 71.369%, and 70.364%, respectively.

7. Conclusions

This paper proposes a DNN-based method to discriminate between inrush currents and internal faults utilizing a data window. The effectiveness of the proposed DNN was assessed through numerical simulations, including inrush currents, internal faults, and cases where the inrush current coincided with internal faults. Despite achieving less accurate results during inrush currents, compared to HAR, DNN performs better in detecting internal faults, even during inrush conditions. Based on graphical illustrations and evaluation metrics, DNN successfully detects internal faults during inrush conditions, enabling the differential relay to operate without delay, regardless of the fault inception angle and residual flux. As DNN does not require a specific threshold to perform the discrimination, it can be applied to different systems to discriminate inrush currents from internal faults.

HAR and UNI are insufficient to deal with both inrush currents and internal faults occurring together. Although EKF can detect internal faults, the effectiveness of EKF is reduced in other systems due to an indecisive threshold. The deficiencies of the prevailing methods, such as reliance on physical parameters and indecisive predefined thresholds, decrease their reliability and generality. In comparison to prevailing methods (HAR, UNI, and EKF), the proposed DNN shows promising results from sample *N*–3, achieving accuracy, sensitivity, and precision values of 100%. It is considered to be one of the promising solutions for discriminating between inrush currents and internal faults. The proposed DNN may produce errors in the presence of CT saturation. Our future work involves developing a discrimination model for the main and backup protections that considers CT saturation and implementing the proposed DNN to discriminate internal faults from inrush currents in real time. The experiment will be based on hardware implementation, which consists of RTDS and EVM boards.

Author Contributions: Conceptualization, S.K. and S.-R.N.; methodology, S.K., G.-W.S. and S.-R.N.; supervision, S.-R.N.; validation, S.K. and G.-W.S.; writing—original draft, S.K. and S.-R.N.; writing—review and editing, S.K., G.-W.S. and S.-R.N. All authors have read and agreed to the published version of the manuscript.

Funding: This research was supported in part by Korea Electric Power Corporation (Grant number: R23XO05-07). This research was also supported in part by the Korea Research Foundation, with funding from the government (Ministry of Education) in 2021 (No. NRF-2021R1F1A1061798).

Data Availability Statement: Data are contained within the article.

Conflicts of Interest: The authors declare no conflicts of interest.

References

1. Gopika, R.; Deepa, S. Study on Power Transformer Inrush Current. *IOSR J. Electr. Electron. Eng.* **2017**, *2*, 59–63.
2. Baoming, G.; Dealmeida, A.; Qionglin, Z.; Xiangheng, W. An Equivalent Instantaneous Inductance-Based Technique for Discrimination between Inrush Current and Internal Faults in Power Transformers. *IEEE Trans. Power Deliv.* **2005**, *20*, 2473–2482. [CrossRef]
3. Guzman, A.; Fischer, N.; Labuschagne, C. Improvements in transformer protection and control. In Proceedings of the 2009 62nd Annual Conference for Protective Relay Engineers, College Station, TX, USA, 30 March–2 April 2009; pp. 563–579.
4. Hamilton, R. Analysis of transformer inrush current and comparison of harmonic restraint methods in transformer protection. *IEEE Trans. Ind. Appl.* **2013**, *49*, 1890–1899. [CrossRef]
5. Zhang, L.; Wu, Q.; Ji, T.; Zhang, A. Identification of inrush currents in power transformers based on higher-order statistics. *Electr. Power Syst. Res.* **2017**, *146*, 161–169. [CrossRef]
6. Mo, C.; Ji, T.; Zhang, L.; Wu, Q. Equivalent statistics based inrush identification method for differential protection of power transformer. *Electr. Power Syst. Res.* **2022**, *203*, 107664. [CrossRef]
7. Hodder, S.; Kasztenny, B.; Fischer, N.; Xia, Y. Low second-harmonic content in transformer inrush currents—Analysis and practical solutions for protection security. In Proceedings of the 2014 67th Annual Conference for Protective Relay Engineers, College Station, TX, USA, 31 March–3 April 2014; pp. 705–722.
8. Lu, Z.; Tang, W.H.; Ji, T.Y.; Wu, Q.H. A Morphological Scheme for Inrush Identification in Transformer Protection. *IEEE Trans. Power Deliv.* **2009**, *24*, 560–568. [CrossRef]
9. Vazquez, E.; Mijares, I.I.; Chacon, O.L.; Conde, A. Transformer differential protection using principal component analysis. *IEEE Trans. Power Deliv.* **2008**, *23*, 67–72. [CrossRef]
10. Afrasiabi, S.; Afrasiabi, M.; Parang, B.; Mohammadi, M.; Samet, H.; Dragicevic, T. Fast GRNN-Based Method for Distinguishing Inrush Currents in Power Transformers. *IEEE Trans. Ind. Electron.* **2021**, *69*, 8501–8512. [CrossRef]
11. Afrasiabi, S.; Afrasiabi, M.; Parang, B.; Mohammadi, M. Integration of Accelerated Deep Neural Network into Power Transformer Differential Protection. *IEEE Trans. Ind. Inform.* **2019**, *16*, 865–876. [CrossRef]
12. Samet, H.; Ghanbari, T.; Ahmadi, M. An Auto-Correlation Function Based Technique for Discrimination of Internal Fault and Magnetizing Inrush Current in Power Transformers. *Electr. Power Compon. Syst.* **2015**, *43*, 399–411. [CrossRef]
13. Mao, P.; Aggarwal, R. A novel approach to the classification of the transient phenomena in power transformers using combined wavelet transform and neural network. *IEEE Trans. Power Deliv.* **2001**, *16*, 654–660. [CrossRef]
14. Key, S.; Ko, C.-S.; Song, K.-J.; Nam, S.-R. Fast Detection of Current Transformer Saturation Using Stacked Denoising Autoencoders. *Energies* **2023**, *16*, 1528. [CrossRef]
15. Key, S.; Kang, S.-H.; Lee, N.-H.; Nam, S.-R. Bayesian Deep Neural Network to Compensate for Current Transformer Saturation. *IEEE Access* **2021**, *9*, 154731–154739. [CrossRef]

16. Sok, V.; Lee, S.-W.; Kang, S.-H.; Nam, S.-R. Deep Neural Network-Based Removal of a Decaying DC Offset in Less Than One Cycle for Digital Relaying. *Energies* **2022**, *15*, 2644. [CrossRef]
17. Silva, K.; Souza, B.; Brito, N. Fault Detection and Classification in Transmission Lines Based on Wavelet Transform and ANN. *IEEE Trans. Power Deliv.* **2006**, *21*, 2058–2063. [CrossRef]
18. Costa, F.B.; de Souza, B.A.; Brito, N.S.D. Detection and Classification of Transient Disturbances in Power Systems. *IEEJ Trans. Power Energy* **2010**, *130*, 910–916. [CrossRef]
19. Gaouda, A.M.; Salama, M.M.A. DSP Wavelet-Based Tool for Monitoring Transformer Inrush Currents and Internal Faults. *IEEE Trans. Power Deliv.* **2010**, *25*, 1258–1267. [CrossRef]
20. Saleh, S.A.; Scaplen, B.; Rahman, M.A. A new implementation method of wavelet-packet-transform differential protection for power transformers. *IEEE Trans. Ind. Appl.* **2011**, *47*, 1003–1012. [CrossRef]
21. Costa, F.B. Fault-induced transient detection based on real-time analysis of the wavelet coefficient energy. *IEEE Trans. Power Deliv.* **2014**, *29*, 140–153. [CrossRef]
22. Medeiros, R.P.; Costa, F.B.; Silva, K.M. Power Transformer Differential Protection Using the Boundary Discrete Wavelet Transform. *IEEE Trans. Power Deliv.* **2015**, *31*, 2083–2095. [CrossRef]
23. Medeiros, R.P.; Costa, F.B. A Wavelet-Based Transformer Differential Protection with Differential Current Transformer Saturation and Cross-Country Fault Detection. *IEEE Trans. Power Deliv.* **2017**, *33*, 789–799. [CrossRef]
24. Medeiros, R.P.; Costa, F.B. A wavelet-based transformer differential protection: Internal fault detection during inrush conditions. *IEEE Trans. Power Deliv.* **2018**, *33*, 2965–2977. [CrossRef]
25. Ruhan, Z.; Mansor, N.N.B.; Illias, H.A. Identification of Inrush Current Using a GSA-BP Network. *Energies* **2023**, *16*, 2340. [CrossRef]
26. Peng, F.; Gao, H.; Huang, J.; Guo, Y.; Liu, Y.; Zhang, Y. Power Differential Protection for Transformer Based on Fault Component Network. *IEEE Trans. Power Deliv.* **2023**, *38*, 2464–2477. [CrossRef]
27. Ali, E.; Helal, A.; Desouki, H.; Shebl, K.; Abdelkader, S.; Malik, O. Power transformer differential protection using current and voltage ratios. *Electr. Power Syst. Res.* **2017**, *154*, 140–150. [CrossRef]
28. Liu, P.; Jiao, B.; Zhang, P.; Du, S.; Zhu, J.; Song, Y. Countermeasure to Prevent Transformer Differential Protection from False Operations. *IEEE Access* **2023**, *11*, 45950–45960. [CrossRef]
29. He, B.; Zhang, X.; Bo, Z. A New Method to Identify Inrush Current Based on Error Estimation. *IEEE Trans. Power Deliv.* **2006**, *21*, 1163–1168. [CrossRef]
30. Elsadd, M.A.; Yousef, W.; Abdelaziz, A.Y. New adaptive coordination approach between generator-transformer unit overall differential protection and generator capability curves. *Int. J. Electr. Power Energy Syst.* **2019**, *118*, 105788. [CrossRef]
31. Rao, J.G.; Pradhan, A.K. Power-Swing Detection Using Moving Window Averaging of Current Signals. *IEEE Trans. Power Deliv.* **2014**, *30*, 368–376. [CrossRef]
32. Hinton, G.E.; Osindero, S.; Teh, Y.-W. A Fast Learning Algorithm for Deep Belief Nets. *Neural Comput.* **2006**, *18*, 1527–1554. [CrossRef]
33. Bengio, Y.; Lamblin, P.; Popovici, D.; Larochelle, H. Greedy Layer-wise Training of Deep Networks. In *Proceedings of the NIPS'06 Proceedings of the 19th International Conference on Neural Information Processing Systems, Vancouver, BC, Canada, 4–7 December 2006*; pp. 153–160.
34. Ken, B.; Normann, F.; Casper, L. Considerations for Using Harmonic Blocking and Harmonic Restraint Techniques on Transformer Differential Relays. *SEL J. Reliab. Power* **2011**, *2*, 1–17.
35. Gunda, S.K.; Dhanikonda, V.S.S.S.S. Discrimination of Transformer Inrush Currents and Internal Fault Currents Using Ex-tended Kalman Filter Algorithm (EKF). *Energies* **2021**, *14*, 6020. [CrossRef]

Disclaimer/Publisher's Note: The statements, opinions and data contained in all publications are solely those of the individual author(s) and contributor(s) and not of MDPI and/or the editor(s). MDPI and/or the editor(s) disclaim responsibility for any injury to people or property resulting from any ideas, methods, instructions or products referred to in the content.

energies

MDPI

Article

Current Sensorless Pole-Zero Cancellation Output Voltage Control for Uninterruptible Power Supply Systems with a Three-Phase Inverter

Hosik Lee [1], Yonghun Kim [2,*] and Seok-Kyoon Kim [3,*]

[1] Green Mobility Team, Tenergy, Suwon 18487, Republic of Korea; hslee@tenergy.co.kr
[2] School of Mechanical Engineering, Chungnam National University, Daejeon 34134, Republic of Korea
[3] Department of Creative Convergence Engineering, Hanbat National University, Daejeon 34158, Republic of Korea
* Correspondence: yonghun.kim@cnu.ac.kr (Y.K.); skkim77@hanbat.ac.kr (S.-K.K.)

Abstract: This article presents a proportional–derivative (PD) type output voltage regulator without the current feedback, taking into account system parameter and load variations. The main advantages are given as follows: First, the first-order output voltage derivative observer is developed without the requirement of system parameter information, which makes it possible to stabilize the system without current sensing. Second, a simple self-tuner implements the feedback-loop adaptation by updating the desired dynamics accordingly. Third, the observer-based active damping injection for the PD-type controller results in the closed-loop system order reduction to 1 by the pole-zero cancellation, including the disturbance observer as a feed-forward term. The prototype uninterruptible power supply system comprised of a 3 kW three-phase inverter, inductors, and capacitors verifies the practical merits of the proposed technique for linear and nonlinear loads.

Keywords: uninterruptible power supply; voltage control; observer; active damping; pole-zero cancellation

1. Introduction

The power conversion between AC and DC is a pivotal task for a wide range of industrial applications, such as home appliances, electric vehicles, factory automation, etc. These applications require of the power converters to secure the high performance in both transient and steady state operations. In particular, the uninterruptible power supply (UPS) systems composed of the three-phase inverter and output *LC*-filter must be carefully designed to meet the specified high-level closed-loop performance under the abrupt power failure scenario (i.e., DC to AC power conversion) [1–5].

The combination of inner-loop current control and outer-loop voltage control has been popularly adopted to regulate the output voltage with its implementation from the proportional–integral (PI) regulator, due to its simple structure and tuning process. The corresponding feedback gains were tuned via the trial-and-error, Zeigler–Nichols, Bode, and Nyquist techniques, conventionally [6,7]. The resultant closed-loop stability and performance were only valid for the fixed operating mode (i.e., given fixed load conditions). The additional gain scheduling mechanism as in [8] successfully enlarged the feasible operating region with an increased computational burden. The feed-forward terms and specific PI gains were introduced with the use of a rotational coordinate transformation to result in the first-order closed-loop transfer functions with desired cut-off frequencies for each loop via pole-zero cancellation (PZC), which required the true system parameters, such as resistance, inductance, and capacitance [9,10]. Novel online identifiers can be used as the solution to the parameter dependence problem [11–13]. The deadbeat-type control provides rapid regulation performance, but is sensitive to high measurement noise sensitivity due to its exact model dependence [14]. This practical limitation was systematically handled by the parameter-dependent observer-based deadbeat controllers [15,16]. The optimal state-feedback controls such as H_∞ and μ-synthesis techniques solved the output voltage control problem considering the external disturbances from the load failure [17,18]. The sliding-mode technique with the conservative discontinuous sign function provides the stabilization and performance recovery property by ensuring the reaching property to the sliding surface [19]. The model predictive control (MPC) generates the optimal duty command for the three-phase inverter considering the input and state constraints with the requirement of parameter dependence and high computational burden [20–22]. To address the computational burden issue, explicit MPC techniques were presented, involving huge offline tasks for partitioning the state space properly; however, this approach could lead to computational burdens from the significant online membership tests [23,24]. To avoid the online optimization and membership tests, the one-step MPC control with full-state feedback was devised, including the disturbance observer (DOB) to replace the regulation error integrators and improving the closed-loop robustness against the load variations [25]. The system parameter, load variations, and one-step time-delay were explicitly considered using a robust-state feedback design technique incorporating regulation error integrators and LMI optimization [26]. Under a similar strategy, the DOB and one-step-ahead state predictor were included for the one-step MPC and finite control set MPC (FCSMPC) as the optimization solution to the output voltage regulation problem with parameter uncertainties [27–29]. There are two practical challenges in the previous results, the parameter dependence of controls and state-observers, at least partially, and fixed closed-loop cut-off frequency, to be handled in this study; it is desirable for the cut-off frequency to be decreased in the steady-state and to be increased in the transient periods.

The aforementioned literature survey identifies the following practical challenges that need to be addressed: (P1) performance inconsistency for load variations, (P2) limited current measurement, (P3) fixed closed-loop bandwidth, and (P4) system model dependence. The proposed solution forms the single-loop feedback structure without the current loop, independent of the exact model information for both the observer and controller. The contributions of the proposed technique are summarized as follows.

- For (P2) and (P4): The output voltage derivative observer is built by combining the Luenberger observer and DOB design techniques without requiring the true system parameter information; this eliminates the need of current feedback.
- For (P3): The online self-tuner adjusts the desired dynamics to implement the feedback-loop adaptation according to the transient and steady-state operation modes.
- For (P1) and (P4): The injection of the active damping term and special form of feedback gain for the DOB-based proportional-derivative (PD) controller result in the stable PZC to render the closed-loop system order to be 1.

A prototype 3 kW UPS system was used to experimentally confirm the effectiveness of the proposed technique under the resistive and rectifier loads, deriving from the beneficial closed-loop properties.

There are five sections to organize this paper. Section 2 introduces the servo motor model with the strategy of handling the parameter and load variations. Section 3 describes the design purpose and presents the proposed solution. Section 4 proves the accomplishment of the design purpose by analyzing the closed-loop dynamics. Section 5 experimentally validates the effectiveness of the proposed solution. Section 6 concludes this paper.

2. Uninterruptible Power Supply System Dynamics

The UPS systems are comprised of a three-phase inverter, inductor, and output capacitor, as shown in Figure 1. The application of Kirchhoff's voltage law to the UPS systems leads to the linear time-invariant system (LTI):

$$C\dot{\mathbf{v}}_{c,abc} = \mathbf{i}_{abc} - \mathbf{i}_{load,abc}, \quad (1)$$

$$L\dot{\mathbf{i}}_{abc} = \mathbf{u}_{abc} - R\mathbf{i}_{abc} - \mathbf{v}_{c,abc}, \ \forall t \geq 0, \quad (2)$$

with the inductor current $\mathbf{i}_{abc} = \begin{bmatrix} i_a & i_b & i_c \end{bmatrix}^T$, capacitor voltage $\mathbf{v}_{c,abc} = \begin{bmatrix} v_{c,a} & v_{c,b} & v_{c,c} \end{bmatrix}^T$, and control signal $\mathbf{u}_{abc} = \begin{bmatrix} u_a & u_b & u_c \end{bmatrix}^T$ defined as

$$\mathbf{u}_{abc} = \frac{1}{6}\begin{bmatrix} 2 & -1 & -1 \\ -1 & 2 & -1 \\ -1 & -1 & 2 \end{bmatrix}\begin{bmatrix} p_a \\ p_b \\ p_c \end{bmatrix} V_{dc}, \tag{3}$$
$$p_x = \begin{cases} 1 \text{ if } S_x : \text{ON and } \bar{S}_x : \text{OFF}, \\ -1 \text{ if } S_x : \text{OFF and } \bar{S}_x : \text{ON}, \end{cases}$$

where $x = a, b, c$. The load current is denoted as $\mathbf{i}_{load,abc} = \begin{bmatrix} i_{load,a} & i_{load,b} & i_{load,c} \end{bmatrix}^T$, to be treated as the unknown disturbance due to the load uncertainties. It should be noted that the capacitor voltage reference is normally given by the three-phase balanced form $\mathbf{v}_{c,abc,ref} = r\begin{bmatrix} \cos(\omega_r t) & \cos(\omega_r - \frac{2\pi}{3}) & \cos(\omega_r t - \frac{4\pi}{3}) \end{bmatrix}^T$ for some $r > 0$ and $\omega_r = 2\pi f$ rad/s.

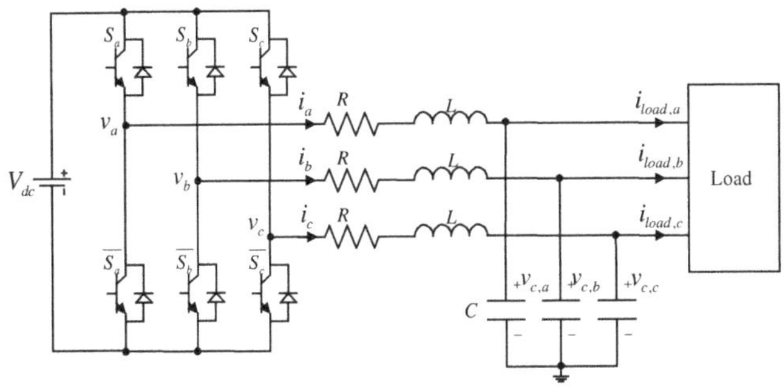

Figure 1. The UPS system topology.

The alternative system behaviors (1) and (2) make the tracking controller design task difficult. The coordinate transformation defined as

$$\mathbf{W}(\theta_r) := \begin{bmatrix} \cos(\theta_r) & \cos(\theta_r - \frac{2\pi}{3}) & \cos(\theta_r - \frac{4\pi}{3}) \\ -\sin(\theta_r) & -\sin(\theta_r - \frac{2\pi}{3}) & -\sin(\theta_r - \frac{4\pi}{3}) \end{bmatrix},$$

synchronized to the phase angle $\theta_r := \omega_r t$ with desired frequency ω_r, eliminates the alternative component of system dynamics as

$$C\dot{\mathbf{v}}_c = \mathbf{i} + C\omega_r \mathbf{J}\mathbf{v}_c - \mathbf{i}_{load}, \tag{4}$$
$$L\dot{\mathbf{i}} = (-R\mathbf{I} + L\omega_r\mathbf{J})\mathbf{i} - \mathbf{v}_c + \mathbf{u},\ \forall t \geq 0, \tag{5}$$

through the calculations resulting in the d-q axis variables $\mathbf{i} = \begin{bmatrix} i_d \\ i_q \end{bmatrix}$, $\mathbf{v}_c = \begin{bmatrix} v_{c,d} \\ v_{c,q} \end{bmatrix}$, $\mathbf{i}_{load} = \begin{bmatrix} i_{load,d} \\ i_{load,q} \end{bmatrix}$, and $\mathbf{u} = \begin{bmatrix} u_d \\ u_q \end{bmatrix}$, such that

$$\mathbf{i} = \frac{2}{3}\mathbf{W}(\theta_r)\mathbf{i}_{abc},\ \mathbf{v}_c = \frac{2}{3}\mathbf{W}(\theta_r)\mathbf{v}_{c,abc},\ \mathbf{i}_{load} = \frac{2}{3}\mathbf{W}(\theta_r)\mathbf{i}_{load,abc},\ \text{and } \mathbf{u} = \frac{2}{3}\mathbf{W}(\theta_r)\mathbf{u}_{abc},$$

$\forall t \geq 0$. The two symbols $\mathbf{I}$ and $\mathbf{J}$ denote the 2×2-dimensional identity and skew-symmetric matrices, respectively, which are $\mathbf{I} = \begin{bmatrix} 1 & 0 \\ 0 & 1 \end{bmatrix}$ and $\mathbf{J} = \begin{bmatrix} 0 & 1 \\ -1 & 0 \end{bmatrix}$.

The passive component (capacitance and inductance) values C and L can be dramatically changed depending on the magnitude of current and voltage, which are assumed to be unknown, but their nominal values C_0 and L_0 are available from the manufacturers. To remove the true system parameter dependence, this study introduces the nominal parameters C_0 and L_0 to systems (4) and (5), yielding the modified system (5) such that

$$C_0\dot{\mathbf{v}}_c = \mathbf{i} + C_0\omega_r\mathbf{J}\mathbf{v}_c + \mathbf{d}_{v,o}, \quad (6)$$
$$L_0\dot{\mathbf{i}} = (-R_0\mathbf{I} + L_0\omega_r\mathbf{J})\mathbf{i} - \mathbf{v}_c + \mathbf{u} + \mathbf{d}_{i,o}, \ \forall t \geq 0, \quad (7)$$

with the unknown lumped disturbances $\mathbf{d}_{v,o}$ and $\mathbf{d}_{i,o}$ modeling the parameter and load variations, which are used for the following controller design and closed-loop analysis sections.

3. Proposed Control Algorithm

The control objective is to render the output voltage $\mathbf{v}_c$ to exponentially behave as the target trajectory $\mathbf{v}_{c,des,0} = \begin{bmatrix} v_{c,d,des,0} & v_{c,q,des,0} \end{bmatrix}^T$, driven by given reference $v_{c,ref} = \begin{bmatrix} v_{c,d,ref} & v_{c,q,ref} \end{bmatrix}^T$, satisfying the desired first-order system

$$\dot{\mathbf{v}}_{c,des,0} = \omega_{v_c}(\mathbf{v}_{c,ref} - \mathbf{v}_{c,des,0}), \ \forall t \geq 0, \quad (8)$$

with the closed-loop specification $\omega_{v_c} > 0$ as the convergence rate of system (8). The next subsection presents the self-tuner through a modification of the desired system (8) to implement the feedback-loop adaptation.

Remark 1. *Taking the Laplace transform to desired system* (8) *(e.g.,* $V_{c,d,des,0}(s) = \mathcal{L}\{v_{c,d,des,0}\}$, $V_{c,q,des,0}(s) = \mathcal{L}\{v_{c,q,des,0}\}$, $V_{c,d,ref}(s) = \mathcal{L}\{v_{c,d,ref}\}$, *and* $V_{c,q,ref}(s) = \mathcal{L}\{v_{c,q,ref}\}$*) derives the transfer function forming the low-pass filter (LPF):*

$$\frac{V_{c,x,des,0}(s)}{V_{c,x,ref}(s)} = \frac{\omega_{vc}}{s + \omega_{vc}}, \ x = d, q, \ \forall s \in \mathbb{C}, \quad (9)$$

with the cut-off frequency ω_{vc} *(rad/s,* $\frac{\omega_{vc}}{2\pi}$ *Hz). Therefore, the closed-loop specification* ω_{v_c} *can be determined as the cut-off frequency of the LPF* (9) *from the reference to the desired output.*

3.1. Self-Tuner

To implement the feedback-loop adaptation, this study slightly modifies the desired system (8) as $\mathbf{v}_{c,des} = \mathbf{v}_{c,des,0}\Big|_{\omega_{v_c} = \hat{\omega}_{v_c}}$, yielding

$$\dot{\mathbf{v}}_{c,des} = \hat{\omega}_{v_c}(\mathbf{v}_{c,ref} - \mathbf{v}_{c,des}), \ \forall t \geq 0, \quad (10)$$

with the proposed self-tuner driving the cut-off frequency $\hat{\omega}_{v_c}$ initiated from $\hat{\omega}_{v_c}(0) = \omega_{v_c}$ as

$$\dot{\hat{\omega}}_{v_c} = \gamma_{v_c}(\|\tilde{\mathbf{v}}_{c,des}\|^2 + \rho_{v_c}\tilde{\omega}_{v_c}), \ \forall t \geq 0, \quad (11)$$

with errors defined as $\tilde{\mathbf{v}}_{c,des} := \mathbf{v}_{c,ref} - \mathbf{v}_{c,des}$, and $\tilde{\omega}_{v_c} := \omega_{v_c} - \hat{\omega}_{v_c}$ and tuning factors $\gamma_{v_c} > 0$ and $\rho_{v_c} > 0$. The self-tuner (11) automatically adjusts the cut-off frequency $\hat{\omega}_{v_c}$ to increase the error decay ratio of $\tilde{\mathbf{v}}_{c,des}$ with the boundedness property of $\hat{\omega}_{v_c} \geq \omega_{v_c}, \forall t \geq 0$. For details, see Section 4. The next issue is designing an observer for the output voltage derivative, which makes it available to inject the active damping term and to feedback the regulation error derivatives.

3.2. Observer

It is undesirable to apply the time derivative operation to the output voltage measurement $\mathbf{v}_c$ to make the signal $\mathbf{a}_c := \dot{\mathbf{v}}_c$ available for feedback, due to the high-frequency measurement noise. To avoid this problem and system parameter dependence, the observer for $\mathbf{a}_c$ is proposed as

$$\dot{\hat{\mathbf{v}}}_c = k_{obs}\mathbf{e}_{v_c} + \hat{\mathbf{a}}_c, \quad (12)$$
$$\dot{\mathbf{z}}_{a_c} = -l_{a_c}\mathbf{z}_{a_c} - l_{a_c}^2\mathbf{e}_{v_c} + l_{a_c}(\hat{\mathbf{a}}_c + k_{obs}\mathbf{e}_{v_c}), \quad (13)$$
$$\hat{\mathbf{a}}_c = \mathbf{z}_{a_c} + l_{a_c}\mathbf{e}_{v_c}, \ \forall t \geq 0, \quad (14)$$

with the estimation error $\mathbf{e}_{v_c} := \mathbf{v}_c - \hat{\mathbf{v}}_c$ and observer gains $k_{obs} > 0$ and $l_{a_c} > 0$. The proposed observer of (12)–(14) drives the estimate $\hat{\mathbf{a}}_c$ to tend to their actual measurement of $\dot{\mathbf{v}}_c$, exponentially, without any use of plant parameters, which corresponds to a feature of this study, unlike the conventional Luenberger observers. The first part (12) is devised by the standard Luenberger observer design technique for the plant dynamics $\dot{\mathbf{v}}_c = \mathbf{a}_c$. The remaining parts (13) and (14) come from the DOB design technique using the observer error dynamics for (12). The estimation error convergence property is provided in Section 4.

3.3. Control Law

To remove the current-loop dependence, this study applies the additional derivative to system (6) to obtain the second-order open-loop system given by

$$\begin{aligned} L_0C_0\ddot{\mathbf{v}}_c &= L_0\dot{\mathbf{i}} + L_0C_0\omega_r\mathbf{J}\dot{\mathbf{v}}_c + L_0\dot{\mathbf{d}}_{v,o} \\ &= (L_0C_0\omega_r^2\mathbf{J}^2 - \mathbf{I})\mathbf{v}_c + \mathbf{u} + \mathbf{d}_v, \ \forall t \geq 0, \end{aligned} \quad (15)$$

with the lumped disturbance defined as $\mathbf{d}_v := L_0\omega_r\mathbf{J}(\mathbf{i} + \mathbf{d}_{v,o}) + (-R_0\mathbf{I} + L_0\omega_r\mathbf{J})\mathbf{i} + \mathbf{d}_{i,o} + L_0\dot{\mathbf{d}}_{v,o}$, where the current dynamics (7) is used for the second equation above. Due to the unavailability of voltage derivative, the observer-based PD-type control law is proposed by involving the estimated variable $\hat{\mathbf{a}}_c$, such that

$$\mathbf{u} = -k_{vc}\hat{\mathbf{a}}_c + L_0C_0\lambda_{vc}(\dot{\mathbf{v}}_{c,des} - \hat{\mathbf{a}}_c) + k_{vc}\lambda_{vc}\tilde{\mathbf{v}}_c - \hat{\mathbf{d}}_v - (L_0C_0\omega_r^2\mathbf{J}^2 - \mathbf{I})\mathbf{v}_c, \ \forall t \geq 0, \quad (16)$$

for the error $\tilde{\mathbf{v}}_c := \mathbf{v}_{c,des} - \mathbf{v}_c$, active damping coefficient $k_{vc} > 0$, feedback gain $\lambda_{vc} > 0$, and observer-based DOB are given by

$$\dot{\mathbf{z}}_v = -l_v\mathbf{z}_v - l_v^2L_0C_0\hat{\mathbf{a}}_c - l_v((L_0C_0\omega_r^2\mathbf{J}^2 - \mathbf{I})\mathbf{v}_c + \mathbf{u}), \quad (17)$$
$$\hat{\mathbf{d}}_v = \mathbf{z}_v + l_vL_0C_0\hat{\mathbf{a}}_c, \ \forall t \geq 0, \quad (18)$$

with the gain $l_v > 0$. It should be noted that the DOB (17) and (18) ensures the convergence of $\hat{\mathbf{d}}_\infty \to \mathbf{d}_\infty$ ($\lim_{t\to\infty} \mathbf{f} = \mathbf{f}_\infty$ for any function $\mathbf{f}$), and that the combination of active damping coefficient k_{vc} and PD gains $k_{vc}\omega_{vc}$ and $L_0C_0\omega_{vc}$ results in the closed-loop system order reduction to 1 by the PZC. For details, see Section 4. Figure 2 shows the control system structure.

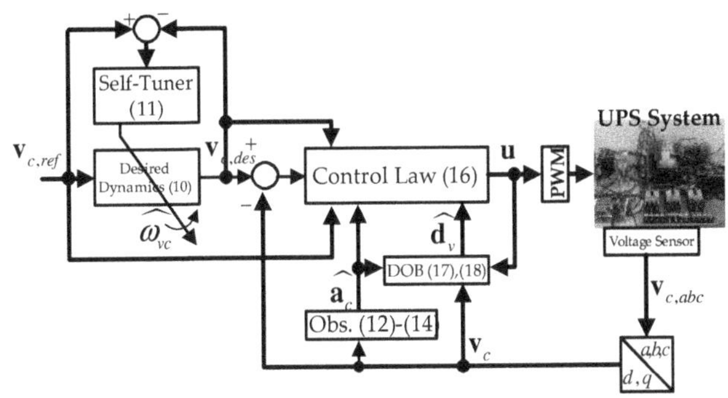

Figure 2. Proposed control system structure.

4. Analysis

This section proves that the proposed control system depicted in Figure 2 accomplishes the control objective (10) without offset errors, by ensuring the performance recovery property. First, Section 4.1 derives the properties of proposed self-tuner (11) to make the closed-loop analysis process simple.

4.1. Self-Tuner

The stability of the desired system (10) remains questionable due to the time-varying nature of $\hat{\omega}_{vc}$ driven by the self-tuner (11). Lemma 1 addresses this issue by analyzing the two dynamics (10) and (11).

Lemma 1. *The cut-off frequency $\hat{\omega}_{v_c}$ from the proposed self-tuner* (11) *ensures that*

$$\lim_{t\to\infty} \mathbf{v}_{c,des} = \mathbf{v}_{c,ref},$$

as $\dot{\mathbf{v}}_{c,ref} \to \mathbf{0}$ exponentially.

Proof. The error $\tilde{\mathbf{v}}_{c,des} = \mathbf{v}_{c,ref} - \mathbf{v}_{c,des}$ yields that

$$\dot{\tilde{\mathbf{v}}}_{c,des} = -\omega_{v_c}\tilde{\mathbf{v}}_{c,des} + \tilde{\omega}_{v_c}\tilde{\mathbf{v}}_{c,des} + \dot{\mathbf{v}}_{c,ref}, \ \forall t \geq 0, \tag{19}$$

using (10). Then, the time derivative of V_{st} defined by

$$V_{st} = \frac{1}{2}||\tilde{\mathbf{v}}_{c,des}||^2 + \frac{1}{2\gamma_{v_c}}\tilde{\omega}_{v_c}^2, \ \forall t \geq 0, \tag{20}$$

is obtained as

$$\begin{aligned} \dot{V}_{st} &= \tilde{\mathbf{v}}_{c,des}^T(-\omega_{vc}\tilde{\mathbf{v}}_{c,des} + \tilde{\omega}_{vc}\tilde{\mathbf{v}}_{c,des} + \dot{\mathbf{v}}_{c,ref}) - \tilde{\omega}_{v_c}(||\tilde{\mathbf{v}}_{c,des}||^2 + \rho_{v_c}\tilde{\omega}_{vc}) \\ &= -\omega_{v_c}||\tilde{\mathbf{v}}_{c,des}||^2 - \rho_{v_c}\tilde{\omega}_{vc}^2 + \dot{\mathbf{v}}_{c,ref}^T\tilde{\mathbf{v}}_{c,des}, \\ &\leq -\alpha_{st}V_{st} + \dot{\mathbf{v}}_{c,ref}^T\tilde{\mathbf{v}}_{c,des}, \ \forall t \geq 0, \end{aligned} \tag{21}$$

with the positive coefficient defined as $\alpha_{st} := \min\{2\omega_{v_c}, 2\gamma_{v_c}\rho_{v_c}\}$, which confirms the result of this theorem, since the strict passivity of $\dot{\mathbf{v}}_{c,ref} \mapsto \tilde{\mathbf{v}}_{c,des}$, as shown in (21), implies the $\mathcal{L}_2$-stability for the same mapping [30]. □

Lemma 2 asserts that the proposed self-tuner (11) renders the convergence rate of the desired system (10) higher than its original version (8), by ensuring the existence of a lower bound of the time-varying gain $\hat{\omega}_{vc}$.

Lemma 2. *The cut-off frequency $\hat{\omega}_{v_c}$ adjusted by self-tuner* (11) *is bounded below by its initial value ω_{v_c} for all time, i.e., $\hat{\omega}_{v_c} \geq \omega_{v_c}, \forall t \geq 0$.*

Proof. The proposed self-tuner (11) can be written as follows in the form of an LTI system:

$$\dot{\hat{\omega}}_{v_c} = -\gamma_{v_c}\rho_{v_c}\hat{\omega}_{v_c} + \gamma_{v_c}\rho_{v_c}\omega_{v_c} + \gamma_{v_c}||\tilde{\mathbf{v}}_{c,des}||^2, \ \forall t \geq 0,$$

whose equivalent form is obtained by the integration of both sides as

$$\begin{aligned} \hat{\omega}_{v_c} &= e^{-\gamma_{v_c}\rho_{v_c}t}\omega_{v_c} + \int_0^t e^{-\gamma_{v_c}\rho_{v_c}(t-\tau)}(\gamma_{v_c}\rho_{v_c}\omega_{v_c} + \gamma_{v_c}||\tilde{\mathbf{v}}_{c,des}||^2)d\tau \\ &\geq \omega_{v_c}, \ \forall t \geq 0. \end{aligned}$$

This completes the proof. □

The inequality (21) is useful to prove the performance recovery property in Theorem 1 as the main result of this section. Lemma 3 derives the state estimation behavior by investigating the observer implementation of (12)–(14).

4.2. Observer and DOB

It is easily seen that the filtering error $\mathbf{e}_{v_c} = \mathbf{v}_c - \hat{\mathbf{v}}_c$ gives the dynamics using (12) as

$$\dot{\mathbf{e}}_{v_c} = -k_{obs}\mathbf{e}_{v_c} + \mathbf{e}_{a_c},\ \forall t \geq 0, \tag{22}$$

with estimation error $\mathbf{e}_{a_c} := \mathbf{a}_c - \hat{\mathbf{a}}_c$, which is obvious due to the Luenberger observer form of (12). The estimation error dynamics (22) make it possible to show the acceleration estimation error dynamics in the first-order LPF form.

Lemma 3. *The proposed observer of* (12)–(14) *yields the following first-order state estimation behaviors:*

$$\dot{\hat{\mathbf{a}}}_c = l_{a_c}(\mathbf{a}_c - \hat{\mathbf{a}}_c),\ \forall t \geq 0. \tag{23}$$

Proof. The observer output (14) gives the dynamics as

$$\dot{\hat{\mathbf{a}}}_c = -l_{a_c}(\hat{\mathbf{a}}_c - l_{a_c}\mathbf{e}_{v_c}) - l_{a_c}^2\mathbf{e}_{v_c} + l_{a_c}(\hat{\mathbf{a}}_c + k_{obs}\mathbf{e}_{v_c}) + l_{a_c}\dot{\mathbf{e}}_{v_c},\ \forall t \geq 0,$$

using the observer dynamics (13), which can be written as

$$\dot{\hat{\mathbf{a}}}_c = l_{a_c}(\dot{\mathbf{e}}_{v_c} + k_{obs}\mathbf{e}_{v_c}) = l_{a_c}\mathbf{e}_{a_c},\ \forall t \geq 0,$$

with the application of (22), which completes the proof. □

The estimation error $\tilde{\mathbf{a}}_c = \mathbf{a}_c - \hat{\mathbf{a}}_c$ satisfies that $\dot{\mathbf{e}}_{a_c} = -l_{a_c}\mathbf{e}_{a_c} + \dot{\mathbf{a}}_c$ because of (23), which implies that $\dot{\mathbf{e}}_{a_c} = -l_{a_c}\mathbf{e}_{a_c}$ as $l_{a_c} \to \infty$. Therefore, it is reasonable to assume that

$$\dot{\mathbf{e}}_{a_c} = -l_{a_c}\mathbf{e}_{a_c},\ \forall t \geq 0, \tag{24}$$

for a sufficiently large value of $l_{a_c} > 0$.

Lemma 4 shows the estimated disturbance behavior from the DOBs, which helps for the proof of performance recovery to be considerably simple with the use of Lemma 3.

Lemma 4. *The DOB of* (17) *and* (18) *yields the first-order disturbance estimation behavior:*

$$\dot{\hat{\mathbf{d}}}_v = l_v(\mathbf{d}_v - \hat{\mathbf{d}}_v) + l_v l_{a_c} L_0 C_0 \mathbf{e}_{a_c},\ \forall t \geq 0. \tag{25}$$

Proof. The differentiation on (18) gives

$$\begin{aligned}
\dot{\hat{\mathbf{d}}}_v &= \dot{\mathbf{z}}_v + l_v L_0 C_0 \dot{\hat{\mathbf{a}}}_c \\
&= -l_v(\hat{\mathbf{d}}_v - l_v L_0 C_0 \hat{\mathbf{a}}_c) - l_v^2 L_0 C_0 \hat{\mathbf{a}}_c - l_v((L_0 C_0 \omega_r^2 \mathbf{J}^2 - \mathbf{I})\mathbf{v}_c + \mathbf{u}) + l_v L_0 C_0 \dot{\mathbf{a}}_c \\
&\quad - l_v L_0 C_0 \dot{\mathbf{e}}_{a_c} \\
&= l_v(\mathbf{d}_v - \hat{\mathbf{d}}_v) + l_v l_{a_c} L_0 C_0 \mathbf{e}_{a_c},\ \forall t \geq 0,
\end{aligned}$$

with the utilization of (17), (18), and (24) for the second and third equations above, sequentially, which completes the proof. □

Defining the disturbance estimation error $\mathbf{e}_{d_v} := \mathbf{d}_v - \hat{\mathbf{d}}_v$, it holds that $\dot{\mathbf{e}}_{d_v} = -l_v\mathbf{e}_{d_v} + l_v l_{a_c} L_0 C_0 \mathbf{e}_{a_c} + \dot{\mathbf{d}}_v$ because of (25), which implies that $\dot{\mathbf{e}}_{d_v} = -l_v\mathbf{e}_{d_v} + l_v l_{a_c} L_0 C_0 \mathbf{e}_{a_c}$ as $l_v \to \infty$. Therefore, it is reasonable to assume that

$$\dot{\mathbf{e}}_{d_v} = -l_v\mathbf{e}_{d_v} + l_v l_{a_c} L_0 C_0 \mathbf{e}_{a_c},\ \forall t \geq 0, \tag{26}$$

for a sufficiently large value of $l_v > 0$. The two results, (24) and (26), can be compactly written as

$$\dot{\mathbf{e}} = \mathbf{A}_e \mathbf{e}, \ \forall t \geq 0, \tag{27}$$

with $\mathbf{e} := \begin{bmatrix} \mathbf{e}_{d_v}^T & \mathbf{e}_{a_c}^T \end{bmatrix}^T$ and stable matrix $\mathbf{A}_e := \begin{bmatrix} -l_v \mathbf{I}_{2\times 2} & l_v l_{a_c} L_0 C_0 \mathbf{I}_{2\times 2} \\ \mathbf{0}_{2\times 2} & -l_{a_c} \mathbf{I}_{2\times 2} \end{bmatrix}$, which is used for the following analysis.

Remark 2. *Assuming $\mathbf{e}_{a_c} \approx \mathbf{0}$ by result* (27)*, the Laplace transform to the system* (25) *(e.g., $\hat{D}_{v,d}(s) = \mathcal{L}\{d_{v,d}\}$, $\hat{D}_{v,q}(s) = \mathcal{L}\{\hat{d}_{v,q}\}$, $D_{v,d}(s) = \mathcal{L}\{d_{v,d}\}$, and $D_{v,q}(s) = \mathcal{L}\{d_{v,q}\}$) derives the transfer function forming the following low-pass filter (LPF):*

$$\frac{\hat{D}_{v,x}(s)}{D_{v,x}(s)} = \frac{l_v}{s + l_v}, \ x = d, q, \ \forall s \in \mathbb{C}, \tag{28}$$

with the cut-off frequency l_v (rad/s, $\frac{l_v}{2\pi}$ Hz). Therefore, the DOB gain l_v can be determined as the cut-off frequency of the LPF (28) from the disturbance to its estimate.

4.3. Control Loop

The combination of active damping and specific form of PD gain occurs in the closed-loop system order reduction to 1 by the stable PZC, which is asserted in Lemma 5.

Lemma 5. *The proposed control law* (16) *forces the output voltage dynamics to be governed by*

$$\dot{\tilde{v}}_c = \lambda_{v_c} \tilde{v}_c + \boldsymbol{B}_{e,F} \boldsymbol{e}_F, \tag{29}$$

with

$$\dot{e}_F = -a_{e_F} e_F + \boldsymbol{B}_e \boldsymbol{e}, \ \forall t \geq 0, \tag{30}$$

where $a_{e_F} := \frac{k_{vc}}{L_0 C_0}$, $\boldsymbol{B}_{e,F} := \begin{bmatrix} \boldsymbol{I}_{2\times 2} & \boldsymbol{I}_{2\times 2} \end{bmatrix}$, and $\boldsymbol{e}_F := \begin{bmatrix} \boldsymbol{e}_{d_v,F}^T & \boldsymbol{e}_{a_c,F}^T \end{bmatrix}^T$ for some $\boldsymbol{B}_e \in \mathbb{R}^{4\times 4}$.

Proof. The substitution of (16) to (15) gives the following closed-loop voltage dynamics:

$$\begin{aligned} L_0 C_0 \ddot{\mathbf{v}}_c &= -k_{vc} \hat{\mathbf{a}}_c + L_0 C_0 \lambda_{vc} (\dot{\mathbf{v}}_{c,des} - \hat{\mathbf{a}}_c) + k_{vc} \lambda_{vc} \tilde{\mathbf{v}}_c + \mathbf{e}_{d_v}, \\ &= -k_{vc} \dot{\mathbf{v}}_c + L_0 C_0 \lambda_{vc} \dot{\tilde{\mathbf{v}}}_c + k_{vc} \lambda_{vc} \tilde{\mathbf{v}}_c + \mathbf{e}_{d_v} + (k_{vc} + L_0 C_0 \lambda_{vc}) \mathbf{e}_{a_c}, \ \forall t \geq 0. \end{aligned}$$

Taking the Laplace transform to both sides above, it follows that

$$\begin{aligned} (L_0 C_0 s^2 + (k_{vc} + L_0 C_0 \lambda_{vc}) s + k_{vc} \lambda_{vc}) \mathbf{V}_c(s) &= \lambda_{vc} (L_0 C_0 s + k_{vc}) \mathbf{V}_{c,des}(s) + \mathbf{E}_{d_v}(s) \\ &\quad + (k_{vc} + L_0 C_0 \lambda_{vc}) \mathbf{E}_{a_c}(s), \end{aligned}$$

which shows that

$$(s + \lambda_{vc}) \mathbf{V}_c(s) = \lambda_{vc} \mathbf{V}_{c,des}(s) + \mathbf{E}_{d_v,F}(s) + \mathbf{E}_{a_c,F}(s), \ \forall s \in \mathbb{C},$$

due to the PZC by the factorization $(L_0 C_0 s^2 + (k_{vc} + L_0 C_0 \lambda_{vc}) s + k_{vc} \lambda_{vc}) = (L_0 C_0 s + k_{vc})(s + \lambda_{vc})$, where $\mathbf{E}_{d_v,F}(s) = \frac{\frac{1}{L_0 C_0}}{s + \frac{k_{vc}}{L_0 C_0}}$ and $\mathbf{E}_{a_c,F} = \frac{\frac{k_{vc} + L_0 C_0 \lambda_{vc}}{L_0 C_0}}{s + \frac{k_{vc}}{L_0 C_0}} \mathbf{E}_{a_c}(s)$. The application of the inverse Laplace transform to both sides above verifies the result of (29) and (30). □

The performance recovery property, as a main result, can be easily proven by the Lyapunov analysis thanks to the results of Lemmas 3–5.

Theorem 1. *The closed-loop system driven by the proposed control law depicted in Figure 2 guarantees that*

$$\lim_{t\to\infty} v_c = v_{c,des},$$

as $\dot{v}_{c,ref} \to \mathbf{0}$ *exponentially.*

Proof. Let us consider the error dynamics from $\tilde{\mathbf{v}}_c = \mathbf{v}_{c,des} - \mathbf{v}_c$ as

$$\begin{aligned} \dot{\tilde{\mathbf{v}}}_c &= \dot{\mathbf{v}}_{c,des} + \dot{\mathbf{v}}_c \\ &= -\lambda_{v_c}\tilde{\mathbf{v}}_c - \mathbf{B}_{e,F}\mathbf{e}_F + \hat{\omega}_{v_c}\tilde{\mathbf{v}}_{c,des}, \ \forall t \geq 0, \end{aligned} \tag{31}$$

with the applications of (11) and (29). The combination of (27), (30), and (31) gives the perturbed linear system

$$\dot{\mathbf{x}}_{cl} = \mathbf{A}_{cl}\mathbf{x}_{cl} + \mathbf{B}_{cl}\hat{\omega}_{v_c}\tilde{\mathbf{v}}_{c,des}, \tag{32}$$

with $\mathbf{x}_{cl} := \begin{bmatrix} \tilde{\mathbf{v}}_c^T & \mathbf{e}_F^T & \mathbf{e}^T \end{bmatrix}^T$, $\mathbf{B}_{cl} := \begin{bmatrix} \mathbf{I}_{2\times 2} & \mathbf{0}_{2\times 4} \end{bmatrix}^T$, and stable matrix $\mathbf{A}_{cl}$ defined as

$$\mathbf{A}_{cl} := \begin{bmatrix} -\lambda_{v_c}\mathbf{I}_{2\times 2} & -\mathbf{B}_{e,F} & \mathbf{0}_{2\times 4} \\ \mathbf{0}_{4\times 2} & -a_1\mathbf{I}_{4\times 4} & \mathbf{B}_e \\ \mathbf{0}_{4\times 2} & \mathbf{0}_{4\times 4} & \mathbf{A}_e \end{bmatrix}.$$

The stability of matrix $\mathbf{A}_{cl}$ makes it solvable for the matrix equation of $\mathbf{A}_{cl}^T\mathbf{P} + \mathbf{P}\mathbf{A}_{cl} = -\mathbf{I}$ with a unique solution $\mathbf{P} = \mathbf{P}^T > \mathbf{0}$. The solution $\mathbf{P} > \mathbf{0}$ defines the positive–definite function as

$$V := \frac{1}{2}\mathbf{x}_{cl}^T\mathbf{P}\mathbf{x}_{cl} + \frac{\kappa}{2}V_{st}, \ \kappa > 0, \ \forall t \geq 0, \tag{33}$$

with the positive–definite function V_{st} given in (20), whose time derivative is obtained using (21) and (32), and Young's inequality (e.g., $\mathbf{x}^T\mathbf{y} \leq \frac{\epsilon}{2}\|\mathbf{x}\|^2 + \frac{1}{2\epsilon}\|\mathbf{y}\|^2$, $\forall \mathbf{x}, \mathbf{y} \in \mathbb{R}^n$, $\forall \epsilon > 0$) as

$$\begin{aligned} \dot{V} &= -\|\mathbf{x}_{cl}\|^2 + \mathbf{x}_{cl}^T\mathbf{P}\mathbf{B}_{cl}\hat{\omega}_{v_c}\tilde{\mathbf{v}}_{c,des} - \kappa\alpha_{st}V_{st} + \kappa\dot{\mathbf{v}}_{c,ref}^T\tilde{\mathbf{v}}_{c,des} \\ &\leq -\frac{1}{2}\|\mathbf{x}_{cl}\|^2 - (\kappa\alpha_{st} - \|\mathbf{P}\|^2\|\mathbf{B}_{cl}\|^2\bar{\omega}_{v_c}^2)V_{st} + \kappa\dot{\mathbf{v}}_{c,ref}^T\tilde{\mathbf{v}}_{c,des}, \ \forall t \geq 0, \end{aligned}$$

with $|\hat{\omega}_{v_c}| \leq \bar{\omega}_{v_c}$. The constant $\kappa = \frac{1}{\alpha_{st}}(\|\mathbf{P}\|^2\|\mathbf{B}_{cl}\|^2\bar{\omega}_{v_c}^2 + \frac{1}{2})$ eliminates the indefinite term so that $\dot{V}$ satisfies

$$\begin{aligned} \dot{V} &\leq -\frac{1}{2}\|\mathbf{x}_{cl}\|^2 - \frac{1}{2}V_{st} + \kappa\dot{\mathbf{v}}_{c,ref}^T\tilde{\mathbf{v}}_{c,des} \\ &\leq -\alpha_{cl}V + \kappa\dot{\mathbf{v}}_{c,ref}^T\tilde{\mathbf{v}}_{c,des}, \ \forall t \geq 0, \end{aligned}$$

with $\alpha_{cl} := \min\{\frac{1}{\lambda_{min}(\mathbf{P})}, \frac{1}{\kappa}\}$, which concludes that $\mathbf{x}_{cl} \to \mathbf{0}$ as $\dot{\mathbf{v}}_{c,ref} \to \mathbf{0}$ exponentially. This completes the proof. □

The proposed PD-type controller shown in Figure 2 does not incorporate any integral action, which may suffer from the offset errors in the actual implementations. The DOB as a feed-forward compensation term makes it possible to remove the offset errors in the absence of regulation error integrators, which are addressed in Theorem 2 as another main result of this study.

Theorem 2. *The proposed control law consisting of* (12)–(14) *and* (16)–(18) *guarantees* $v_{c,\infty} = v_{c,ref,\infty}$ *in the actual implementations, where* f_∞ *denotes the steady-state value of* f *for any convergent function* f.

Proof. The closed-loop dynamics of (23), (25), and (29)–(30) result in the steady-state equations

$$\begin{aligned}
\mathbf{0} &= l_{a_c}\mathbf{e}_{a_c,\infty}, \\
\mathbf{0} &= l_v\mathbf{e}_{d_v,\infty} + l_v l_{a_c} L_0 C_0 \mathbf{e}_{a_c,\infty}, \\
\mathbf{0} &= -a_1\mathbf{e}_{d_v,F,\infty} + a_2\mathbf{e}_{d_v,\infty}, \\
\mathbf{0} &= -a_1\mathbf{e}_{a_c,F,\infty} + a_3\mathbf{e}_{a_c,\infty}, \\
\mathbf{0} &= \lambda_{v_c}\tilde{\mathbf{v}}_{c,\infty} + \mathbf{e}_{d_v,F,\infty} + \mathbf{e}_{a_c,F,\infty},
\end{aligned}$$

which show that $\tilde{\mathbf{v}}_{c,\infty} = \mathbf{0}$ is equivalent to $\mathbf{v}_{c,\infty} = \mathbf{v}_{c,des,\infty} = \mathbf{v}_{c,ref,\infty}$, due to the steady-state equation of self-tuner $\mathbf{0} = \hat{\omega}_{v_c,\infty}\tilde{\mathbf{v}}_{c,des,\infty}$ obtained from (10). Therefore, the claim is true. □

Remark 3. *The closed-loop analysis results in this section reveal a design parameter tuning process as follows:*

1. *Observer:*
   - *(Lemma 3) Choose l_{a_c} and k_{obs} such that $l_{a_c} >> k_{obs}$ satisfies the desired observer error dynamics $\dot{\mathbf{e}}_{a_c} = -l_{a_c}\mathbf{e}_{a_c}$ and $\dot{\mathbf{e}}_{v_c} \approx -k_{obs}\mathbf{e}_{v_c}$.*
2. *DOB:*
   - *(Lemma 4 and Remark 2) Choose l_v for a given specification $\frac{\hat{D}_{v,x}(s)}{D_{v,x}(s)} = \frac{l_v}{s+l_v}$, $x = d, q$.*
3. *Controller:*
   - *(Lemma 5 and Theorem 1) For a given specification $\omega_{vc}(= \hat{\omega}_{vc}(0))$ in the nominal system (8), increase k_{vc} to obtain the error dynamics $\dot{\tilde{v}}_c \approx -\lambda_{vc}\tilde{v}_c$ for some choice $\lambda_{vc} >> \omega_{vc}$.*
4. *Self-tuner:*
   - *(Lemma 2) After specifying $\rho_{st} = \frac{\beta_{st}}{\gamma_{st}}$ with $\beta_{st} > 0$, increase ρ_{st} and β_{st} considering the maximum closed-loop bandwidth ($\bar{\omega}_{vc} \geq \hat{\omega}_{vc}$) from the hardware specification.*

This process obtains the closed-loop tuning results for Section 5.

5. Experimental Results

Figure 3 visualizes the experimental setup comprised of the prototype 3 kW three-phase inverter, inductor, output capacitor, and digital signal processor (DSP; Texas Instrument (TI) DSP28377, Dallas, TX, USA) board. The passive component (e.g., inductance and capacitance) values were given as $R = 0.038\ \Omega$, $L = 1$ mH, and $C = 80$ µF. The DC-Link level was set to $V_{dc} = 90$ V with the bi-directional power supply. The synchronization for control and pulse-width modulation (PWM) periods was done to 0.1 ms of internal interrupt service routine. An additional laptop PC using MATLAB/Simulink (2022a) was used to observe and collect the real-time system responses, such as current and voltage, under the controller area network environment.

The control algorithms were realized using the nominal parameter values $R_0 = 0.8$ R, $L_0 = 1.3$ L, and $C_0 = 0.9$ C for taking into account the parameter variations in this experimental study. The design parameter tuning result was given as follows: (observer) $k_{obs} = 20$, $l_{a_c} = 628$, (DOB) $l_v = 942$, (self-tuner) $\gamma_{st} = 20$, $\rho_{st} = 10/\gamma_{st}$, and (active damping) $k_v = 5 \times 10^{-3}$ with the cut-off frequency $\omega_{vc} = 12.56$ rad/s ($f_{vc} = \omega_{vc}/2\pi = 2$ Hz).

The conventional PZC technique comprised of proportional–integral control and feed-forward compensators was chosen for comparison and given by the following specifications [10]:

- (Outer-loop)

$$\mathbf{i}_{ref} = -b_{d,v}\mathbf{v}_c + L_0\omega_{vc}\tilde{\mathbf{v}}_c + b_{d,v}\omega_{vc}\int_0^T \tilde{\mathbf{v}}_c d\tau - C_0\omega_r \mathbf{J}\mathbf{v}_c$$

- (Inner-loop)

$$\mathbf{u} = L_0\omega_{cc}\tilde{\mathbf{i}} + R_0\omega_{cc}\int_0^t \tilde{\mathbf{i}}d\tau - L_0\omega_r\mathbf{J}\mathbf{i},\ \tilde{\mathbf{i}} = \mathbf{i}_{ref} - \mathbf{i},\ \forall t \geq 0,$$

The specifications also included with the additional inclusion of active damping term $-b_{d,v}\mathbf{v}_c$ for a better performance and tuned parameters $b_{d,v} = 0.5$ and $\omega_{cc} = 1885$ rad/s ($f_{cc} = \omega_{cc}/2\pi = 300$ Hz); this control also tries to assign the desired cut-off frequency ω_{vc} to the closed-loop system, which is identical to the control objective of proposed controller.

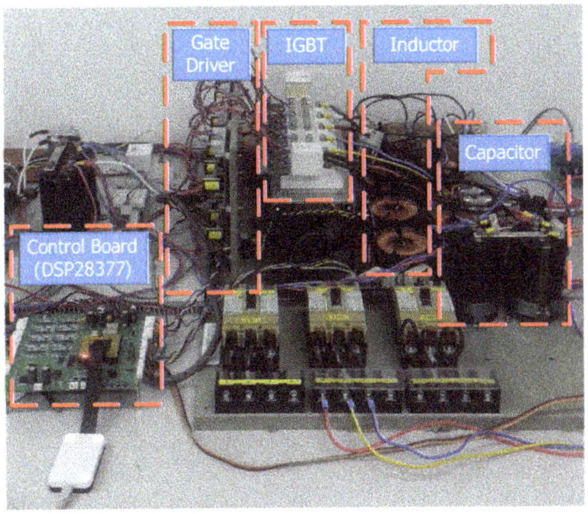

Figure 3. Experimental setup.

The next sections clarify the practical advantages of the proposed PD-type controller through the comparison study with the PZC controller in the output voltage tracking and regulation tasks under the linear (resistive and resistive–inductive) and nonlinear (rectifier) load variations.

5.1. *Performance Comparison for Linear Load Variation*

5.1.1. Tracking Task

This section observes the tracking performance variations under the use of three resistive loads $R_L = 2, 4, 10\ \Omega$. The output voltage reference was suddenly increased from its initial value $r = 15$ V to $r = 30$ V. Figure 4 shows that there were no tracking performance variations under the proposed control, but they were present under the PZC control, in the absence of current feedback for the proposed control system. Moreover, the proposed self-tuner effectively improved the closed-loop performance in the transient periods by increasing and restoring the feedback gain (e.g., feedback-loop adaptation). This beneficial point comes from the observer, online self-tuner, active damping, and DOBs embedding in the proposed controller. The resultant a-phase voltage responses are depicted in Figure 5 with the consistent closed-loop performance with the proposed controller in the presence of operating condition changes. Figures 6 and 7 show the d-q current responses to be driven more rapidly by the proposed controller than the PZC controller. The load current waveform for each controller is presented in Figure 8, resulting in the total harmonic distortion (THD) calculation result of 1.1% (approximately the same with both controllers), which is acceptable for the actual applications. Figure 9 presents the observer error behaviors in this task with the successful state estimation results by rendering the estimation error to be convergent to zero by the proposed observer. The DOB and self-tuner responses are shown in Figures 10 and 11.

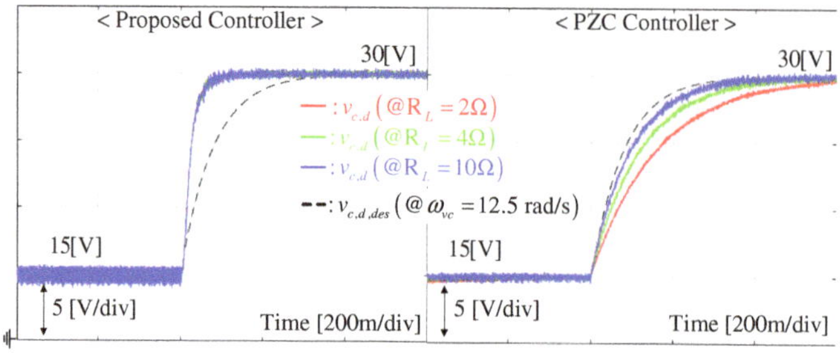

Figure 4. *d*-axis output voltage tracking performance variation comparison for three resistive loads $R_L = 2, 4, 10\ \Omega$.

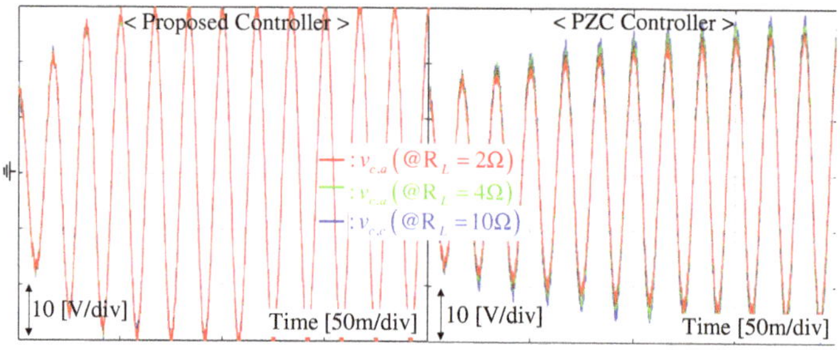

Figure 5. *a*-phase output voltage ($V_{c,a}$) tracking performance variation comparison for three resistive loads $R_L = 2, 4, 10\ \Omega$.

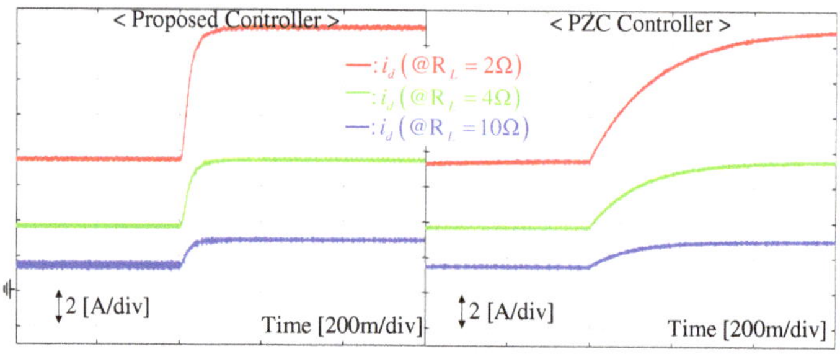

Figure 6. *d*-axis current response comparison under tracking task for three resistive loads $R_L = 2, 4, 10\ \Omega$.

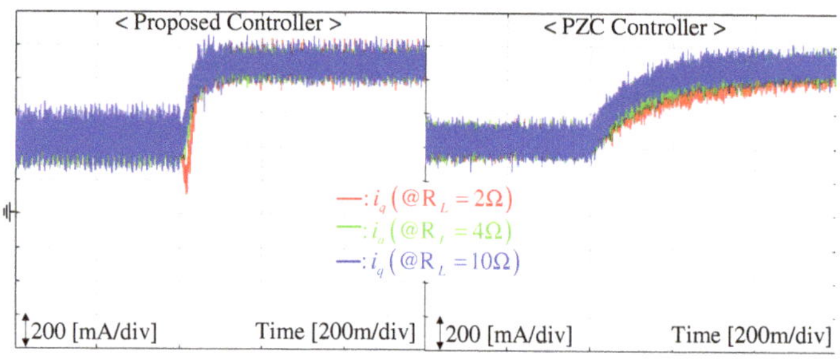

Figure 7. *q*-axis current response comparison under tracking task for three resistive loads $R_L = 2, 4, 10\ \Omega$.

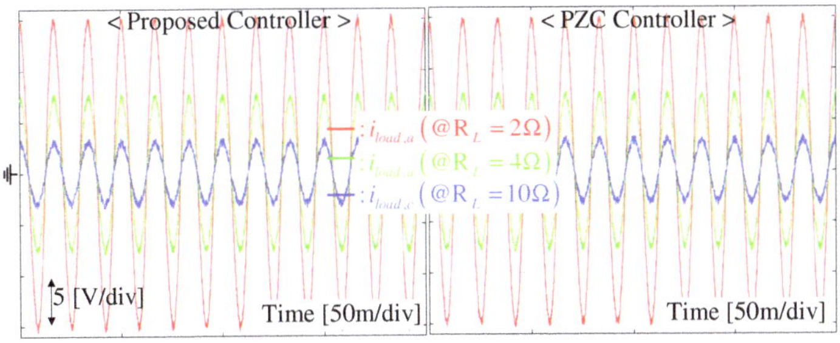

Figure 8. *a*-phase load current response comparison under tracking task for three resistive loads $R_L = 2, 4, 10\ \Omega$.

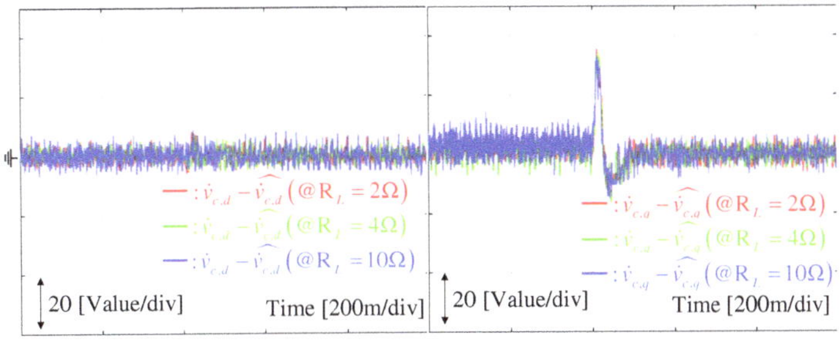

Figure 9. Observer error responses under tracking task for three resistive loads $R_L = 2, 4, 10\ \Omega$.

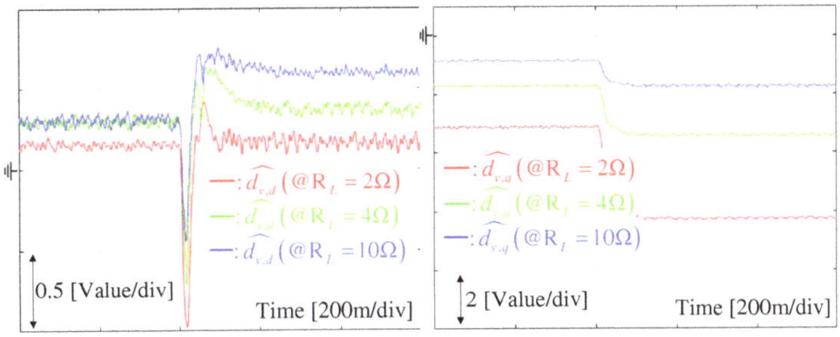

Figure 10. DOB responses under tracking task for three resistive loads $R_L = 2, 4, 10\ \Omega$.

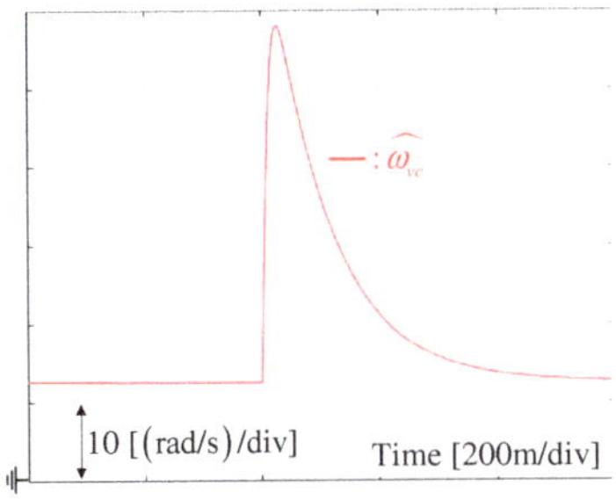

Figure 11. Self-tuner response under tracking task for resistive load $R_L = 10\ \Omega$.

5.1.2. Tracking Task under Resistive–Inductive Load

This section demonstrates the tracking performance variations for three resistive–inductive load pairs of ($R_L = 2\ \Omega$, $L_L = 1$ H), ($R_L = 4\ \Omega$, $L_L = 1$ H), and ($R_L = 10\ \Omega$, $L_L = 1$ H) under the same settings as in Section 5.1.1. Comparing the output voltage responses in Figure 4, Figure 12 indicates the consistent output voltage responses by the proposed controller for resistive and resistive–inductive load operations, unlike the conventional PZC controller. The corresponding a-phase voltage responses are presented in Figure 13 with considerable rapidity and consistency, compared with the PZC controller.

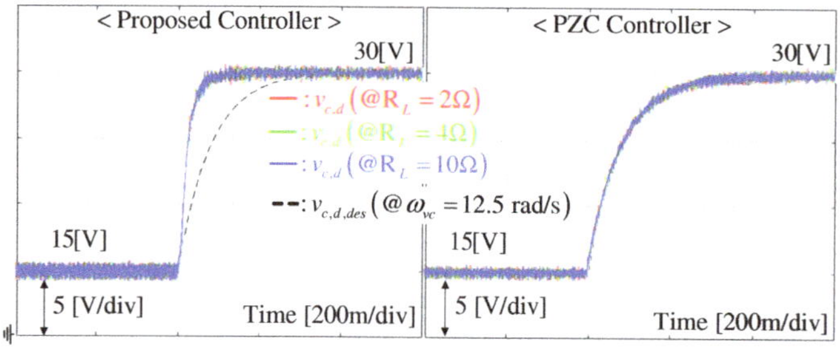

Figure 12. d-axis output voltage tracking performance variation comparison for three resistive–inductive load pairs of ($R_L = 2\,\Omega$, $L_L = 1$ H), ($R_L = 4\,\Omega$, $L_L = 1$ H), and ($R_L = 10\,\Omega$, $L_L = 1$ H).

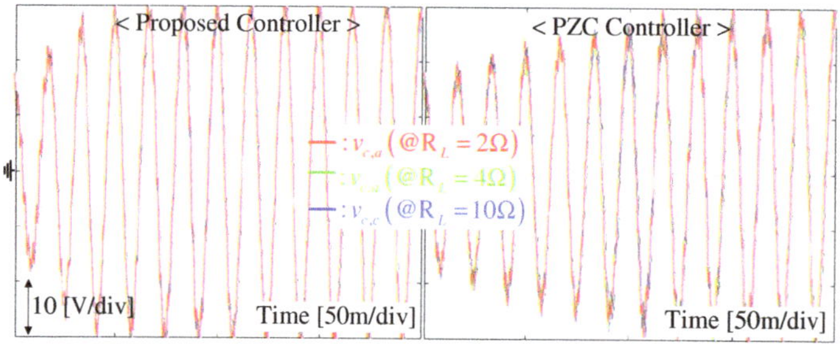

Figure 13. a-phase output voltage tracking performance variation comparison for three resistive–inductive load pairs of ($R_L = 2\,\Omega$, $L_L = 1$ H), ($R_L = 4\,\Omega$, $L_L = 1$ H), and ($R_L = 10\,\Omega$, $L_L = 1$ H).

5.1.3. Regulation Task under Resistive Load

This section demonstrates the output voltage regulation performance improvement at the fixed 30 V operation mode with three sudden resistive load change scenarios from the initial resisitve load $R_L = 10\,\Omega$ to $1.6\,\Omega$, $3.3\,\Omega$, and $5\,\Omega$. Figure 14 depicts the regulation performances from the proposed and PZC controllers by showing the d-axis voltage responses. There were significant performance improvements rather than the tracking task under the three load variation scenarios with the considerable reduction of transient periods by the proposed controller. This improvement was also presented in a-phase voltage responses shown in Figure 15. Figures 16 and 17 depict the d-q current whose rapid behaviors result in this output voltage's regulation performance improvement.

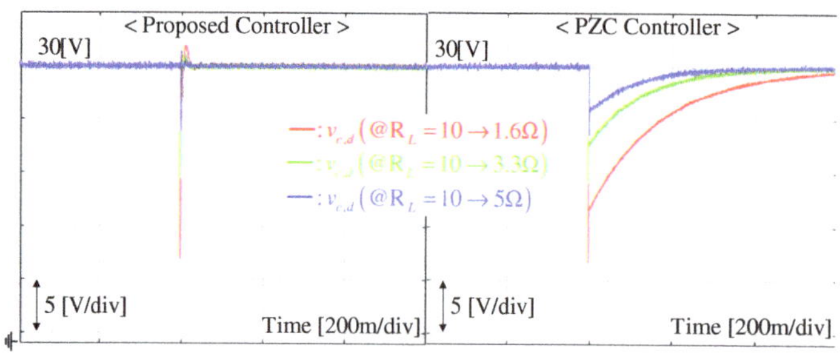

Figure 14. d-axis output voltage regulation performance comparison under three resistive load change scenarios as $R_L = 10 \rightarrow 1.6, 3.3, 5\,\Omega$.

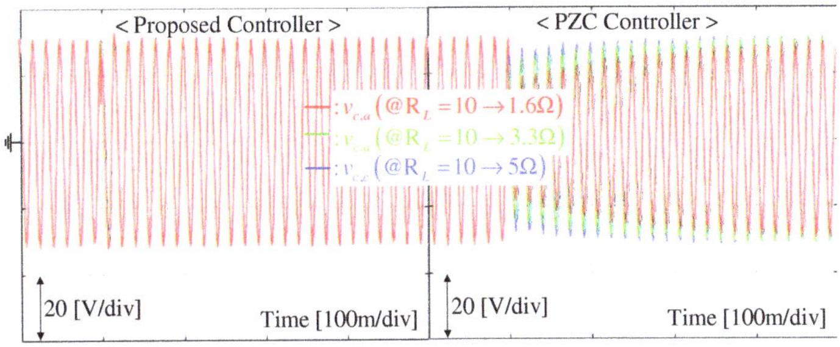

Figure 15. *a*-phase output voltage regulation performance comparison under three resistive load change scenarios as $R_L = 10 \rightarrow 1.6, 3.3, 5\,\Omega$.

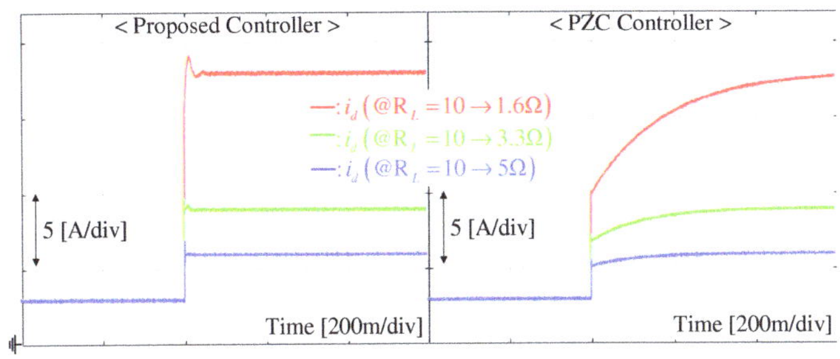

Figure 16. *d*-axis current response comparison under regulation task for three resistive load change scenarios as $R_L = 10 \rightarrow 1.6, 3.3, 5\,\Omega$.

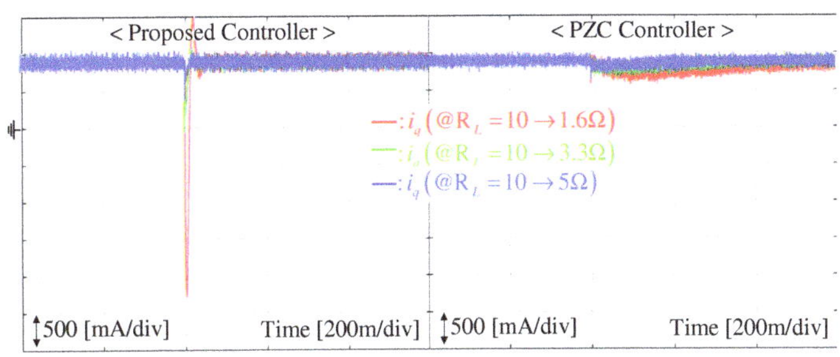

Figure 17. *q*-axis current response comparison under regulation task for three resistive load change scenarios as $R_L = 10 \rightarrow 1.6, 3.3, 5\,\Omega$.

5.2. Performance Comparison for Nonlinear Load Variation

5.2.1. Tracking Task

This section uses the nonlinear loads comprised of the three-phase diode rectifier and single resistor to evaluate the output voltage tracking performance improvement. This experiment was conducted three times for three single resistors of 2, 4, and 10 Ω. The same output voltage reference for the resistive load case was used. Figure 18 shows similar result as the resistive load case, except for the voltage ripple magnification from the three-phase diode rectifier switching actions. There were no tracking performance variations under the proposed control, but they were present under the classical control, in the absence of current feedback for the proposed control system.

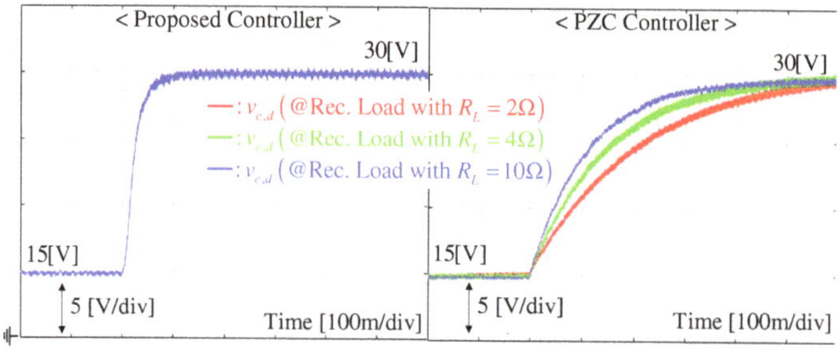

Figure 18. d-axis output voltage tracking performance variation comparison for rectifier load with three resistors of $R_L = 2, 4, 10\ \Omega$.

5.2.2. Regulation Task

This section verifies the output voltage regulation performance improvement under the nonlinear load where the resistor value attached in the three-phase diode rectifier was abruptly changed from 10 Ω to 1.6, 3.3, and 5 Ω. The output voltage reference was fixed to 30 V. Figure 19 also shows the similar result as the resistive load case, except for the voltage ripple magnification. The proposed controller provided a considerable better regulation performance with the reduction of undershoot magnitude and transient periods.

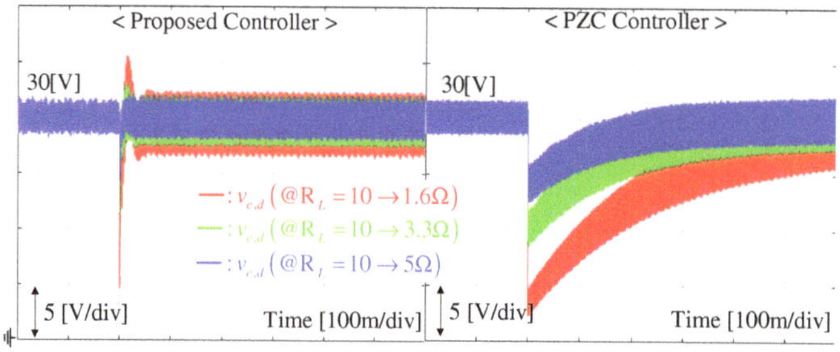

Figure 19. d-axis output voltage regulation performance comparison for rectifier load with three resistor value change scenarios as $R_L = 10 \rightarrow 1.6$, 3.3, and 5 Ω.

5.3. Tracking Task: Self-Tuner Effect

This section presents the merit of variable cut-off frequency from self-tuner under the tracking task with the resistive load $R_L = 2\ \Omega$. The self-tuner design parameter γ_{st} was increased to 0, 20, and 50 with the step reference from 15 V to 30 V. As intended, Figure 20 shows that the increased cut-off frequency excitation level resulted in the improved tracking performance without any over- or undershoots due to the performance recovery property proved in Theorem 1.

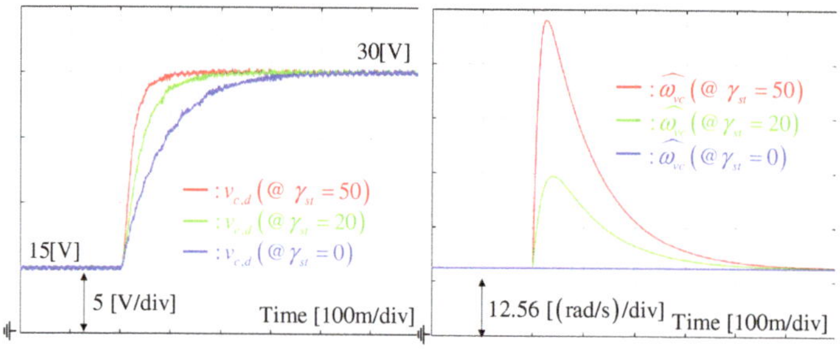

Figure 20. d-axis output voltage tracking performance variations and cut-off frequency behaviors as increasing $\gamma_{st} = 0, 20$, and 50 under resistive load $R_L = 2\ \Omega$.

5.4. Summary

This section concludes the experimental studies in Sections 5.1 and 5.2 by showing the quantitative comparison results obtained from the metric function $J := \sqrt{\int_0^\infty \|\mathbf{v}_{c,des} - \mathbf{v}_c\|^2 dt}$. This comparison additionally included the case of a conventional multi-loop PI controller tuned for the base bandwidth ω_{v_c}. Figure 21 shows the table summarizing the performance comparison results indicating an improvement of 31 % at least by the proposed technique.

J	Linear Load		Nonlinear Load		Average
	Tracking	Regulation	Tracking	Regulation	
Proposed Controller	2593	1815	2812	2733	2488
PZC Controller	3127	3472	3715	4113	3606
PI Controller	4882	4255	5133	5739	5002

Figure 21. Performance comparison results.

6. Conclusions

The proposed output voltage regulator was designed without the current feedback with the consideration of practical constraints, system parameter and load variations. The output voltage derivative observer was proposed to estimate the actual state without dependence on system parameters, which removes the need for current feedback. The improved closed-loop robustness was secured by the combination of active damping and specific forms of PD gains, leading to the closed-loop system order reduction to 1. Moreover, the beneficial closed-loop properties, named performance recovery and offset-free, were guaranteed by the closed-loop analysis. The various experimental data verified the practical advantages of the proposed controller. As a future work, the proposed technique will be expanded as a solution to the phase voltage and current synchronization problem for multiple power converters, providing an optimization process to automatically determine the design parameters for the controller, observer, DOB, etc.

Author Contributions: Conceptualization and methodology, S.-K.K.; software, validation, formal analysis, investigation, writing—original draft preparation, and writing—review and editing, Y.K.; resources, supervision, project administration, and funding acquisition, H.L. All authors have read and agreed to the published version of the manuscript.

Funding: This work is supported by the Korea Agency for Infrastructure Technology Advancement (KAIA) grant funed by the Ministry of Land, Infrastructure and Transport (Grant RS-2020-KA158067) and this research was financially supported by the Ministry of Small and Medium-sized Enterprises (SMEs) and Startups(MSS), Korea, under the "Regional Specialized Industry Development Plus Program (R&D, S3364188)" supervised by the Korea Technology and Information Promotion Agency for SMEs.

Data Availability Statement: The data presented in this study are available on request from the corresponding author. The data are not publicly available due to a legal issue.

Conflicts of Interest: Author Hosik Lee was employed by the company Green Mobility Team, Tenergy. The remaining authors declare that the research was conducted in the absence of any commercial or financial relationships that could be construed as a potential conflict of interest.

References

1. Saeed, M.; Fernández, D.; Guerrero, J.M.; Díaz, I.; Briz, F. Insulation Condition Assessment in Inverter-Fed Motors Using the High-Frequency Common Mode Current: A Case Study. *Energies* **2024**, *17*, 470. [CrossRef]
2. Patel, M.; Zhou, Z. An Interleaved Battery Charger Circuit for a Switched Capacitor Inverter-Based Standalone Single-Phase Photovoltaic Energy Management System. *Energies* **2023**, *16*, 7155. [CrossRef]

3. Luo, Z.; Zhang, B.; Li, L.; Tang, L. A Decentralized Control Strategy for Series-Connected Single-Phase Two-Stage Photovoltaic Grid-Connected Inverters. *Energies* **2023**, *16*, 7099. [CrossRef]
4. Liao, Z.; Peng, T.; Liu, J.; Guo, T. Multi-Adjustment Strategy for Phase Current Reconstruction of Permanent Magnet Synchronous Motors Based on Model Predictive Control. *Energies* **2023**, *16*, 5694. [CrossRef]
5. Kim, S.K. Performance-recovery proportional-type output-voltage tracking algorithm of three-phase inverter for uninterruptible power supply applications. *IET Circuits Devices Syst.* **2019**, *13*, 185–192. [CrossRef]
6. Loh, P.C.; Newman, M.J.; Zmood, D.N.; Holmes, D.G. A comparative analysis of multiloop voltage regulation strategies for single and three-Phase UPS systems. *IEEE Trans. Power Electron.* **2003**, *18*, 1176–1185.
7. Kassakian, J.C.; Schlecht, M.; Verghese, G.C. *Principles of Power Electronics*; Addison-Wesley: Reading, MA, USA, 1991.
8. Matausek, M.R.; Jeftenic, B.I.; Miljkovic, D.M.; Bebic, M.Z. Gain scheduling control of DC motor drive with field weakening. *IEEE Trans. Ind. Electron.* **1996**, *43*, 153–162. [CrossRef]
9. Sul, S.K. *Control of Electric Machine Drive Systems*; Wiley: Hoboken, NJ, USA, 2011; Volume 88.
10. Kazmierkowski, M.P.; Krishnan, R.; Blaabjerg, F. *Control in Power Electronics—Selected Problems*; Academic Press: Cambridge, MA, USA, 2002.
11. Bustos, R.; Gadsden, S.; Biglarbegian, M.; AlShabi, M.; Mahmud, S. Battery State of Health Estimation Using the Sliding Interacting Multiple Model Strategy. *Energies* **2024**, *17*, 536. [CrossRef]
12. Xin, C.; Li, Y.X.; Ahn, C.K. Adaptive Neural Asymptotic Tracking of Uncertain Non-Strict Feedback Systems With Full-State Constraints via Command Filtered Technique. *IEEE Trans. Neural Netw. Learn. Syst.* **2023**, *34*, 8102–8107. [CrossRef]
13. Liu, D.; Zhao, H. Affine Projection Sign Subband Adaptive Filter Algorithm With Unbiased Estimation Under System Identification. *IEEE Trans. Circuits Syst. II Express Briefs* **2023**, *70*, 1209–1213. [CrossRef]
14. Kawabata, T.; Miyashita, T.; Yamamoto, Y. Deadbeat control of three phase PWM inverter. *IEEE Trans. Power Electron.* **1990**, *5*, 21–28. [CrossRef]
15. Ito, Y.; Kawauchi, S. Microprocessor-based robust digital control for UPS with three-phase PWM inverter. *IEEE Trans. Power Electron.* **1995**, *10*, 196–204. [CrossRef]
16. Cho, J.S.; Lee, S.Y.; Mok, H.S.; Choe, G.H. Modified deadbeat digital controller for UPS with 3-phase PWM inverter. In Proceedings of the Industry Applications Conference, Thirty-Fourth IAS Annual Meeting, Phoenix, AZ, USA, 3–7 October 1999.
17. Willmann, G.; Coutinho, D.F.; Pereira, L.F.A.; Libano, F.B. Multiple-loop H_∞ control design for uninterruptible power supplies. *IEEE Trans. Ind. Electron.* **2007**, *54*, 1591–1602. [CrossRef]
18. Lee, T.; Tzeng, K.; Chong, M. Robust controller design for a single-phase UPS inverter using μ-synthesis. *IEE Proc. Electr. Power Appl.* **2004**, *151*, 334–340. [CrossRef]
19. Pichan, M.; Rastegar, H. Sliding-Mode Control of Four-Leg Inverter With Fixed Switching Frequency for Uninterruptible Power Supply Applications. *IEEE Trans. Ind. Electron.* **2017**, *64*, 6805–6814. [CrossRef]
20. Vargas, R.; Cortes, P.; Ammann, U.; Rodriguez, J.; Pontt, J. Predictive control of a three-phase neutral-point-clamped inverter. *IEEE Trans. Ind. Electron.* **2007**, *54*, 2697–2705. [CrossRef]
21. Cortes, P.; Rodriguez, J.; Vazquez, S.; Franquelo, L.G. Predictive control of a three-phase UPS inverter using two steps prediction horizon. In Proceedings of the 2010 IEEE International Conference on Industrial Technology (ICIT), Via del Mar, Chile, 14–17 March 2010.
22. Cortes, P.; Ortiz, G.; Yuz, J.I.; Rodriguez, J.; Vazquez, S.; Franquelo, L.G. Model Predictive Control of an Inverter With Output LC Filter for UPS Applications. *IEEE Trans. Ind. Electron.* **2009**, *56*, 1875–1883. [CrossRef]
23. Bemporad, A.; Borrelli, F.; Morari, M. Model predictive control based on linear programming; The explicit solution. *IEEE Trans. Autom. Control* **2002**, *47*, 1974–1985. [CrossRef]
24. Borrelli, F. *Constrained Optimal Control of Linear and Hybrid Systems*; Springer: New York, NY, USA, 2003.
25. Kim, S.K.; Park, C.; Yoon, T.W.; Lee, Y. Disturbance-observer-based model predictive control for output voltage regulation of three-phase inverter for uninterruptible-powersupply applications. *Eur. J. Control* **2015**, *23*, 71–83. [CrossRef]
26. Lim, J.; Park, C.; Han, J.; Lee, Y. Robust tracking control of a three-phase DC-AC inverter for UPS applications. *IEEE Trans. Ind. Electron.* **2014**, *61*, 4142–4151. [CrossRef]
27. Danayiyen, Y.; Lee, K.; Choi, M.; Lee, Y. Model Predictive Control of Uninterruptible Power Supply with Robust Disturbance Observer. *Energies* **2019**, *12*, 2871. [CrossRef]
28. Nam, N.; Choi, M.; Lee, Y. Model Predictive Control of a Grid-Connected Inverter with LCL Filter using Robust Disturbance Observer. In Proceedings of the IFAC Workshop on Control of Smart Grid and Renewable Energy Systems CSGRES 2019, Jeju, Republic of Korea, 10–12 June 2019.
29. Danayiyen, Y.; Altaş, I.; Lee, Y. Robust Discrete Time Disturbance Observer with Finite Control Set Model Predictive Control in UPS System. In Proceedings of the IFAC Workshop on Control of Smart Grid and Renewable Energy Systems CSGRES 2019, Jeju, Republic of Korea, 10–12 June 2019.
30. Khalil, H.K. *Nonlinear Systems*, 3rd ed.; Prentice Hall: Hoboken, NJ, USA, 2002.

Disclaimer/Publisher's Note: The statements, opinions and data contained in all publications are solely those of the individual author(s) and contributor(s) and not of MDPI and/or the editor(s). MDPI and/or the editor(s) disclaim responsibility for any injury to people or property resulting from any ideas, methods, instructions or products referred to in the content.

energies

MDPI

Article

A Flexible Envelope Method for the Operation Domain of Distribution Networks Based on "Degree of Squareness" Adjustable Superellipsoid

Kewei Wang [1], Yonghong Huang [1,*], Junjun Xu [2] and Yanbo Liu [1]

1 School of Electrical and Information Engineering, Jiangsu University, Zhenjiang 212013, China; 2222207122@stmail.ujs.edu.cn (K.W.); 2222307102@stmail.ujs.edu.cn (Y.L.)
2 School of Automation, Nanjing University of Posts and Telecommunications, Nanjing 210023, China; jjxu@njupt.edu.cn
* Correspondence: hyh@ujs.edu.cn; Tel.: +86-138-6139-8265

Abstract: The operation envelope of distribution networks can obtain the independent *p*-*q* controllable range of each active node, providing an effective means to address the issues of different ownership and control objectives between distribution networks and distributed energy resources (DERs). Existing research mainly focuses on deterministic operation envelopes, neglecting the operational status of the system. To ensure the maximization of the envelope operation domain and the feasibility of decomposition, this paper proposes a modified hyperellipsoidal dynamic operation envelopes (MHDOEs) method for distribution networks based on adjustable "Degree of Squareness" hyperellipsoids. Firstly, an improved convex inner approximation method is applied to the non-convex and nonlinear model of traditional distribution networks to obtain a convex solution space strictly contained within the original feasible region of the system, ensuring the feasibility of flexible operation domain decomposition. Secondly, the embedding of the adjustable "Degree of Squareness" maximum hyperellipsoid is used to obtain the total *p*-*q* operation domain of the distribution network, facilitating the overall planning of the distribution network. Furthermore, the calculation of the maximum inscribed hyperrectangle of the hyperellipsoid is performed to achieve *p*-*q* decoupled operation among the active nodes of the distribution network. Subsequently, a correction coefficient is introduced to penalize "unknown states" during the operation domain calculation process, effectively enhancing the adaptability of the proposed method to complex stochastic scenarios. Finally, Monte Carlo methods are employed to construct various stochastic scenarios for the IEEE 33-node and IEEE 69-node systems, verifying the accuracy and decomposition feasibility of the obtained *p*-*q* operation domains.

Keywords: distributed power supply; distribution network; the voltage exceeds the limit; convex inner approximation method; operational envelope; hyperellipsoid; unknown state; operational domain

1. Introduction

In response to the national dual carbon goals, in recent years, there has been a high penetration of distributed energy resources (DERs) in distribution networks, accompanied by a significant increase in flexible and controllable resources [1,2]. Consequently, the operational risks of the systems have risen sharply. However, there are differences in the focus of control between the dispatch center and DERs. The dispatch center primarily coordinates system operations while satisfying constraints for safe operation, while DERs primarily aim to maximize their own benefits. There is some conflict between the two parties during operation. To achieve safe and reliable operation of the distribution network and flexible control of controllable resources, it is particularly important to carry out precise calculations of the operating domains [3] of each active node (i.e., nodes connected to flexible and controllable resources such as DERs) in the distribution network, and to achieve decoupled operation of each active node by controlling the output of flexible resources, such as distributed generation at each node.

In order to achieve the safe and reliable operation of the distribution network and the flexible regulation of controllable resources, the operation domain [3] of each active node in the distribution network (that is, nodes that flexibly regulate resources such as access to DERs) is accurately calculated, and the decoupling operation of each active node is realized by controlling the output of each node's flexible resources, such as distributed power supply.

The relevant literature proposes reactive voltage [4], active voltage [5], sag control curves, and reactive–active cooperative control strategies [6] for the direct control of DERs, enabling them to satisfy the safe operation constraints of the distribution network. However, it is challenging to monitor the operation status of DERs in real time [7].

The authors of [8] proved for the first time the existence of all distribution network security operation domains and provided a strict definition thereof. The calculation of the safe operating domain for distribution networks relies on the precise localization of stable operating boundaries. The authors of [9] gradually observed the system's operating status to obtain the local boundaries of the safe operating domain in distribution networks, while [10] obtained a collection of flexible operating points for DERs in terms of active and reactive power through extensive simulations of practical application scenarios using the Monte Carlo method. In [11], the problem of determining the critical point of stable operation of the system is transformed into the problem of finding the optimal solution, and the piecewise linearization method is used to fit the security domain boundary and generate the observable security domain space.

However, the above study cannot fit the boundary of the operation domain more accurately, and the related study [12] further proposes using the dynamic operation envelopes (DOEs) method of the distribution network operation domain to compute the safe operation domain of the distribution network. The solution of the flexible operation domain of distribution networks needs to follow two principles: one is to maximize the volume of the operating domain under the premise of approaching the actual solution, in order to improve the utilization rate of system capacity and ensure the flexible operation of the system; the other is to ensure the decomposition feasibility of the resulting operation domain, i.e., without violating the operation constraints of the network or the DERs, a complete trajectory of the total power in the feasible region can be obtained by appropriately scheduling the DERs.

The papers [13,14] are based on the exact unbalanced three-phase power flow method. On the premise of ensuring safe system operation, this method gradually adjusts the input or output power of fixed loads according to the system's operating status [15]. However, the method involves multiple iterations, the safe operating domain enveloped is conservative, and the conservative use of the network capacity leads to the waste of system resources [16–18]. Some scholars have proposed a method based on the unbalanced three-phase optimal power flow (UTOPF) computation [19], which takes the maximum of the scheduling region as the objective function to obtain the allowable regulation range of DERs. However, this method sacrifices the accuracy of running domain calculations. To solve the problem that UTPF and UTOPF methods are too conservative or have low solution accuracy, [20] collected a large amount of system data and used machine learning algorithms to fit a black-box model of the system for predicting DERs. However, this type of method relies on the accuracy and real-time nature of the data.

Existing studies mainly use conventional parameterized convex sets to fit the flexible operation domain of distribution networks; however, due to the randomness of DERs and the diversity of distribution network equipment, this method has poor adaptability to complex and stochastic scenarios, leading to the intensification of the system security risk during network operation. Considering that the ellipse has high adaptability and can dynamically adjust its parameters to change its size and shape, adapting to different operating states of the system, [21] proposes the *p-q* optimal elliptic safe operation domain envelope method for distribution networks with time decoupling, but this method only provides the total safe operation domain for the system operation, and it cannot provide the safe operation range for each active node separately. The authors of [22] constructed a rotating rectangular model for the operating envelope for each active node to obtain the *p-q* flexible operating domain between each active node, but the *p-q* of each active node cannot be independently regulated and cannot provide a specific scheduling scheme for the distribution network.

The operating domains of the active nodes in the distribution network exist in different spaces, and mapping from high-dimensional feasible domains to low-dimensional ones can result in the loss of depth information. However, most current research focuses on low-dimensional operating domains, which can easily lead to system violations [23–25] even when all active nodes follow the calculated flexible operating domains.

To solve the above problems, this paper first adopts the improved convex inner approximation method to obtain the convex solution space strictly contained in the original feasible region of the system, so as to ensure the feasibility of the flexible operation domain decomposition of the distribution network under the high proportion of new energy penetration. In order to avoid the loss of depth information accompanied by the mapping of the high-dimensional feasible domain to the low-dimensional one, a superellipsoid is proposed to obtain the maximum inner connected super-rectangle of the multidimensional feasible domain, and to realize *p-q* decoupling of the operation between each active node of the distribution network. A "Degree of Squareness" adjustable superellipsoid with higher adaptability is further proposed to envelope the flexible operation domain of the distribution network, which is closer to the optimal solution while achieving a pre-specified level. In view of the randomness of DERs and the diversity of distribution network equipment, a penalty term is added to the objective function considering the running state of each node to improve the applicability of the model to calculate multiple scenarios. Finally, for IEEE 33-node and IEEE-69 node systems, the Monte Carlo method is used to construct a variety of random scenarios, and the accuracy and decomposition feasibility of the *p-q* running domain are verified.

The remainder of this paper is structured as follows: Section 2 examines the use of an improved convex inner approximation method to obtain a convex solution space within the original feasible region of the distribution network, ensuring the feasibility of decomposing the flexible operation domain of the distribution network. Section 3 proposes a new flexible envelope method for the operation domain of distribution networks based on adjustable "Degree of Squareness" hyperellipsoids. Section 4 presents the simulations and analysis, while Section 5 provides the conclusions.

2. Feasible Domain Modeling Based on Network Operational Constraints

The active distribution network model encompasses multiple sets of constraints, with the power flow constraints exhibiting non-convex nonlinearity [26]. To ensure the feasibility of decomposing the flexible operation domain of the distribution network, it is crucial to rewrite the original distribution network model.

2.1. Rewriting of the Power Flow Model of the Distribution Network

Although the traditional distribution network model convex relaxation method [27–29] provides a good solution space for distribution network optimization, it cannot guarantee that the model is strictly contained in the original feasible region of the system, and the boundary conditions are often difficult to meet. Therefore, this paper introduces an improved convex inner approximation method to obtain a convex solution space strictly contained in the original feasible region of the system.

A feeder line of a distribution network is generally expressed, as shown in Figure 1.

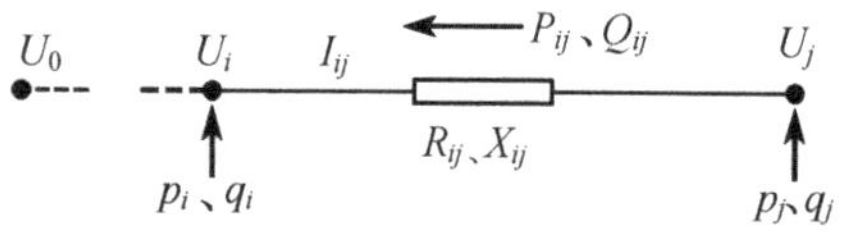

Figure 1. Radial network equivalent model.

In Figure 1, U_0 is the voltage of the first node; U_i and U_j are the voltage of node i and j, respectively; p_i and q_i are the net injected active and reactive power, respectively, at node i; P_{ij} and Q_{ij} are the active and reactive power of branch ij, respectively; I_{ij} is the current of branch ij; and $Z_{ij} = R_{ij} + jX_{ij}$ is the impedance of branch ij.

In this paper, $l_{ij} = I_{ij}^2$, $\boldsymbol{P} = [P_{ij}]^{\mathrm{T}}$, $\boldsymbol{Q} = [Q_{ij}]^{\mathrm{T}}$, and reference [30] were used to process the original distribution network model, and the following results were obtained:

$$\begin{cases} \boldsymbol{P} = \boldsymbol{Cp} - \boldsymbol{D}_R\boldsymbol{l} \\ \boldsymbol{Q} = \boldsymbol{Cp} - \boldsymbol{D}_X\boldsymbol{l}' \end{cases} \tag{1}$$

where $\boldsymbol{C} = (\boldsymbol{I}-\boldsymbol{A})^{-1}$; $\boldsymbol{p} = [p_i]^{\mathrm{T}}$; $\boldsymbol{D}_X = (\boldsymbol{I}-\boldsymbol{A})^{-1}\boldsymbol{AX}$; $\boldsymbol{D}_R = (\boldsymbol{I}-\boldsymbol{A})^{-1}\mathrm{AR}$; $\boldsymbol{C} = (\boldsymbol{I}-\boldsymbol{A})^{-1}$; $\boldsymbol{l} = [I_{ij}^2]_n$; $\boldsymbol{A}$ is the association matrix of nodes and branches; and the matrix $(\boldsymbol{I} - \boldsymbol{A})$ is invertible [30].

The variables $\boldsymbol{P}$ and $\boldsymbol{Q}$ are coupled to each other [31,32], and the rest of the variables except for the branch currents are decision or state variables, using $\boldsymbol{I}$ as an intermediate quantity to denote the other variables.

Using $\boldsymbol{p}$ and $\boldsymbol{l}$ as decision quantities, the upper and lower bounds for each proxy variable can be obtained:

$$\begin{cases} \boldsymbol{P}^+ = \boldsymbol{Cp} - \boldsymbol{D}_R\boldsymbol{l}_{\min} \\ \boldsymbol{P}^- = \boldsymbol{Cp} - \boldsymbol{D}_R\boldsymbol{l}_{\max} \\ \boldsymbol{Q}^+ = \boldsymbol{Cq} - \boldsymbol{D}_{X+}\boldsymbol{l}_{\min} - \boldsymbol{D}_{X-}\boldsymbol{l}_{\max} \\ \boldsymbol{Q}^- = \boldsymbol{Cq} - \boldsymbol{D}_{X+}\boldsymbol{l}_{\max} - \boldsymbol{D}_{X-}\boldsymbol{l}_{\min} \end{cases}, \tag{2}$$

where $\boldsymbol{D}_{X+}$ is a non-negative element in matrix $\boldsymbol{D}_X$; $\boldsymbol{q} = [q_i]^{\mathrm{T}}$; $\boldsymbol{D}_{X-}$ is a negative element in $\boldsymbol{D}_X$; $q^+ = q(p^+)$; $\boldsymbol{l}_{\max}$ and $\boldsymbol{l}_{\min}$ are the maximum and minimum values of $\boldsymbol{l}(\boldsymbol{p})$, respectively; and $\boldsymbol{P}^- \le \boldsymbol{P} \le \boldsymbol{P}^+, \boldsymbol{Q}^- \le \boldsymbol{Q} \le \boldsymbol{Q}^+$.

The state variables P and Q are functions of the branch currents l. Therefore, the accuracy of the relaxation depends on the values of the upper and lower bounds of l. The relaxation bounds of l will be illustrated below.

For the branch ij, the second-order Taylor expansion of the branch power flow rate based on the rated operating point $x_{ij}^0 = col\left\{P_{ij}^0, Q_{ij}^0, \left(U_j^0\right)^2\right\}$ of the system is given by the following expression:

$$l \approx l^0 + \boldsymbol{J}^{\mathrm{T}}\delta + \frac{1}{2}\delta^{\mathrm{T}}\boldsymbol{H}\delta, \tag{3}$$

where $l^0 = l(x_{ij}^0)$ indicates l when the system is at the rated operating point, while the matrices $\boldsymbol{\delta}$, $\boldsymbol{J}$, and $\boldsymbol{H}$ are defined in reference [33]:

$$\delta = \begin{bmatrix} P_{ij} - P_{ij}^0 \\ Q_{ij} - Q_{ij}^0 \\ v_j - v_j^0 \end{bmatrix}, \boldsymbol{J} = \begin{bmatrix} \frac{\partial l_{ij}}{\partial P_{ij}} \\ \frac{\partial l_{ij}}{\partial Q_{ij}} \\ \frac{\partial l_{ij}}{\partial v_j} \end{bmatrix}, \boldsymbol{H} = \begin{bmatrix} \frac{\partial^2 l_{ij}}{\partial P_{ij}^2} & \frac{\partial^2 l_{ij}}{\partial P_{ij}\partial Q_{ij}} & \frac{\partial^2 l_{ij}}{\partial P_{ij}\partial v_j} \\ \frac{\partial^2 l_{ij}}{\partial Q_{ij}\partial P_{ij}} & \frac{\partial^2 l_{ij}}{\partial Q_{ij}^2} & \frac{\partial^2 l_{ij}}{\partial Q_{ij}\partial v_j} \\ \frac{\partial^2 l_{ij}}{\partial v_j\partial P_{ij}} & \frac{\partial^2 l_{ij}}{\partial v_j\partial Q_{ij}} & \frac{\partial^2 l_{ij}}{\partial v_j^2} \end{bmatrix}.$$

The derivation of the upper and lower bound expressions for l is shown below:

$$\begin{aligned} l = |l| &\approx \left| l^0 + J^{\mathrm{T}}\delta + \tfrac{1}{2}\delta^{\mathrm{T}} H \delta \right| \\ &\leq \left| l^0 \right| + \left| J^{\mathrm{T}}\delta \right| + \left| \tfrac{1}{2}\delta^{\mathrm{T}} H \delta \right| \\ &\leq l^0 + \max\left\{ 2\left| J^{\mathrm{T}}\delta \right|, \left| \delta^{\mathrm{T}} H \delta \right| \right\}, \end{aligned} \tag{4}$$

$$l \leq l^0 + \max\left\{ 2\left| J_+^{\mathrm{T}}\delta_+, J_-^{\mathrm{T}}\delta_- \right|, \left| \delta^{\mathrm{T}} H \delta \right| \right\} = l_{\max}, \tag{5}$$

$$l \geq l^0 + J_+^{\mathrm{T}}\delta_- + J_-^{\mathrm{T}}\delta_+ = l_{\min}, \tag{6}$$

where J_+ and J_- are matrices consisting of non-negative and negative elements in J, respectively, while δ_+ and δ_- are matrices consisting of non-negative and negative elements in δ, respectively. $l_{\min}$ may be negative, but $l_{ij} = I_{ij}^2 \geq 0$, so the lower bound of l_{ij} should take the value of $\max\{0, l_{\min}\}$.

After defining the upper and lower bounds of l, P and Q can be replaced by proxy variables to obtain a convex solution space inside the initial non-convex region.

2.2. Distribution Network Constraints

The following constraints need to be met for distribution network operation:

$$\begin{cases} I_{ij} \leq I_{ij,\max} \\ U_{i,\min} \leq U_i \leq U_{i,\max} \\ P_{ij,\min} \leq P_{ij} \leq P_{ij,\max} \\ P_{ij}^2 + Q_{ij}^2 \leqslant S_{ij,\max}^2 \\ P_{g,i} - P_{L,i} = U_i \sum\limits_{j=1}^{n} U_j (G_{ij}\cos\theta_{ij} + B_{ij}\sin\theta_{ij}) \\ Q_{g,i} - Q_{L,i} = U_i \sum\limits_{j=1}^{n} U_j (G_{ij}\sin\theta_{ij} - B_{ij}\cos\theta_{ij}) \end{cases}, \tag{7}$$

where: $I_{ij,\max}$ is the upper limit of the branch current amplitude; U_i and U_j are the voltage amplitudes at nodes i and j, respectively; $U_{i,\max}$ and $U_{i,\min}$ are the upper and lower limits of the node voltage amplitudes, respectively; P_{ij} and Q_{ij} are the active and reactive powers of the branch ij, respectively; $P_{ij,\min}$ and $P_{ij,\max}$ are the upper and lower bounds of the active power of the branch ij, respectively; G_{ij} is the conductance of the line ij; θ_{ij} is the phase-angle difference between the voltages at nodes i and j; $S_{ij,\max}$ is the apparent power; $P_{g,i}$ is the active power injected by the power supply at node i, and $P_{L,i}$ is the active power consumed by the load at node i; $Q_{g,i}$ is the reactive power injected by the power supply at node i, and $Q_{L,i}$ is the reactive power consumed by the load at node i.

3. Construction and Improvement of Flexible Operational Domain Models

Due to the loss of depth information accompanying the mapping from high-dimensional feasible regions to lower dimensions, even if all active nodes are located within the calculated flexible operation domain, the network may still face operational safety risks. Therefore, when using polygon or ellipse approximation methods to solve for the flexible operation domain of the distribution network, the obtained capacity allocation schemes may not all be practically feasible.

3.1. Distribution Network Operating Envelope Definition

The feasibility region (FR) model developed above, denoted as $\mathcal{F}(p,q)$, is represented as follows:

$$\begin{aligned} \mathcal{F}(p,q) &= \{(p,q) \mid \tau_1 p + \tau_2 q + \tau_3 \Lambda = \ell_1, \tau_4 \Lambda \leq \ell_2\} \\ &\left\{(p,q) \mid -\tau_4 \tau_3^{-1} \tau_1 p \leq \ell_2 + -\tau_4 \tau_3^{-1} (\ell_1 - \tau_2 q)\right\}, \end{aligned} \tag{8}$$

where: $\boldsymbol{p} = \{P_1, \ldots, P_n\}$, and $\boldsymbol{q} = \{q_1, \ldots, q_n\}$; p and q are the variables corresponding to the active and reactive power to be optimized, respectively; Λ is the vector consisting of all the variables of the distribution network (including state and control variables); $\boldsymbol{\tau}_1$, $\boldsymbol{\tau}_2$, $\boldsymbol{\tau}_3$, $\boldsymbol{\tau}_4$, ℓ_1, and ℓ_2 are the constant parameter matrices of suitable size, and $\overline{\tau}_4 = -\tau_4\tau_3^{-1}$; $\tau_4\Lambda \leq \ell_2$ indicates all distribution network operation constraints, including the voltage amplitude and current amplitude constraints of distribution lines.

$\mathcal{F}(p,q)$ provides a guideline for the safe operation [34,35] of the distribution network; however, the high penetration rate of resources such as DERs exacerbates the risks associated with active distribution network operations. Issues like voltage overruns and line overloads become prominent, and to achieve practicality within this feasible domain, it is necessary to consider the coupling between controllable resources at each node of the distribution network [36,37].

In this paper, we aim to find a decoupled feasibility region (DFR) within the convex solution space of the original feasible region of the system, which allows p-q independent scheduling among active nodes, and each active node can control its power independently within the boundary of the DFR without violating the network operation constraints, so as to achieve the decoupled operation of each active node in the active distribution network. The DFR is represented as a hyperrectangle in multidimensional space, and Figure 2 shows the schematic diagram of the decoupled p-q operation domain between active node 1 and active node 2, where $\boldsymbol{H_F}$ is the high-dimensional feasible domain space. From the perspective of spatial geometry, the power planes of each active node should be orthogonal to each other [22].

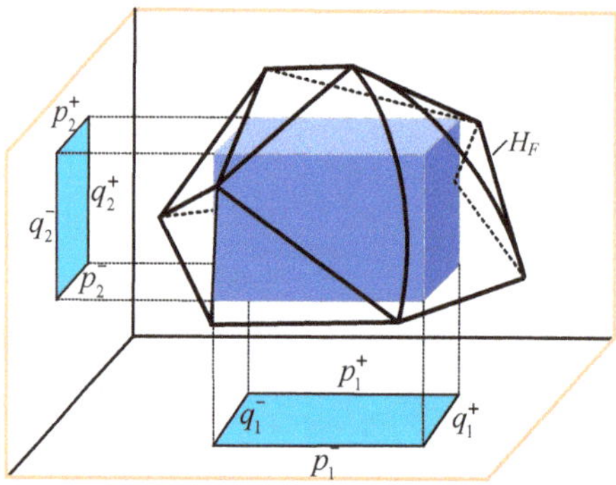

Figure 2. A p-q decoupling diagram between active nodes.

The mathematical expression for the hyperrectangle is as follows:

$$F_c(p_v, q_v) = \begin{cases} p_v^- \leq p_v \leq p_v^+ \\ q_v^- \leq q_v \leq q_v^+ \end{cases}, \tag{9}$$

where v is the vth active node of the distribution network; p_v^- and p_v^+ are the upper and lower bounds of active power after decoupling of node v, respectively; q_v^- and q_v^+ are the upper and lower bounds of reactive power after decoupling of node v, respectively.

3.2. Tunable Superellipsoid Running Envelope Solution Model Based on "Degree of Squareness"

Due to the loss of depth information associated with the mapping from high-dimensional to low-dimensional feasible domains, the network may still face the risk of safe operation even if all active nodes are located in the calculated flexible operation domain. Therefore, the capacity allocation schemes obtained by using polygon or ellipse approximation methods to solve the flexible operation domain of the distribution network may not all be practical and feasible.

As shown in Figure 3, the blue "*" indicates the baseline operating point, and the red "*" indicates the actual operating point. it is assumed that active nodes 1 and 2 have operation points Θ' and Θ'', respectively, at a certain moment, both of which are located in their respective flexible operation domains, but the risk of voltage overrun still occurs for the system. The process of mapping the high-dimensional model to the low-dimensional model is irreversible with missing depth information.

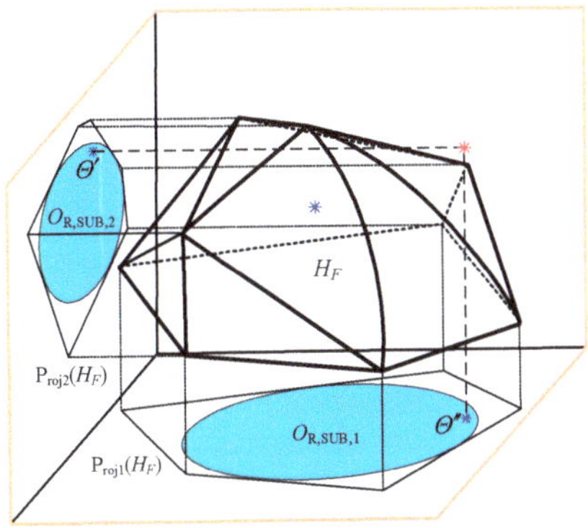

Figure 3. Run domain assignment: non-feasible example.

In Figure 3, P_{roj1} and P_{roj2} are its mappings on different low-dimensional running domain spaces for three-dimensional examples, while $O_{R,SUB}$ are the ellipsoidal running domains obtained from the low-dimensional feasible domain approximation.

Aiming at the above problems, this paper proposes a flexible operation domain envelopment method based on hyperellipsoids to avoid the loss of depth information due to dimensionality reduction; by embedding the maximum hyperellipsoid in the polyhedron FR in order to obtain the maximum hyperrectangle, we achieve the *p*-*q* decoupled operation among the active nodes of the power distribution network.

The hyperellipsoid is essentially a stretching transformation of the unit sphere, with the following formula:

$$\mathcal{E} = \begin{Bmatrix} \Theta = \Theta_c + L\omega \\ \omega : \|\omega\|_2 \leqslant 1 \end{Bmatrix}, \tag{10}$$

where ω is the coordinate vector of the point inside the superellipsoid, Θ_c is the center of the superellipsoid, and L is a positive definite $\nu - 1$-dimensional diagonal matrix representing the lengths of all axes of the ellipsoid.

Figure 4 shows the schematic diagram of the flexible operation domain envelope method based on the superellipsoid, which first searches for a maximal superellipsoid within the convex solution space of the original feasible region of the system to obtain the total *p*-*q* operation domain of the distribution network, which is convenient for the overall planning and operation of the distribution network, and further calculates the maximal internally connected super-rectangles of the superellipsoid, which achieves the decoupling of the *p*-*q* operation among the active nodes.

Assuming that the coupling of multiple active nodes is not considered in this paper, the ellipsoid approximation method has a good envelope for the two-dimensional flexible operation domain projected by an active node, but the actual operation of the system needs to take into account the coupling of multiple active nodes in order to determine the feasibility of the capacity allocation scheme; therefore, the construction of a high-dimensional *p*-*q* operation domain is particularly important. According to the existing studies, the three-dimensional running domain derived from multiple sets of two-dimensional *p*-*q* elliptic running domains is not necessarily a standard ellipsoid but, rather, an ellipsoid with "square degree". This conclusion can be extended to higher dimensions.

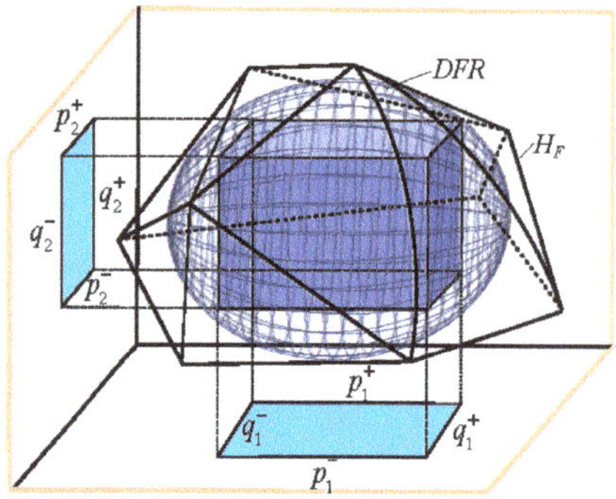

Figure 4. Dynamic operation envelope method based on hyperellipsoids.

From the above analysis, the dynamic envelope of the flexible operation domain of the distribution network considering the "Degree of Squareness" adjustable superellipsoid can be closer to the optimal solution based on reaching the pre-specified level. Figure 5 shows the power boundary curves of the active nodes with different squareness parameters β. The power curves are convex when $\beta \geq 1$ and are closer to the rectangle as β increases.

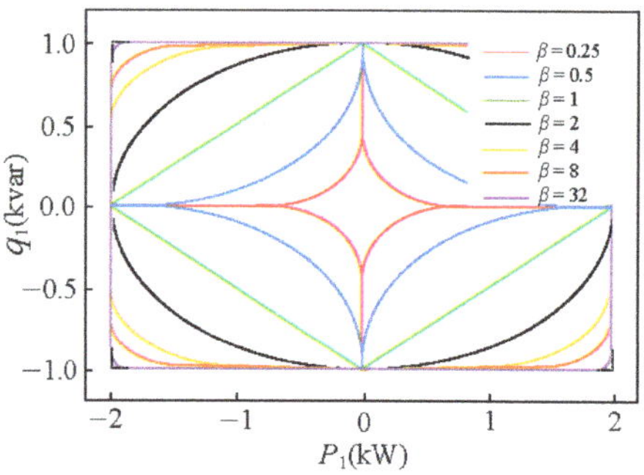

Figure 5. The concept of a "square" adjustable hyperellipsoid.

Consider the "Degree of Squareness" adjustable hyperellipsoid, which can be expressed as follows:

$$\mathcal{E} = \begin{Bmatrix} \Theta = \Theta_c + L\omega \\ \omega : \|\omega\|_\beta^\beta \leqslant 1 \end{Bmatrix}, \tag{11}$$

In this paper, the squareness parameter β is adjusted in order to expand the volume of the superellipsoid, but the volume of the superellipsoid changes slowly as the parameter becomes larger, so in this paper we take $\beta = 2^K$ (K is a positive integer), and the superellipsoid equation is further expressed as follows:

$$\mathcal{E} = \begin{Bmatrix} \Theta = \Theta_c + L\omega \\ \omega : \omega \in \gamma_h \end{Bmatrix}, \tag{12}$$

where $\gamma_h = \left\{ y_1 \mid y_{k,i}^2 \leq y_{k+1,i} \; \forall i, \; \forall k \leq K-1; \|y_k\|_2 \leq 1 \right\}$ is the set of constraints, where y_k is an intermediate vector variable,$y_{k,i}$ is the ith element of y_k, and γ_h contains $mK + 1$ quadratic constraints.

Using the "Degree of Squareness" adjustable hyperellipsoid parameters, the formula for the volume of the tangent hyperrectangle is obtained as follows:

$$V_h = 2^K \prod |L_{ii}\omega_i| = 2^K \det(L) \prod_i \omega_i, \tag{13}$$

where ω_i is the ith element of ω.

The optimal solution ω_i^* is denoted when the volume of the interior hyperrectangle is maximized:

$$\omega_i^* = argmax_{\omega:\|\omega\|_\beta^\beta=1} 2^K \det(L) \sum \log(\omega_i) = v^{-1/n}, \tag{14}$$

where $\omega :\| \ \omega \ \|_\beta^\beta = 1$ is the surface of the "square" adjustable hyperellipsoid.

The hyperellipsoid's internal tangent hyperrectangle can be re-expressed as follows:

$$V_h = \left(2v^{-1/n}\right)^K \det(L), \tag{15}$$

Most of the existing studies ignore the operation state of each active node, and the solved operation domain is fixed. In order to reflect the flexibility of the operation domain solution and weaken the unfavorable effects of the uncertainty and complexity of resources such as DERs, this paper presents the following discussion:

(1) Disregarding node operational states

The problem of computing the maximum internally connected hyperrectangle of a hyperellipsoid can be expressed as follows:

$$\begin{aligned} V_{h,\max} &= \max_{p,q,L,\Theta_c} \log(\det(\boldsymbol{L})) \\ &= \textstyle\sum_i \log(\boldsymbol{L}_{ii}) \\ &= \max \prod_i \left(p_i^+ + p_i^-\right)\left(q_i^+ + q_i^-\right) \\ &= \max \sum_i \log\left(p_i^+ + p_i^-\right)\left(q_i^+ + q_i^-\right) \\ &= \max \sum_i \left[\log\left(p_i^+ + p_i^-\right)\right] \\ &\quad + \log\left(q_i^+ + q_i^-\right), \end{aligned} \tag{16}$$

$$\begin{aligned} s.t.\ \ & \underline{p} \le p \le \overline{p}, \underline{q} \le q \le \overline{q}, \\ & \left\| \tau_4 \tau_3^{-1} \tau_1 L \right\|_2 - \tau_4 \tau_3^{-1} \tau_1 \Theta_c \\ & \le \boldsymbol{\ell}_2 - \tau_4 \tau_3^{-1} (\boldsymbol{\ell}_1 - \tau_2 q), \end{aligned} \tag{17}$$

where $\overline{p}$ and $\underline{p}$ are the upper and lower bounds of the active power constraints, respectively, while $\overline{q}$ and $\underline{q}$ are the upper and lower bounds of the reactive power constraints, respectively.

(2) Consideration of node operational state

When the operating state of the node is unknown, introduce correction coefficients η, penalizing $\sum_i |\Theta_c(i)|$ in the objective function to ensure that the inputs and outputs are as equal in magnitude as possible and V_h can be maximized as follows:

$$\begin{aligned} V_{h,\max} &= \max_{p,q,L,\Theta_c} \log(\det(L)) - \eta \sum_i \xi_i \\ &= \sum_i \log(L_{ii}) - \eta \sum_i \xi_i \\ &= \max \prod_i (p_i^+ + p_i^-)(q_i^+ + q_i^-) - \eta \sum_i \xi_i \\ &= \max \sum_i \log(p_i^+ + p_i^-)(q_i^+ + q_i^-) - \eta \sum_i \xi_i \\ &= \max \sum_i \left[\log(p_i^+ + p_i^-)\right] \\ &\quad + \log(q_i^+ + q_i^-) - \eta \sum_i \xi_i, \end{aligned} \tag{18}$$

$$\begin{aligned}&s.t.\ \ \underline{p} \le p \le \overline{p}, \underline{q} \le q \le \overline{q},\\&\overline{\tau}_4\tau_1 L y_1 + \overline{\tau}_4\tau_1\Theta_c \le \ell_2 + \overline{\tau}_4(\ell_1 - \tau_2 q)\ ,\forall y_1 \in \gamma_h,\\&-\xi_i \le \Theta_c(i) - \lambda_i L_{ii} v^{-1/2^K} \le \xi_i,\ \forall i,\end{aligned} \tag{19}$$

where η is the correction coefficient, ξ_i is the slack variable introduced by considering the operating state of the node, $\Theta_c(i)$ is the ith element of the vector Θ_c, λ_i is a 0–1 variable denoting the operating state of node i, $\lambda_i = 1$ denotes the input power, $\lambda_i = -1$ denotes the output power, and $\Theta_c(i) - \lambda_i L_{ii}/\sqrt{n}$ is the proposed input/output limit to be assigned to each node.

Formula (19) Constraint (3) ensures that the limit of output power is as close to 0 kW as possible when the node inputs power, and the limit of input power is as close to 0 kW as possible when the node outputs power, so that the node can freely vary its power between 0 kW and the assigned capacity limit. For example, if the node is outputting power, it will be penalized as its input limit $\Theta_c(i) + L_{ii} v^{-1/2^K}$ in the objective function. Then, consider the operational state of the customer after replacing n with 2^K.

The upper and lower limits of the capacity of the active node are denoted as follows:

$$\Theta_c(i) - L_{ii} v^{-1/2^K}, \tag{20}$$

$$\Theta_c(i) + L_{ii} v^{-1/2^K}, \tag{21}$$

where $\Theta_c(i)$ and L_{ii} need to be optimized.

For the uncertain $y_1 \in \gamma_h$ in (19), a generalized formula for the maximum value of $\overline{\tau}_4\tau_1 L y_1$ is derived in order to enable it to express the general case of distribution network operation:

$$\begin{aligned}&\mathbf{max}_{(y_1,\ldots,y_k)} x^T y_k\\&s.t.\ y_{k,i}^2 \le y_{k+1,i}\ \ \forall k \le K-1, \forall i(\alpha_{k,i}),\\&\|y_k\|_2 \le 1(\alpha_k)\end{aligned} \tag{22}$$

where x is a known vector, a row of $\overline{\tau}_4\tau_1 L$; α_k and $\alpha_{k,i}$ are Lagrange multipliers.

The Lagrange function of Equation (22) is

$$\begin{aligned}&W(\boldsymbol{\alpha}_{k,i}, \boldsymbol{\alpha}_K, y_k) = x^T y_k - \sum_{2\le k\le K-1}\sum_i [\boldsymbol{\alpha}_{k,i} y_{k,i}^2 - \boldsymbol{\alpha}_{k-1,i} y_{k,i}]\\&-\sum_i\left[\boldsymbol{\alpha}_{1,i} y_{1,i}^2 - x_i y_{1,i}\right] - \boldsymbol{\alpha}_k(\|y_k\|_2 - 1)\\&\le x^T y_k + \sum_i \frac{x_i^2}{4\boldsymbol{\alpha}_{1,i}} + \sum_{2\le k\le K-1}\sum_i \frac{\boldsymbol{\alpha}_{k-1,i}^2}{4\boldsymbol{\alpha}_{k,i}}\\&-\boldsymbol{\alpha}_K(\|y_k\|_2 - 1),\end{aligned} \tag{23}$$

Then, $\min\limits_{\alpha_{k,i},\alpha_K}\max\limits_{y_K} W(\alpha_{k,i}, \alpha_K, y_k)$ is an equivalent problem of Formula (22), which can be further expressed as follows:

$$\begin{aligned}&\min_{\alpha_{k,i},\alpha_K} \alpha_K + \sum_i \frac{x_i^2}{4\alpha_{1,i}} + \sum_{2\le k\le K-1}\sum_i \frac{\alpha_{k-1,i}^2}{4\alpha_{k,i}},\\&s.t.\ \|\alpha_{k-1}\| \le \alpha_k,\end{aligned} \tag{24}$$

By introducing the intermediate variable $t_{m,k,i}$, replacing x with $[\overline{\tau}_4\tau_1 L]_m^T$, and removing the minimum operator in the objective function of Formula (24), Formulas (18) and (19) can be reformulated as follows:

$$\begin{aligned}
&\max_{p,q,L,\Theta_c} \log(\det(L)) - \eta \sum_i \xi_i, \\
&\text{s.t. } \underline{p} \le p \le \overline{p}, \underline{q} \le q \le \overline{q}, \\
&\alpha_{m,k} + \sum_{k \le K-1} \sum_i t_{m,k,i} + [\overline{\tau}_4 \tau_1 \Theta_c]_m \\
&\le [\ell_2 + \overline{\tau}_4 \ell_1]_m - [\overline{\tau}_4 \tau_2 q]_m), \forall m, \\
&\left([\overline{\tau}_4 \tau_1 L]_{m,i}\right)^2 \le 4\alpha_{m,1,i} t_{m,1,i}, \forall m, \forall i, \\
&\alpha_{m,k-1,i}^2 \le 4\alpha_{m,k,i} t_{m,k,i}, \forall m, \forall k \in \{2, \ldots, K-1\}, \forall i, \\
&\|\alpha_{m,k-1}\|_2 \le \alpha_{m,k}, \forall m, \\
&-\xi_i \le \Theta_c(i) - \Lambda_i L_{ii} v^{-1/2^K} \le \xi_i, \forall i,
\end{aligned} \tag{25}$$

where $[\cdot]_m$ denotes the mth row of the matrix or the mth element of the vector, and $[\cdot]_{m,i}$ denotes the (m, i)th element of the matrix.

The corresponding maximum internally connected super-rectangle M is obtained by solving the above formulas:

$$M = \begin{cases} \frac{-L_{ii}^*}{\sqrt{n}} + \Theta_{c,i}^* \le p_i \le \frac{L_{ii}^*}{\sqrt{n}} + \Theta_{c,i}^* \\ \frac{-L_{\varphi\varphi}^*}{\sqrt{n}} + \Theta_{c,\varphi}^* \le q_i \le \frac{L_{\varphi\varphi}^*}{\sqrt{n}} + \Theta_{c,\varphi}^* \end{cases}, i = 1, \ldots, v, \tag{26}$$

where L^* and Θ_c^* are the maximal hyperrectangles L and Θ_c, respectively, and $\varphi = i + 1$, corresponding to the maximal hyperellipsoid at its maximum.

4. Experimental Verification

In this paper, IEEE 33-node and IEEE 69-node systems were used for simulation and analysis, the system topology and related parameter settings were derived from the literature [38,39], and the Monte Carlo method was used to construct a variety of stochastic scenarios to verify the accuracy of the resulting *p-q* operation domain of each active node and the feasibility of decomposition. The IEEE 33-node system used in this paper selects nodes 6, 13, 9, 24, 30, and 33 as active nodes, while the IEEE 69-node system selects nodes 4, 8, 21, 25, 36, 40, 45, 48, 52, and 60 as active nodes. The specific configurations of the IEEE 33-bus system and the IEEE 69-bus system are detailed in Appendix A.

Line loss is not considered in this study, so the line power flow constraint depends on the network structure. The reference capacity of all examples was set to 10 MVA, and the node voltage constraint ranged from 0.95 p.u. to 1.05 p.u. The CONOPT solver in GAMS 31.1.1 software was used to obtain the security domain boundary points. The hardware environment of the test system was an AMD A10-8700P 1.80 GHz processor by AMD (Santa Clara, CA, USA), and the memory capacity was 8 GB.

4.1. Total Operational Domain Envelope Security Assessment

Taking the IEEE 33-node system as an example, 1000 actual operating points were randomly selected during the whole time period of the distribution network to analyze the coverage of the actual operating points by the superellipsoid enveloping method and the adjustable superellipsoid enveloping method with the "Degree of Squareness" compared with the traditional convex dynamic operation envelopes (CDOEs) [40]. Considering that there are errors in the prediction of the actual operation scenarios of the system, this paper improves the prediction accuracy requirement by setting the maximum of hazardous scenarios to not exceed 5%.

Figure 6 shows the actual operating point coverage of the IEEE 33-node system based on the CDOEs approach. Most of the actual operating points in the analyzed high percentage penetration of distributable power are located inside and on the boundaries of the operating domain of the envelope, where 14% of the actual operating points are located outside the operating domain, exceeding the upper limit of the hazardous scenario setup.

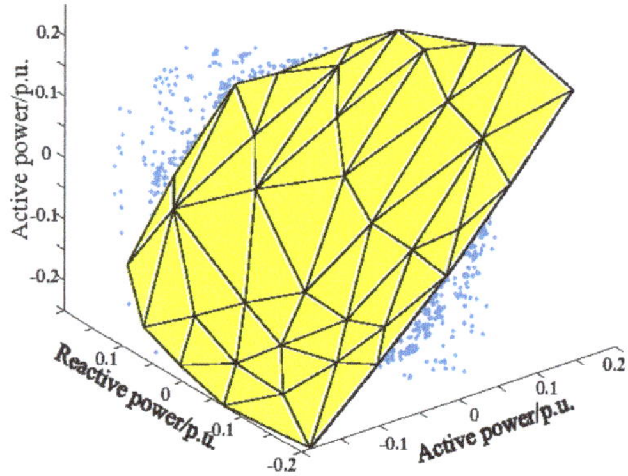

Figure 6. Envelope of actual running points for the CDOEs method.

Figure 7 shows the running domain covering the actual running points enveloped by the HDOEs method adopted in this paper, and the analysis shows that the running domain established by the HDOEs method can cover 91% of the actual running points, which is an overall improvement of about 5% compared with the effect of the CDOEs method.

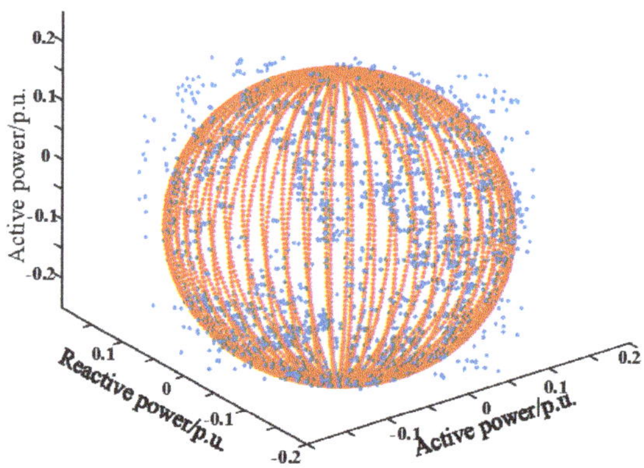

Figure 7. Envelope of actual running points for the HDOEs method.

In order to obtain higher envelope accuracy and a larger envelope range, we added the "Degree of Squareness" on the basis of the HDOEs method, and Figure 8 shows the HDOEs method with an adjustable "Degree of Squareness" on the envelope of actual operating points, which is able to cover 97% of the selected actual operating points, while the envelope range is further improved by about 6% to meet the maximum upper limit of hazardous scenarios set by the system.

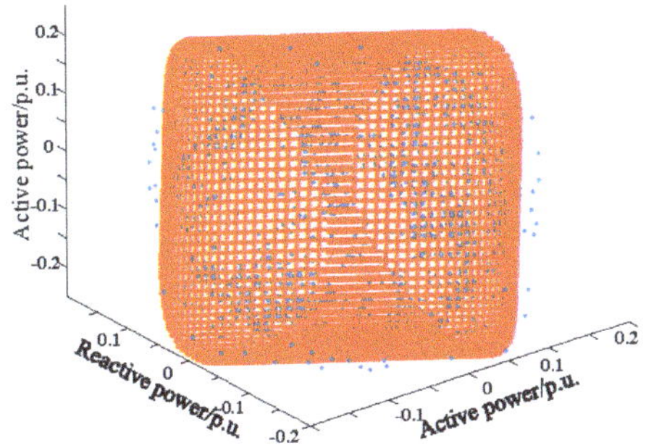

Figure 8. The envelope of actual operating points for the MHDOEs method.

4.2. Active Node Independent Regulatory Scope Analysis

The envelope of the distribution network operation domain needs to maximize the operation domain in close proximity to the actual solution in order to improve the utilization of the system capacity as well as the flexibility of the operation. Figure 9 compares the *p-q* operation domains of each active node generated by the rectangular DOEs (RDOEs) [22] method with the method proposed in this paper.

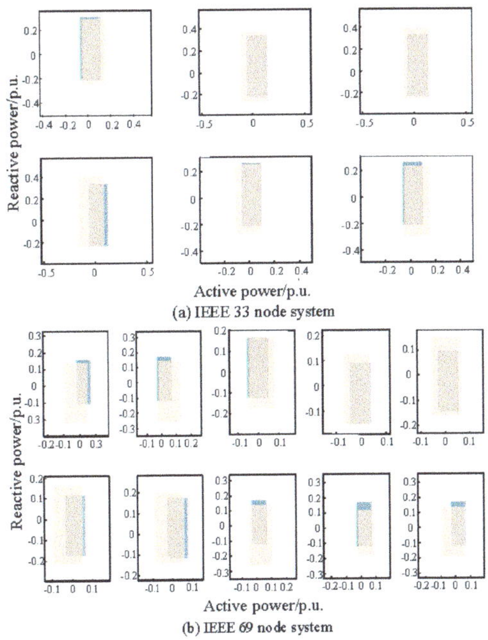

Figure 9. Comparison of *p-q* operating domains of active nodes generated by different methods.

The blue color in Figure 9 shows the area covered by the rectangular envelopment method, while the light orange color shows the area covered by the method proposed in this paper. Figure 9a shows the *p-q* operation domain range of 6 active nodes of the IEEE 33-node system, while Figure 9b shows the *p-q* operation domain range of 10 active nodes of the IEEE 69-node system, and both methods ensure that the benchmark operation point is within the feasible domain of the enveloped area.

Figure 10 shows the area analysis of the operating domains enveloped by different methods. Compared with the conventional RDOEs method, the HDOEs method proposed in this paper increases the operation envelope area, with an average enhancement effect of 144.82% for the IEEE 33-node system and 131.21% for the IEEE 69-node system. By using MHDOEs to correct the operation domains encompassed by the HDOEs method, the IEEE 33- and 69-node systems' envelope areas are further improved by 9.87% and 1.37%, respectively.

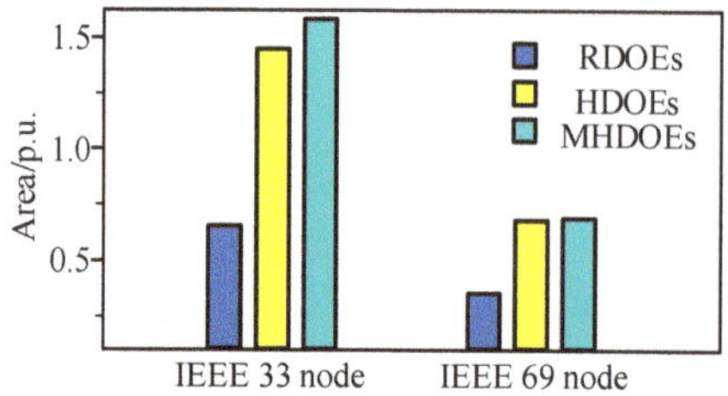

Figure 10. Different methods to run the envelope area comparison.

The above analysis verifies that the *p*-*q* operation domain envelopment method based on the "Degree of Squareness" adjustable superellipsoid can expand the operation domain on the basis of the existing method, providing more operable *p*-*q* adjustment ranges for the active nodes, and the operation domain areas of the active nodes enveloped by the method in this paper do not differ much, which is in line with the actual operation situation.

Table 1 shows the average errors generated by the RDOEs method, HDOEs method, and MHDOEs method applied to the operating domain envelope of the IEEE 33-node system and the IEEE 69-node system.

Table 1. Comparison of envelope mean error of different methods.

	RDOEs	HDOEs	MHDOEs
IEEE 33	0.21	0.09	0.01
IEEE 69	0.23	0.13	0.02

As can be seen from Table 1, compared with the conventional RDOEs method, the HDOEs method proposed in this paper improves the accuracy of the operational envelope, where the average error of the IEEE 33-node system is reduced by 0.12, and that of the IEEE 69-node system is reduced by 0.1. Using MHDOEs to modify the operating domain enclosed by the HDOEs method, the envelope area of the IEEE 33- and 69-node systems is further reduced by 0.08 and 0.11, respectively, so the security domain constructed by the method in this paper is more accurate.

4.3. Operational Domain Envelope Time Analysis

Table 2 shows the analysis of the runtime domain computation time for different methods.

Table 2. Different methods to run envelope time analysis.

	RDOEs	HDOEs	MHDOEs
IEEE 33	0.039 s	16.154 s	18.674 s
IEEE 69	0.055 s	24.564 s	28.389 s

By analyzing Table 2, it can be seen that the HDOEs method and MHDOEs method proposed in this paper directly envelope the operational domain in high-dimensional space; thus, compared with the conventional RDOEs method, the calculation time of the operational domain is longer, and the calculation time is slightly increased compared with the RDOEs method but still remains at a small order of magnitude. The calculation time of the operational domain meets the needs of the calculation of the operational domain of the distribution network. Moreover, for the IEEE 33- and 69-node systems, the operating domain can be maximized on the premise of ensuring model accuracy, and the distribution network resources can be flexibly regulated under the high proportion penetration of DERs.

4.4. Operational Domain Envelope Security Validation

While verifying the accuracy of the envelope method in this paper, it is also necessary to verify the security of each actual operation point in the operation domain; in this paper, the Monte Carlo method is used to simulate the actual operation of the distribution network, the flexible resources (such as DERs) of each active node are controlled to satisfy the active–reactive power of randomly selected operation points, the power of the baseline operation point is used as the injected power of the nodes (except for the active node), and we analyze whether the voltage of each node overruns the limits.

Figure 11 shows the node multi-scenario voltage fluctuation curves for the IEEE 33-node and IEEE 69-node systems.

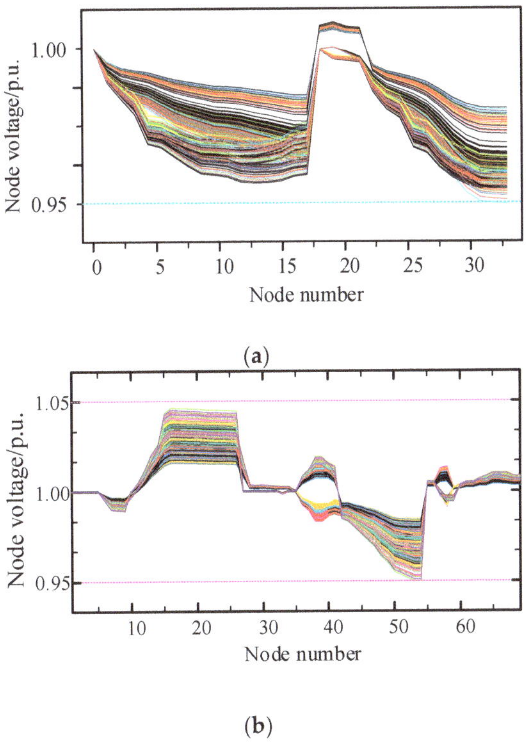

Figure 11. Run envelope security verification: (**a**) IEEE 33-node system voltage fluctuation curve; (**b**) IEEE 69-node system voltage fluctuation curve.

The results show that there is no overrun in the voltage of each node within the flexible operation domain of the DOEs methodology envelope, as shown in Figure 11. It is able to ensure the safe operation of the distribution network.

5. Conclusions

This paper proposes a dynamic envelope method for the operation domain of distribution networks using an adjustable "degree of squareness" superellipsoid, which achieves p-q decoupling operation among various active nodes in the distribution network. This provides an effective means to address the issues of different ownership between distribution networks and distributed energy resources (DERs) and discrepancies in control objectives. Furthermore, considering the operating states of the distribution network, a margin measure with a penalty term was added to the model to enhance the network applicability of the proposed method. Finally, simulation analyses were conducted on the IEEE 33-node and IEEE 69-node systems, leading to the following conclusions:

(1) The method adopted in this paper uses an improved convex inner approximation approach, providing a convex solution space that is strictly contained within the original feasible region of the system for subsequent envelope construction of the distribution network operation domain.
(2) This paper employs a "degree of squareness" adjustable superellipsoid envelope method with high network adaptability, capable of dynamically adjusting parameters to change its size and adapt to different operating states of the system, achieving p-q decoupling operation among various active nodes in the distribution network. Compared with traditional methods, the proposed method significantly enhances the range of the operation envelope while ensuring the feasibility of solution decomposition.
(3) In response to the high penetration of DERs, compared with traditional convex envelope methods, this paper's method adds a penalty term to the model to penalize unknown states of each node during the calculation of the operation domain, effectively mitigating the adverse effects brought by the uncertainty and complexity of resources such as DERs.

The method proposed in this paper can adjust the shape and size of the envelope by changing the envelope parameters according to the operating state of the system, and it introduces the margin measure with a penalty term. This flexibility is an important embodiment of the extensibility of the method. In the future, the design of envelope parameters and margin metrics could be further strengthened to consider more types of system disturbances and uncertainties, such as weather changes, equipment failures, etc., to enhance the robustness and practicability of the method.

Author Contributions: Investigation, Y.H.; Project Administration, Y.H.; GAMS 31.1.1 software, K.W., J.X., and Y.L.; Supervision, Y.H.; Visualization, J.X.; Writing—Original Draft, K.W. All authors have read and agreed to the published version of the manuscript.

Funding: This work was sponsored by the National Natural Science Foundation of China (52107101), Priority Academic Program Development of Jiangsu Higher Education Institutions (PAPD2024), and Science and Technology Project of State Grid Jiangsu Electric Power Co., Ltd. (J2020014).

Data Availability Statement: The original contributions presented in the study are included in the article.

Conflicts of Interest: The authors declare no conflict of interest. The funder was not involved in the study design, collection, analysis, interpretation of data, the writing of this article or the decision to submit it for publication.

Nomenclature

The following are some of the symbols and abbreviations used in this text:

Symbol

$\mathcal{F}(p,q)$	Mathematical expression of feasible domain
$\tau_4 \Lambda \leq \ell_2$	The set of all run constraints
$F_c(p_v, q_v)$	Mathematical expression of hyperrectangle
L	The length of all axes of a hyperellipsoid
$\mathcal{E}$	Hyperellipsoid with adjustable "square"
V_h	Built-in hyperrectangular volume
ξ_i	Introduced relaxation variable
$\Theta_c(i) - \lambda_i L_{ii} / \sqrt{n}$	The input/output quota to be allocated by each node

Abbreviations

DERs	Distributed energy resources
DFR	Decoupled feasibility region
UTOPF	Unbalanced three-phase optimal power flow
DOEs	Dynamic operation envelopes
CDOEs	Convex DOEs
RDOEs	Rectangular DOEs
HDOEs	Hyperellipsoidal DOEs
MHDOEs	Modified HDOEs

Appendix A

Figure A1 is the schematic diagram of the IEEE 33-node system, and Table A1 shows the distributed power configuration of the IEEE 33-node system.

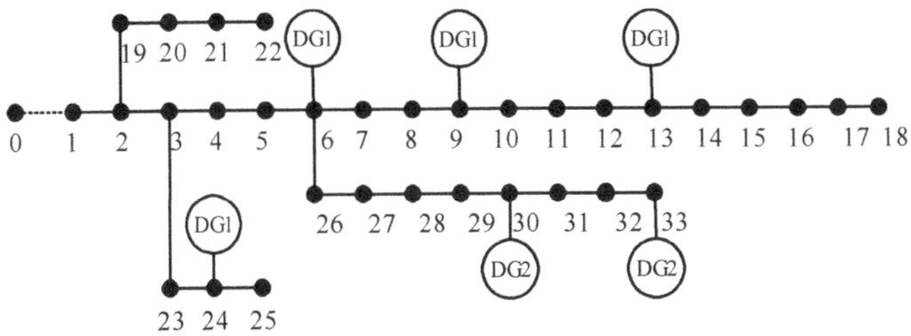

Figure A1. Distributed power supply configuration of the IEEE 33-node system.

As shown in Table A1, the DG1 active power output of nodes 6, 9, 13, and 24 is 0.5 MW, and the DG2 active power output of nodes 30 and 33 is 0.35 MW and 0.65 MW, respectively.

Table A1. Parameters of added DGs of 33-bus systems.

No.	DG Type	Quantity	Location	Active Power/MW
DG1	PV cell	1	6, 9, 13, 24	0~0.5
DG2	WTG	1	30	0~0.35
DG2	WTG	1	33	0~0.65

Figure A2 is the schematic diagram of the IEEE 69-node system, and Table A2 shows the distributed power configuration of the IEEE 69-node system.

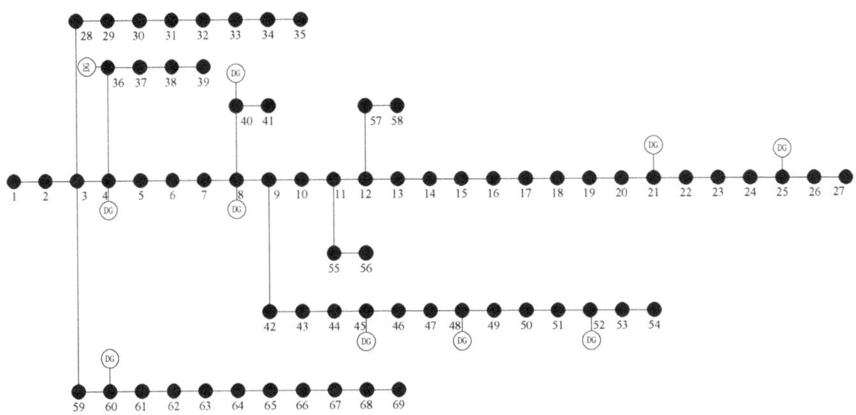

Figure A2. Distributed power supply configuration of the IEEE 69-node system.

As shown in Table A2, the DG1 active power output of nodes 4, 8, 21, 25, and 36 is 0.35 MW, and the DG2 active power output of nodes 40, 45, 48, 52, and 60 is 0.25 MW.

Table A2. Parameters of added DGs of 69-bus systems.

No.	DG Type	Quantity	Location	Active Power/MW
DG1	PV cell	1	4, 8, 21, 25, 36,	0~0.35
DG2	WTG	1	40, 45, 48, 52, 60	0~0.25

Considering that the distribution area covers a small area, all loads, photovoltaic (PV) cells, and wind turbine (WT) units adopt the same time-series curve, using a typical summer curve of a certain place, as shown in Figure A3.

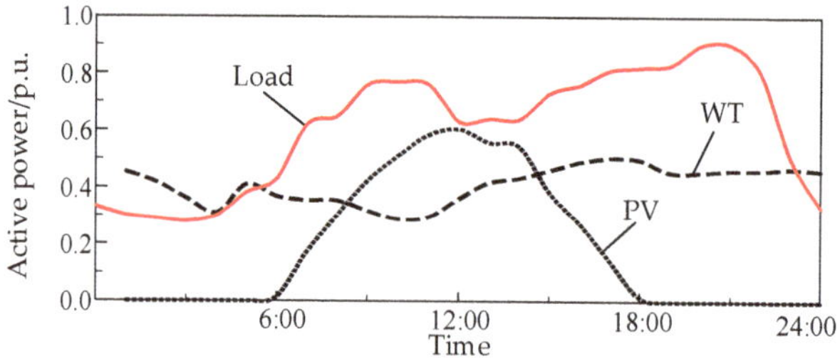

Figure A3. Daily load curve and daily operation curve of wind turbines and photovoltaic cells.

References

1. Zhang, Z.; Chen, Z. Transformation mechanism and characterization method of multi-agent flexibility and randomness in active distribution Network. *Autom. Electr. Power Syst.* **2018**, *48*, 116–129. (In Chinese)
2. Koutsoukis, N.C.; Georgilakis, P.S.; Hatziargyriou, N.D. Multistage coordinated planning of active distribution networks. *IEEE Trans. Power Syst.* **2017**, *33*, 32–44. [CrossRef]
3. Jiang, X. Feasible Operation Region of an Electricity Distribution Network with SOPs. Ph.D. Thesis, Cardiff University, Cardiff, UK, 2024.
4. Sun, X.; Qiu, J.; Zhao, J. Optimal Local Volt/Var Control for Photovoltaic Inverters in Active Distribution Networks. *IEEE Trans. Power Syst.* **2021**, *36*, 5756–5766. [CrossRef]
5. Vandoorn, T.L.; Meersman, B.; De Kooning, J.D.M.; Vandevelde, L. Transition From Islanded to Grid-Connected Mode of Microgrids With Voltage-Based Droop Control. *IEEE Trans. Power Syst.* **2013**, *28*, 2545–2553. [CrossRef]
6. Samadi, A.; Eriksson, R.; Söder, L.; Rawn, B.G.; Boemer, J.C. Coordinated Active Power-Dependent Voltage Regulation in Distribution Grids With PV Systems. *IEEE Trans. Power Deliv.* **2014**, *29*, 1454–1464. [CrossRef]
7. Xing, Q.; Chen, Z.; Zhang, T.; Li, X.; Sun, K. Real-time optimal scheduling for active distribution networks: A graph reinforcement learning method. *Int. J. Electr. Power Energy Syst.* **2023**, *145*, 108637. [CrossRef]
8. Xiao, J.; Zu, G.; Bai, G.; Zhang, M.; Wang, C.; Zhao, J. Mathematical Definition and Existence Proof of Security Domain of Distribution System. *Proc. CSEE* **2016**, *36*, 4828–4836. (In Chinese)
9. Xiao, J.; Zuo, L.; Zu, G.; Liu, S. Security Domain Model of Distribution System Based on Power Flow Calculation. *Proc. CSEE* **2017**, *37*, 4941–4949. (In Chinese)
10. Ageeva, L.; Majidi, M.; Pozo, D. Analysis of Feasibility Region of Active Distribution Networks. In Proceedings of the 2019 International Youth Conference on Radio Electronics, Electrical and Power Engineering (REEPE), Moscow, Russia, 14–15 March 2019; pp. 1–5.
11. Pei, L.; Wei, Z.; Chen, S.; Sun, G.; Lv, S. Security Domain of AC-DC Hybrid Distribution Network based on Convex envelope. *Power Syst. Autom.* **2021**, *45*, 45–51. (In Chinese)
12. Petrou, K.; Procopiou, A.T.; Gutierrez-Lagos, L.; Liu, M.Z.; Ochoa, L.F.; Langstaff, T.; Theunissen, J.M. Ensuring Distribution Network Integrity Using Dynamic Operating Limits for Prosumers. *IEEE Trans. Smart Grid* **2021**, *12*, 3877–3888. [CrossRef]
13. Blackhall, L. *On the Calculation and Use of Dynamic Operating Envelopes*; Australian National University: Canberra, Australian, 2020.
14. Liu, B.; Braslavsky, J.H. Sensitivity and robustness issues of operating envelopes in unbalanced distribution networks. *IEEE Access* **2022**, *10*, 92789–92798. [CrossRef]
15. Rayati, M.; Bozorg, M.; Cherkaoui, R.; Carpita, M. Distributionally robust chance constrained optimization for providing flexibility in an active distribution network. *IEEE Trans. Smart Grid* **2022**, *13*, 2920–2934. [CrossRef]

16. Karimianfard, H.; Haghighat, H. Generic resource allocation in distribution grid. *IEEE Trans. Power Syst.* **2018**, *34*, 810–813. [CrossRef]
17. Wang, R.; Ji, H.; Li, P.; Yu, H.; Zhao, J.; Zhao, L.; Zhou, Y.; Wu, J.; Bai, L.; Yan, J.; et al. Multi-resource dynamic coordinated planning of flexible distribution network. *Nat. Commun.* **2024**, *15*, 4576. [CrossRef] [PubMed]
18. Karthikeyan, N.; Pillai, J.R.; Bak-Jensen, B.; Simpson-Porco, J.W. Predictive control of flexible resources for demand response in active distribution networks. *IEEE Trans. Power Syst.* **2019**, *34*, 2957–2969. [CrossRef]
19. Liu, M.Z.; Ochoa, L.F.; Wong, P.K.C.; Theunissen, J. Using opf-based operating envelopes to facilitate residential der services. *IEEE Trans. Smart Grid* **2022**, *13*, 4494–4504. [CrossRef]
20. Bassi, V.; Ochoa, L.N. *Deliverables 1-2-3a: Model-Free Voltage Calculations and Operating Envelopes*; The University of Melbourne: Melbourne, Australia, 2022.
21. Chen, X.; Li, N. Leveraging two-stage adaptive robust optimization for power flexibility aggregation. *IEEE Trans. Smart Grid* **2021**, *12*, 3954–3965. [CrossRef]
22. Shi, L.; Xu, X.; Yan, Z.; Tan, Z. Envelop Calculation Method for Active-reactive Power Operation of Distribution Network. *Proc. CSEE* **2024**, 1–15. (In Chinese)
23. Yang, T.; Yu, Y. Steady-state security region-based voltage/var optimization considering power injection uncertainties in distribution grids. *IEEE Trans. Smart Grid* **2018**, *10*, 2904–2911. [CrossRef]
24. Oladeji, I.; Makolo, P.; Abdillah, M.; Shi, J.; Zamora, R. Security impacts assessment of active distribution network on the modern grid operation—A review. *Electronics* **2021**, *10*, 2040. [CrossRef]
25. Gu, C.; Wang, Y.; Wang, W.; Gao, Y. Research on Load State Sensing and Early Warning Method of Distribution Network under High Penetration Distributed Generation Access. *Energies* **2023**, *16*, 3093. [CrossRef]
26. Hu, D.; Peng, Y.; Wei, W.; Xiao, T.; Cai, T.; Xi, W. Reactive power optimization strategy of deep reinforcement learning for distribution network with multiple time scales. *Proc. CSEE* **2022**, *42*, 5034–5045. (In Chinese)
27. Zhang, J.; Cui, M.; Yao, X.; He, Y. Dual time-scale coordinated optimization of active distribution network based on data-driven and physical model. *Automation of Electr. Power Syst.* **2023**, *47*, 64–71. (In Chinese)
28. Lin, Z.; Hu, Z.; Song, Y. Review of Convex relaxation Techniques for optimal power flow Problems. *Proc. CSEE* **2019**, *39*, 3717–3728.
29. Ju, Y.; Huang, Y.; Zhang, R. Optimal power flow of three-phase AC-DC hybrid active distribution Network based on second-order cone programming convex relaxation. *Trans. China Electrotech. Soc.* **2021**, *36*, 1866–1875. (In Chinese)
30. Heidari, R.; Seron, M.M.; Braslavsky, J.H. Non-local approximation of power flow equations with guaranteed error bounds. In Proceedings of the 2017 Australian and New Zealand Control Conference (ANZCC), Gold Coast, QLD, Australia, 17–20 December 2017; pp. 83–88.
31. Borghetti, A.; Bosetti, M.; Grillo, S.; Massucco, S.; Nucci, C.A.; Paolone, M.; Silvestro, F. Short-term scheduling and control of active distribution systems with high penetration of renewable resources. *IEEE Syst. J.* **2010**, *4*, 313–322. [CrossRef]
32. Pilo, F.; Pisano, G.; Soma, G.G. Optimal coordination of energy resources with a two-stage online active management. *IEEE Trans. Ind. Electron.* **2011**, *58*, 4526–4537. [CrossRef]
33. Huang, Y.; Wang, Y.; Kong, W.; Cao, C.; Su, J.; Wang, K. Active power and reactive power coordination optimization of active distribution network based on improved convex inner approximation method. *Proc. CSEE* **2024**, 1–11. (In Chinese)
34. Li, H.; He, H. Learning to operate distribution networks with safe deep reinforcement learning. *IEEE Trans. Smart Grid* **2022**, *13*, 1860–1872. [CrossRef]
35. Kou, P.; Liang, D.; Wang, C.; Wu, Z.; Gao, L. Safe deep reinforcement learning-based constrained optimal control scheme for active distribution networks. *Appl. Energy* **2020**, *264*, 114772. [CrossRef]
36. Mu, Y.; Jin, S.; Zhao, K.; Dong, X.; Jia, H.; Qi, Y. "people-car-pile-road-net" depth under the coupling of the collaborative planning and operation optimization distribution network. *Autom. Electr. Power Syst.* **2018**, *48*, 24–37. (In Chinese)
37. Lu, S.; Mo, Y. Distributed dispatching method of active distribution network considering multiple regulation resources. *J. Phys. Conf. Ser.* **2022**, *2237*, 012004. [CrossRef]
38. Sun, H.; Ding, X.; Wu, Z.; Zheng, S.; Xu, Z. Voltage regulation strategy of distribution network with automatic voltage control system limit optimization. *Electr. Power Syst. Autom.* **2024**, 1–13. (In Chinese)
39. Guo, Q.; Wu, J.; Mo, C.; Xu, H. Collaborative optimization model of voltage and reactive power for new energy distribution Network based on mixed integer second-order cone programming. *Proc. CSEE* **2018**, *38*, 1385–1396. (In Chinese)
40. Chen, S.; Wei, Z.; Sun, G.; Wei, W.; Wang, D. Convex hull based robust security region for electricity-gas integrated energy systems. *IEEE Trans. Power Syst.* **2018**, *34*, 1740–1748. [CrossRef]

Disclaimer/Publisher's Note: The statements, opinions and data contained in all publications are solely those of the individual author(s) and contributor(s) and not of MDPI and/or the editor(s). MDPI and/or the editor(s) disclaim responsibility for any injury to people or property resulting from any ideas, methods, instructions or products referred to in the content.

energies

MDPI

Article

Analysis of Underground Distribution System Models for Secondary Substations

Boohyun Shin [1,2], Hyeseon Lee [2] and Sungyun Choi [1,*]

[1] School of Electrical Engineering, Korea University, Seoul 02841, Republic of Korea; vectorist@korea.ac.kr
[2] Distribution Power Laboratory, KEPCO Research Institute, Daejeon 34056, Republic of Korea; leehyeseon@kepco.co.kr
* Correspondence: sungyun@korea.ac.kr

Abstract: In Korea, the demand for complete underground installation of power distribution equipment installed on roads and green areas is increasing. In addition, KEPCO is making efforts to build a more reliable system for the underground distribution system. To meet these needs, this paper proposes the S-substation. In the S-substation, an RMU, a large power transformer, and an LV-Board (including ATCB and MCCB) are installed within the underground structure. This paper proposes three models to apply the S-substation to the underground distribution system. Power flow analysis is conducted for each model by simulating a variety of loads and DERs, and the frequency fluctuations are also examined under different distribution system events. An economic analysis is also conducted to select the optimal model. The economic analysis focuses on VOLL and construction costs. Based on power flow and economic analysis, one model is selected, and the underground distribution system that the model is applied is presented.

Keywords: secondary substation; power flow; underground distribution system; underground compact station; ring main unit; power quality; MCCB; MATLAB/SIMULINK; value of lost load

1. Introduction

This paper analyzes the optimal Low Voltage (LV) distribution system for introducing the Secondary Substation (S-substation) in underground distribution systems. The S-substation has been implemented in the underground distribution system in Korea for the following reasons:

- The need for complete undergrounding of Power Distribution Equipment (PDE) for underground Distribution Lines (D/Ls) supply.
- A desire to improve the aesthetics of PDE installed on sidewalks.
- The introduction of a highly reliable underground distribution system.
- The expansion of the LV supply range.

To meet these requirements, the S-substation installs a larger and more reliable PDE than the existing PDE for supplying underground D/Ls. The Ring Main Unit (RMU) is installed in place of the pad-mounted switchgear, and the oil-filled transformer replaces the pad-mounted transformer. Transformers in S-substations can be installed with capacities ranging from 1 to 2 MVA, which is larger than conventional transformers. This paper concentrates on the LV distribution system rather than the High Voltage (HV) distribution system. In the S-substation, the LV-board (including Automatic Transfer Circuit Breaker (ATCB) and Mold Case Circuit Breaker (MCCB)) is installed to reduce the fault duration time in the LV distribution system and to prevent cascading failures. Additionally, all PDE are waterproof so that the structures can operate under flood conditions [1].

Unlike the HV distribution system, the LV distribution system operates as a radial system, and in the event of a fault on the transformer, all connected loads experience a power outage. In contrast, during a power outage in the HV distribution systems, only a short power outage is experienced due to switching HV D/Ls. However, in case of a transformer fault, the recovery time is prolonged, so the customers experience a long power outage.

Only now has the LV distribution system received less attention than the HV distribution system. The HV distribution system is equipped with an automated system known as the Distribution Automation System (DAS), and continuous efforts are made to reduce fault duration and sections through protection coordination analysis. In contrast, the LV distribution system lacks such automation and coordination efforts. In particular, the underground LV distribution system should be of greater interest due to the following characteristics.

- The capacity of transformers in underground systems is typically larger than those in overhead distribution systems.
- Three-phase cables are installed under the assumption of three-phase loads.
- The fuses within the transformer protect the entire LV distribution system.
- Fault sections are challenging to locate due to the underground installation.

Therefore, this paper focuses more on the LV distribution system than on the HV distribution system, as installing the S-substation requires modifications to the existing LV distribution system model.

This paper also introduces the concept of the S-substation and presents a case study of its installation by KEPCO. The current underground distribution system is introduced, and three types of LV distribution system models suitable for S-secondary are presented. Power Flow (PF) and economic analysis are conducted to compare each model. The "Result & Discuss" section presents the example model and considerations when the model selected as the optimal LV model is currently applied to the underground distribution system.

The main keywords in this paper are power flow, the LV distribution system, and secondary substation. Previous studies on these topics include a paper on PF in distribution systems, which primarily analyzed the influence of the Distributed Energy Resources (DERs) connection. The algorithm for PF focuses on the economic dispatch of energy and ancillary services based on market prices at the system level. Simulation results show the effectiveness of the approach [2]. The main aim of the second paper is to enhance solar performance distribution and increase photovoltaic (PV), providing an overview of unbalanced optimal power flow [3]. The third paper provides a comprehensive understanding of the optimal power flow problem and proposes potential solutions. This paper proposes a comprehensive overview of prospective advancements and compares the diverse methodologies employed in optimal power flow [4].

Few studies specifically examine the LV distribution system. One paper examined the control of electric currents from renewable energy sources (RESs) in Microgrid systems [5], while another analyzed algorithms that control voltage and reactive power in LV distribution networks [6].

Finally, one of the studies on S-substations includes that S-substations devote autonomous management of the connected LV network and provide grid flexibility services [7]. With the digitalization of these nodes, S-substation allows the development of innovative solutions for predictive maintenance and life extension of the assets [8]. Other studies primarily focused on testing the electrical characteristics of its structure or internal PDE (such as dielectric tests, internal arc tests, and temperature rise tests) [9–11]. However, previous studies have not examined the optimal LV distribution system model for applying S-substations.

2. Secondary Substation

2.1. S-Substation

The distribution station improves the aesthetics of the city by collecting and installing many PDEs installed on the ground in one space. In addition, RMUs, large power transformers, and LV boards are installed to enable more reliable operation of the distribution systems. However, applying distribution stations (inside large Buildings) in all areas is challenging due to difficulties in securing and maintaining buildings. In particular, securing a building or a place where LV loads range from 500 to 2000 kW can be difficult; otherwise, using S-substations is increasingly attracting attention as a solution in such cases. This method involves installing and supplying PDE in underground structures without requiring the construction or purchase of a separate building. Figure 1 shows the circuit diagram of the station and Figure 2 depicts actual pictures of the S-substation installed in Korea.

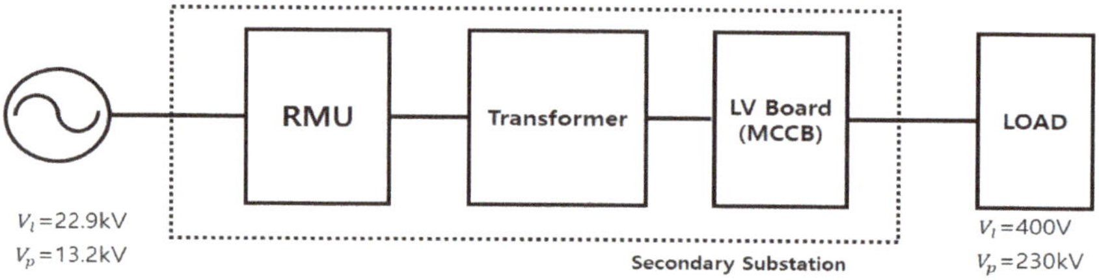

Figure 1. S-substation concept figure.

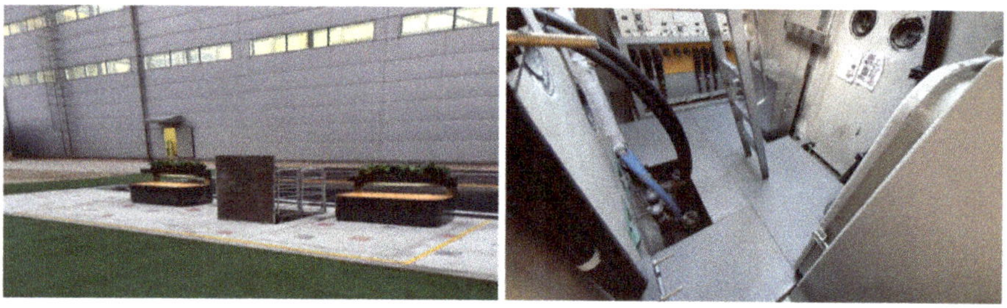

Figure 2. S-substation in Korea.

2.2. Underground Distribution Systems

For LV distribution systems, both overhead and underground D/Ls are supplied as radial systems. On overhead D/Ls, single-phase transformers ranging from 10 to 150 kVA are used, while for underground D/Ls, three-phase transformers ranging from 150 to 500 kVA are currently used. Large power transformers are also expected to be used in the future (Currently, individual load is limited to less than 1000 kW). A figure of the current underground distribution system is shown below [12].

In this figure, solid lines indicate HV D/Ls, whereas dotted lines represent LV D/Ls. The switchgear mainly uses four circuits, and the HV D/Ls comprise an open-loop system. Conversely, the LV D/Ls connected to the transformer are connected in a radial system. Figure 3 illustrates an example of a system currently operating as an underground distribution system in Korea. In the system shown in the figure, LV systems of pad-mounted transformers are operated independently. Thus, when a transformer experiences a fault, it is impossible to connect the LV distribution system to other transformers. As a result, LV customers may experience prolonged power outages until the transformer is repaired [13].

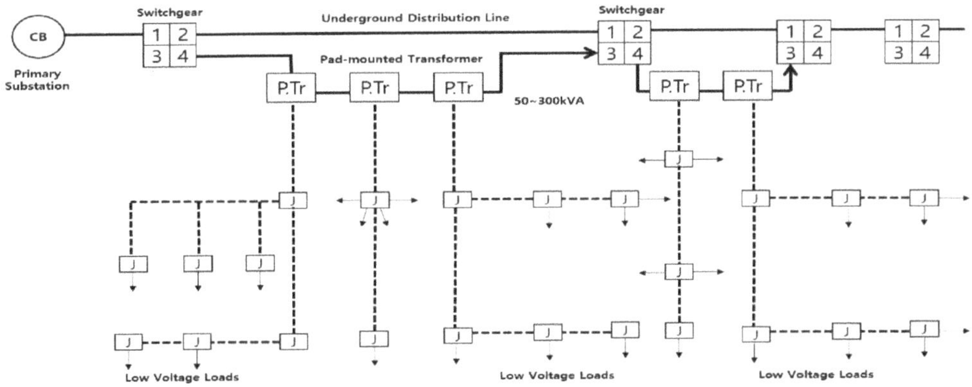

Figure 3. Underground distribution system (Current).

2.3. RMU Method for Strengthening Reliability

The introduction of an RMU-based method is being considered in the underground distribution system. The RMU is a PDE for constructing a ring-type distribution system. It includes a circuit breaker function to protect the PDE connected to the RMU and isolate the D/Ls in case of faults. This method can be applied when building a more reliable system [14].

Figure 4 illustrates the configuration of the S-substation where the RMU method is applied. The S-substation is not connected to the main D/L connected to other D/Ls or the HV Loads but only to the branch D/L. This S-substation has an RMU, a transformer, and an LV board in the same underground structure [12].

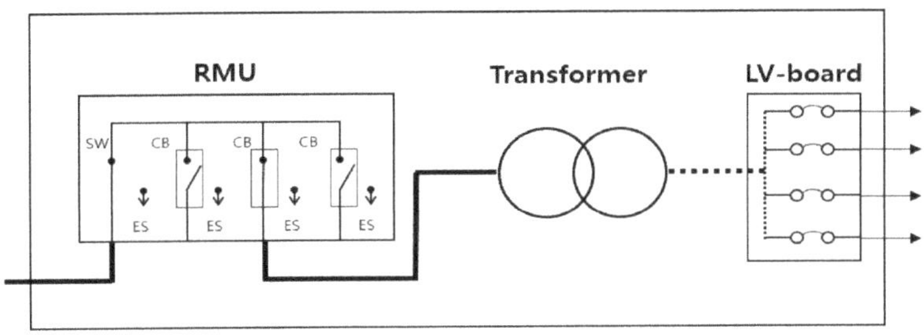

Figure 4. S-Substation with PDE.

3. Three Types of Distribution System Models

This paper proposes three types of distribution system models for the S-substation.

3.1. Model A (Dual S-Substation)

The first model is composed of supplying power to each S-substation by installing two S-substations. For instance, if one transformer experiences a fault, the power can be supplied by replacing the transformer installed in another S-substation. In this case, after detecting a fault in the transformers installed in the two S-substations, power cut-off/power-on through ATCB should be performed to minimize fault. This model allows loads to be shared between S-substations. Therefore, in the event of a fault, the transformer installed in one S-substation shares all the loads, so it is important to consider the load factor of the transformers. Model A is shown in Figure 5.

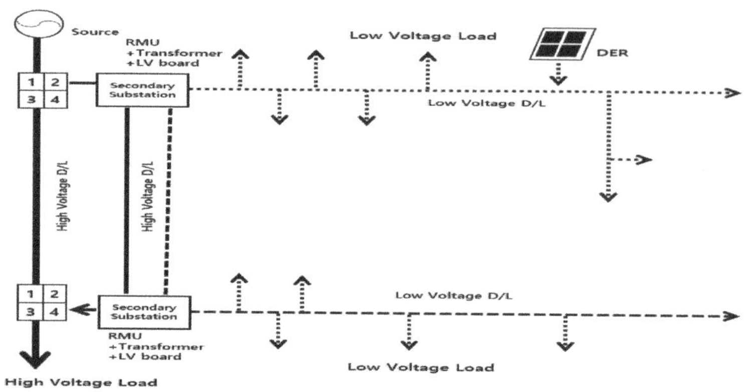

Figure 5. Underground distribution systems based on model A.

3.2. Model B (Auxiliary Transformer)

Model B involves installing two transformers in the S-substation. In this approach, power is primarily supplied by one transformer, and in the case of a transformer fault, the power is switched and supplied with an auxiliary transformer installed in the same S-substation. In this model, since two transformers are housed simultaneously in the S-substation, it has the advantage of reducing the replacement time in the case of a peripheral voltage fault compared to Model A, thus improving reliability. Model B is shown in Figure 6.

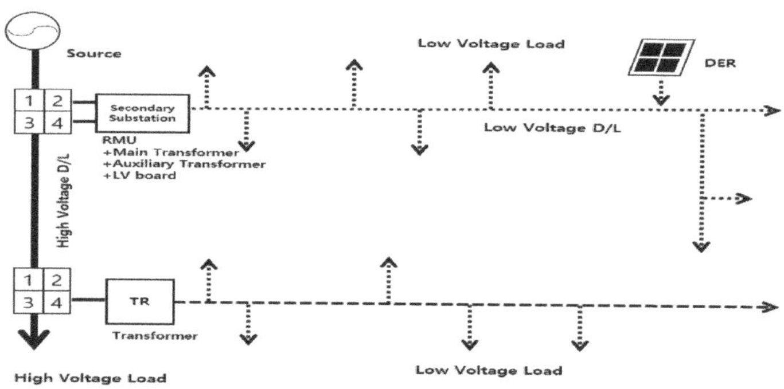

Figure 6. Underground distribution system based on model B.

However, since two transformers are installed in a single S-substation, not only does it take up significant space, but due to space constraints, it can also cause problems such as ventilation and waterproofing.

3.3. Model C (Pad Transformer and S-Substation Connection)

In the case of Model C, only one S-substation is applied to the existing underground distribution system. Model A installs two substations, whereas this model utilizes only one. It is considered that this model is suitable for existing subordinate consumers outside of housing site development districts due to the limitations associated with the installation of underground structures. The current underground D/Ls are a type in which one to four transformers are connected to one terminal of a four-circuit switchgear, and the switchgear does not have a circuit-breaking function, as shown in Figure 3. Model C switches the power source during a power outage via the ATCB upon detecting a fault of a transformer installed in the S-substation, reducing power outage time. Model C is shown in Figure 7.

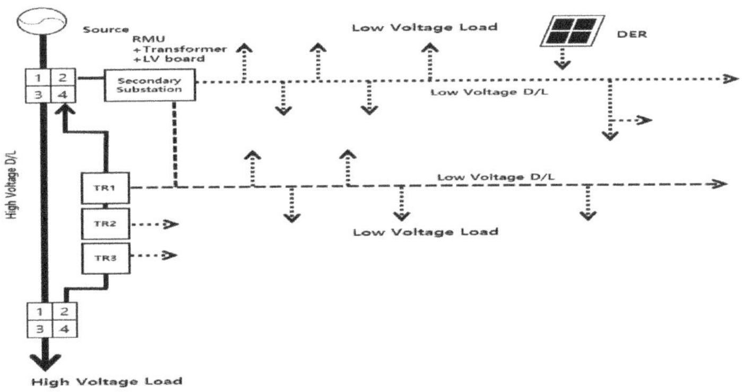

Figure 7. Underground distribution system based on model C.

4. Power Flow and Parameters

4.1. Power Flow

4.1.1. Definition

PF is a numerical analysis of the flow of electric power in interconnected power systems. In this paper, the PF is analyzed using MATLAB/SIMULINK, which calculates the PF using the Newton–Raphson method [15].

4.1.2. Direction of Power Flow

The key parameters of interest in the PF results in MATLAB/SIMULINK are voltage and phase. Each load is connected in parallel and must be supplied with the specified voltage. The phase information can also tell the direction of the tide. In MATLAB/SIMULINK, the phase is displayed in degree units, and one can observe the difference in the direction and time of the tide by looking at the difference in phase.

$$I_R = \frac{V_S - V_R}{Z'} - \frac{Y'}{2}V_R = \frac{V_s e^{-j\delta} - V_R}{jX'} - \frac{j\omega C' l}{2}V_R \quad (1)$$

$$S_R = V_R I_R^* = V_R\left(\frac{V_S e^{j\delta} - V_R}{jX'}\right)^* + \frac{j\omega C' l}{2}V_R^2 = V_R\left(\frac{V_S e^{-j\delta} - V_R}{-jX'}\right) + \frac{j\omega C' l}{2}V_R^2 = \frac{jV_R V_S \cos\delta + V_R V_S \sin\delta - jV_R^2}{X'} + \frac{j\omega C' l}{2}V_R^2 \quad (2)$$

$$P = P_S = P_R = R_e(S_R) = \frac{V_R V_S}{X'}\sin\delta[W] \quad (3)$$

Figure 8 depicts the distribution system as a π-type equivalent circuit, with V_S and I_S representing the sending-end voltage and current and V_R and I_R representing the receiving-end voltage and current. Z' denotes the total series impedance, while Y' represents the total shunt admittance. Additionally, δ represents the voltage-phase angle between the sending and receiving ends. From the above equations, it can be seen that, in AC, the active PFs come from the point where the phase is leading to the point where it is lagging [16]. In MATLAB/SIMULINK, the slack value is typically set as $1\angle 0$, so the load side can be expected to come out as a negative value.

Figure 9 illustrates a waveform for active power. For example, if the phase angle at Bus 1-2 is 0 deg but there is a difference of −0.01 deg in Bus 1-3, which is the subsequent load, it would be necessary to verify whether power is supplied after a few seconds.

$$\mathrm{T} = \frac{1}{\mathrm{f}} = \frac{2\pi}{w} = \frac{1}{60\ \mathrm{Hz}} = 0.01666\ [\mathrm{s}] \fallingdotseq 16.7\ \mathrm{ms}\ (= -360\ \mathrm{deg}) \rightarrow \frac{16.7\ \mathrm{ms} \times -0.01\ \mathrm{deg}}{-360\ \mathrm{deg}} = 0.46388\ \mu\mathrm{s} \quad (4)$$

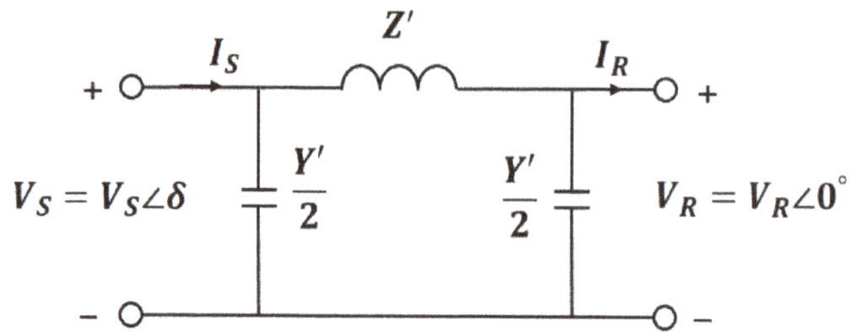

Figure 8. Simple power system equivalent circuit.

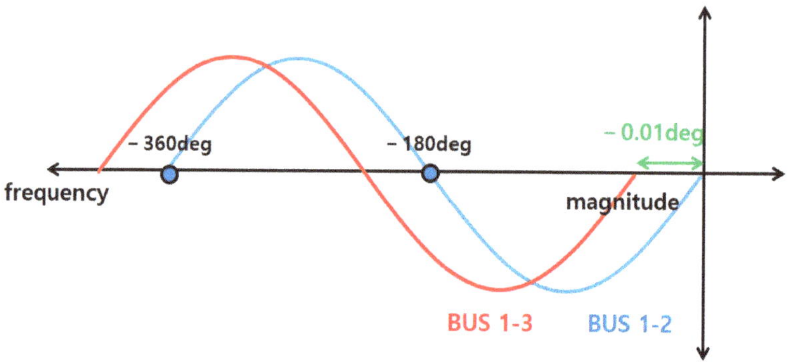

Figure 9. Example of the phase difference of load bus.

As shown in the above equation, since the power system in Korea uses 60 Hz, 1 cycle can be regarded as 16.7 ms; therefore, based on the above formula, a phase angle difference of 0.01 deg corresponds to 0.4638 μs.

4.2. Standard Voltage

In the distribution system, the loads are connected in parallel. A parallel load is a connection method in which the voltage is the same and the current varies depending on the load. The standard voltage ranges are listed in Table 1.

Table 1. Standard voltage ranges in Korea.

Standard Voltage	Range
110 V	110 ± 6 V
220 V	220 ± 13 V
380 V	380 ± 38 V

From the above standard, it can be seen that the voltages must be maintained from 207 V to 238 V for single-phase 220 V and from 342 V to 418 V for three-phase 380 V.

Proper arrangement of the loads is required to maintain the standard voltage ranges to the loads connected at the end of the D/Ls. Even if the impedances of the D/Ls are small, a voltage drop occurs when the current is large, so it is preferable to connect heavy LV loads directly to a source (transformer). Figures 10 and 11 illustrate this concept.

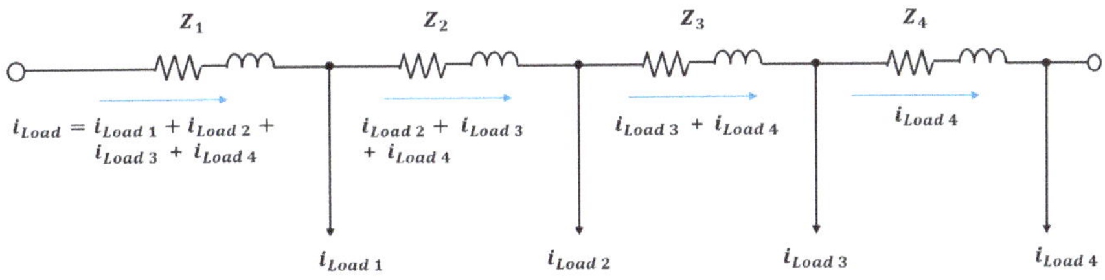

Figure 10. Calculation formula for voltage drop D/L.

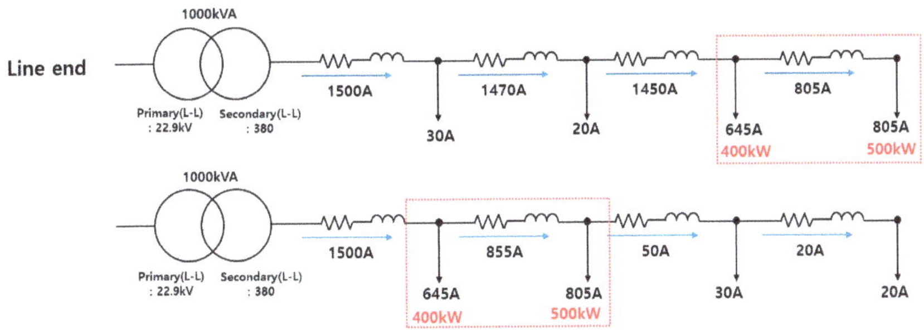

Figure 11. Comparison of load displacement.

In MATLAB/SIMULINK, one can set the voltage value of the slack bus to 1 [pu] and calculate the PF, which allows one to determine the amount of voltage drop that occurred as a result of the calculation.

4.3. Load Placement

This paper analyzes in detail the load placement that was briefly addressed in "B. Standard Voltage." The kVA ratings of the transformer installed in the S-substation are 500, 1000, and 2000. Additionally, heavy LV loads are connected. In the LV, i.e., on the transformer's secondary side, the current varies according to the capacity of the loads, even though the voltage connected to the loads is the same.

For optimal load placement, this paper analyzes how the voltage drop changed for three cases: ① direct connection to the transformer, ② end of the D/L, ③ separation connection. To calculate the voltage drop for each of the three cases, the configuration in Model A is modified in the proposed system types. The first case is that heavy LV loads (250~500 kW) are directly connected to the transformer's secondary side (directly under the transformer), and the second case is that the heavy LV loads are connected to the end of the D/L. The third is separately connected to other LV loads. Additionally, in the third case, the branch D/L is long, so the line impedances are set to be large [12]. Figure 12 describes the simulation results.

The voltage is expressed as a PU value and the specified voltage maintenance range must be 0.855 [pu] or higher; however, when connected with heavy LV loads at the end of the D/L, there is a significant voltage drop from Bus 1-3/2-3, making it difficult to maintain Power Quality (PQ). When heavy LV loads are directly connected to the transformer, the specified voltage is maintained (0.855 pu) in all buses. In contrast, when connected to the end of the D/L, it is observed that the voltage drops rapidly from Bus 1-3. In the case of a separate line configuration, Bus 1-2 is configured as a bus on a separate line rather than on the existing line. The simulation results indicate that the optimal arrangement is to connect heavy LV loads directly under the transformer's secondary side. The transformer's secondary side current and voltage drop can be expressed using the following formula.

$$\text{Rated Current[A]} = \frac{Transformer\ capacity\ [\text{kVA}]}{\sqrt{3} \times rated\ voltage_{l-l}\ [\text{V}]} = \frac{1000\ [\text{kVA}]}{\sqrt{3} \times 380\ [\text{V}]} \cong 1500\ [\text{A}] \quad (5)$$

A rated current of 1500 A is calculated while assuming the transformer capacity is 1000 kVA. As indicated by the above equations, when heavy LV loads are connected to the end of the D/L, the current of heavy loads flows through the D/L, thus resulting in a voltage drop due to line impedance, which shows that the specified voltage cannot be maintained; therefore, in the system model presented in this paper, the PF is analyzed while assuming that all heavy LV loads are placed directly under the transformer.

$$Load_{total} = \sum Load_G + \sum Load_R + \sum Load_E + \sum Load_I + \sum Load_A \quad (6)$$

Several LV loads are connected to the bus. These are General ($Load_G$), Residential ($Load_R$), Educational ($Load_E$), Industrial ($Load_I$), and Agricultural ($Load_A$) Loads. These loads are connected to each bus [17].

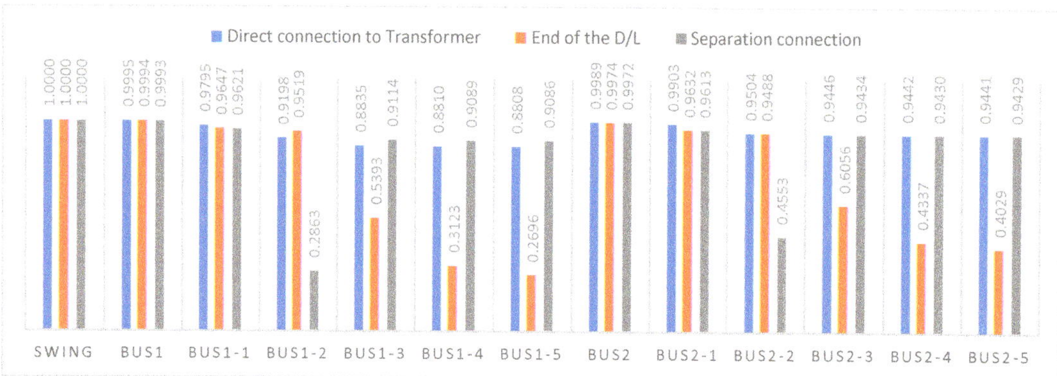

Figure 12. Bus voltage drop cases.

5. Analysis of Power Flow by Model

5.1. Model A

Figure 13 shows the equivalent circuit of Model A, and Table 2 lists the results of PF when it is operated in a normal state (steady state) without any fault or conversion. As shown in Table 2, it can be observed that there are cases where it does not reach standard 1 [PU]. In this simulation, since the transformer's secondary side is set to 380 V, if it is 0.8808 of Bus 1-5, which is the lowest value, 380 × 0.8808 = 334 V, which is lower than the standard voltage range; however, since the actual transformer supplies the secondary at 400 V, it can be seen that it is necessary to maintain up to 342/400 V = 0.855 [PU] to maintain 342 V (380 ± 38 V), which is the minimum standard voltage. That is, the simulation results show that the load connected to each BUS is receiving power within the standard voltage range. To compare the PF results for the three models, it is necessary to examine not only the steady-state operation (Normal state) but also the operating conditions under other scenarios. In this paper, we review the following conditions: (1) Normal state, (2) Line switching, and (3) Dual Line. Additionally, changes in voltage and phase due to the connection of DERs are compared.

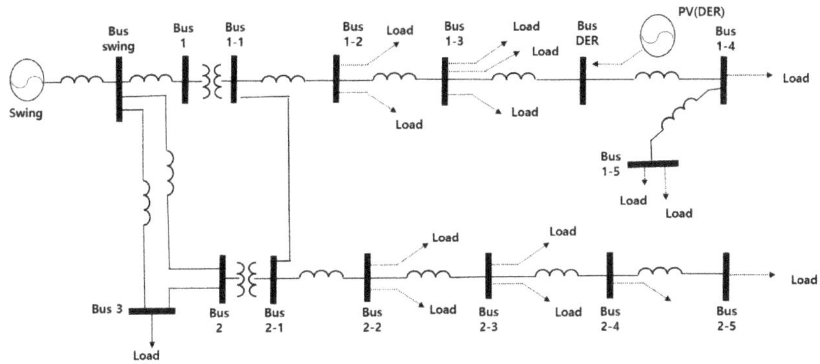

Figure 13. Underground distribution system based on model A.

Table 2. Comparison of power flow result by model A.

BUS	Normal State		Normal + DER		Line Switching		Line Switching + DER		Dual Line		Dual Line + DER	
	V	Deg	V	Deg	V	Deg	V	Deg	V	Deg	V	Deg
swing	1	0	1	0	1	0	1	0	1	0	1	0
Bus 1	0.9995	−0.13	0.9996	−0.12	0.9999	0	1	0	0.9996	−0.12	0.9997	−0.11
Bus 1-1	0.9795	−7.87	0.9855	−7.03	0.9998	−0.01	0.9998	−0.01	0.9839	−7.08	0.9882	−6.63
Bus 1-2	0.9198	−9.01	0.9326	−8.12	0.8204	−14.08	0.8403	−13.16	0.9455	−8.32	0.953	−7.79
Bus1-3	0.8835	−23.2	0.9137	−8.92	0.788	−28.27	0.8172	−10.58	0.9082	−22.5	0.9349	−9.2
Bus 1-4	0.881	−26.27	0.9087	−1.58	0.7858	−31.34	0.8099	−0.65	0.9056	−25.57	0.9305	−2.33
Bus 1-5	0.8808	−26.93	0.9085	−2.24	0.7856	−32	0.8096	−1.3	0.9053	−26.23	0.9302	−2.99
Bus 2	0.9989	−0.9	0.999	−0.9	0.9959	−1.35	0.9966	−1.29	0.9985	−0.96	0.9987	−0.94
Bus 2-1	0.9903	−6.14	0.9904	−6.14	0.9619	−13.16	0.9685	−12.26	0.9854	−7.13	0.9872	−6.72
Bus 2-2	0.9504	−6.96	0.9504	−6.96	0.8732	−13.82	0.8859	−13	0.9374	−7.9	0.9418	−7.51
Bus 2-3	0.9446	−12.06	0.9446	−12.06	0.8679	−18.92	0.8805	−18.1	0.9317	−13	0.9361	−12.61
Bus 2-4	0.9442	−12.94	0.9442	−12.94	0.8675	−19.8	0.8802	−18.98	0.9313	−13.88	0.9356	−13.49
Bus 2-5	0.9441	−13.16	0.9441	−13.16	0.8675	−20.02	0.8801	−19.2	0.9312	−14.1	0.9356	−13.71
Bus 3	0.9991	−0.81	0.9992	−0.81	0.9968	−1.15	0.9973	−1.11	0.9988	−0.86	0.9989	−0.84

The table above lists the PF results for all cases [12]. As a result of the simulation, it can be confirmed that most of the methods of supplying power from an auxiliary source after the main power source is cut off cause voltage drop and the phase delay. In the case of parallel operation of transformers, dual power is supplied to the loads. Moreover, when DER is connected, a power source is present midway along the D/L other than the Slack (main source). Therefore, when DER is operational, it can be understood that the voltage of the D/L increases and the phase becomes faster. To facilitate understanding, the paper presents the following comparative graphs (Refer to Figure 14).

The upper graph shows the voltage value for each bus for line No. 1 to which the DER is connected, while the lower graph shows the voltage value for each bus to which the DER is not connected. The simulation for the second type of Line Switching sets the route where the power supply of the transformer supplying line 1 is cut off, so it can be confirmed that power is supplied when it is supplied from the transformer of line 2 after the power is cut off, but the voltage drop, in this case, is very large. Although the power in line 2 is not cut off, it can be confirmed that loads of line 2 are also affected because it also divides the line 1 loads. The graph on the right of the upper/lower graph shows how the voltage changes when DER is connected. It can be seen that the voltage increases when DER is applied to both lines 1 and 2. In this paper, the DER is configured to 99 kW. If additional DERs are connected, the voltage drop can be further compensated.

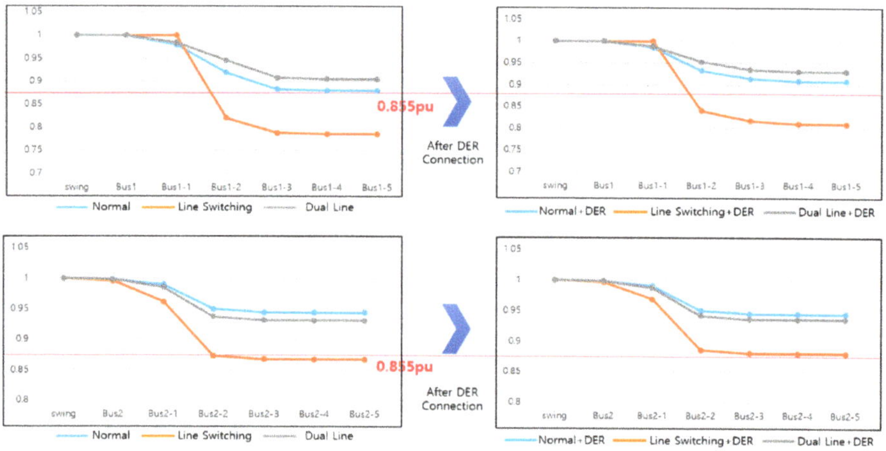

Figure 14. Power flow comparison graphs based on model A.

From the results, it can be seen that parallel operation is the most effective configuration for maintaining PQ, and when DER is connected, the phase becomes faster. Buses 1 and 2 represent the HV terminal on the primary side of the transformer. Buses 1-1 and 2-1 represent the transformer's secondary side, while loads are connected from Buses 1-2 and 2-2.

As shown in the table and graph, if we compare buses 1-2 and 2-2 with buses 1-3 and 2-3, it can be seen that the phase is significantly left behind. This is because heavy LV loads are connected to the 1-2 and 2-2 buses, and the phase from the next bus is delayed; that is, when heavy LV loads are connected, it can be seen that the flow of the current is delayed.

Figure 15 demonstrates that the first line is bus 1, followed by buses 1-2 and 2-2, and the last line is bus 2-5; that is, Bus 1 > 2 >> 1-1 > 2-1 > 2-2 > 1-2 >> 2-3 > 2-4 > 2-5 >> 1-3 > 1-4 > 1-5. This sequence highlights the differences in phase angles across the buses.

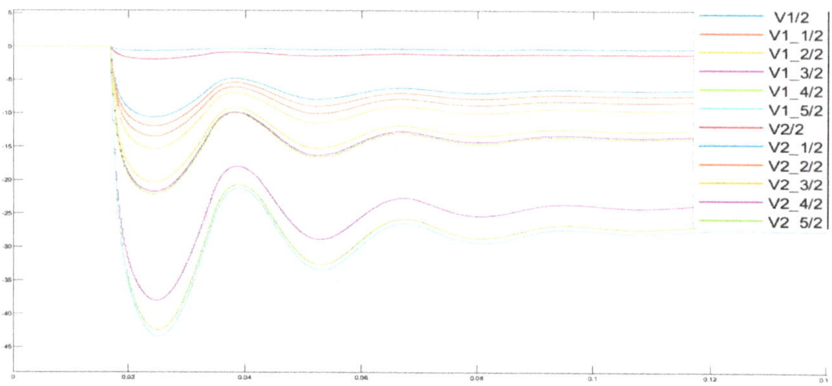

Figure 15. Phase change over time based on the bus.

To determine the frequency change according to major events, (1) in 20 cycles, main transformer fault, (2) in 40 cycles, Line switching (to reduce power outage), (3) in 60 cycles, connecting DER, (4) in 80 cycles, recovery fault (with dual line state), (5) in 100 cycles, 1 phase ground fault (Refer to Figure 16). During most of these events, the frequency deviates from the specified frequency range, but it returns to normal immediately. The frequency drops when power is lost or the load factor increases; the frequency rises when power is supplied or DER is applied.

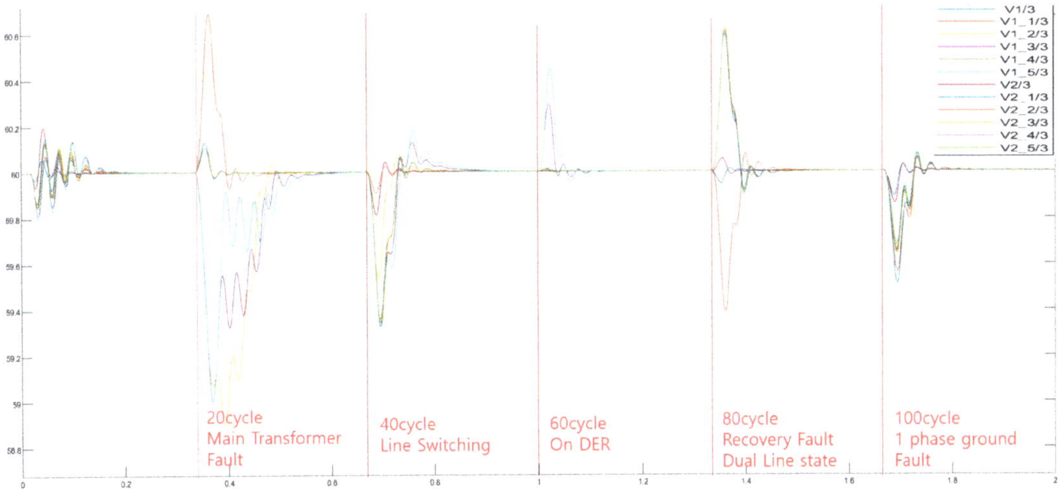

Figure 16. Frequency changes according to specific orders.

The above simulation is conducted with each transformer operating at 70~90% of its rated load; therefore, it is noticeable that the voltage drop is large when switching power to another transformer due to a transformer fault. It is necessary to examine how the PF changes according to the transformer load factor. To this end, the load factor is adjusted to 50% or less and the simulation is performed again.

The results of the analyses are presented in the following Figure 17.

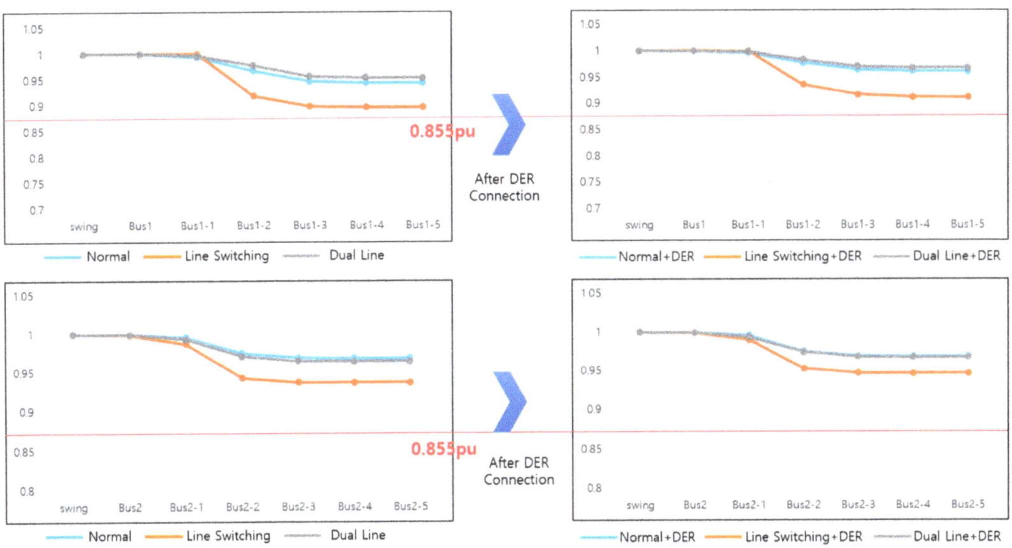

Figure 17. Change in power flow after adjusting the loads.

The above graph shows that when the loads are maintained within 50%, the standard voltage range is maintained without a significant decrease in voltage even in the case of Line switching.

5.2. Model B

In the second model, B, the main and auxiliary transformers are installed simultaneously in the same S-substation. The Figure 18 is represented as an example circuit for easy understanding.

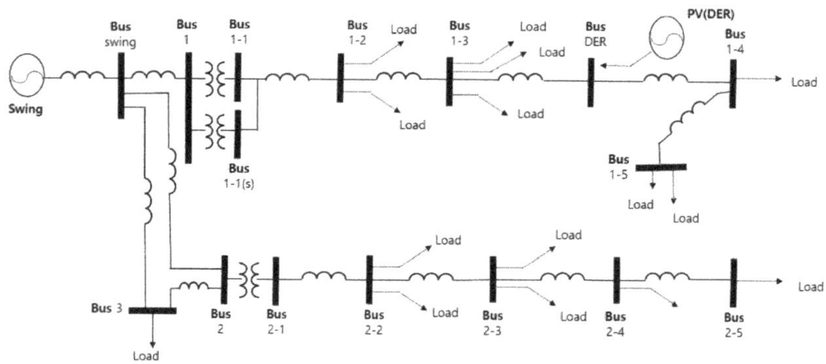

Figure 18. Underground distribution system based on model B.

In this model, the 1000 kVA transformer for the upper line is supplied under normal conditions, and in the event of a fault in D/Ls due to a transformer fault, the loads are switched and supplied to the auxiliary transformer; in other words, the auxiliary transformer is normally operated with no load, and it is only operated with loads applied when there is a transformer fault.

To maintain the standard voltage in Model A in this paper, a value of 0.855 [pu] should be maintained. As a result of the simulation, the loads connected to each bus do not satisfy 0.855 [pu]. Of course, this sample drawing also satisfies the standard because the load arrangement and impedance are considered such that 0.855 [pu] or more is obtained in the normal state.

Next, let us identify the situation where the main transformer fault, which is not in the normal state, and the parallel operation in which the main/auxiliary transformer is operated at the same time using the Table 3 below.

Table 3. Comparison of power flow results by case.

BUS	Normal		Normal + DER		Line Switching		Line Switching + DER		Dual Line		Dual Line + DER	
	V	Deg	V	Deg	V	Deg	V	Deg	V	Deg	V	Deg
swing	1	0	1	0	1	0	1	0	1	0	1	0
Bus 1	0.9995	−0.13	0.9996	−0.12	0.9995	−0.13	0.9996	−0.12	0.9996	−0.14	0.9997	−0.12
Bus 1-1	0.9795	−7.87	0.9855	−7.03	0.9795	−7.87	0.9855	−7.03	0.9912	−4.18	0.994	−3.8
Bus 1-2	0.9198	−9.01	0.9326	−8.12	0.9198	−9.01	0.9326	−8.12	0.9629	−5.43	0.9688	−4.96
Bus 1-3	0.8835	−23.2	0.9137	−8.92	0.8835	−23.2	0.9137	−8.92	0.9248	−19.62	0.9537	−7.93
Bus 1-4	0.881	−26.27	0.9087	−1.58	0.881	−26.27	0.9087	−1.58	0.9223	−22.69	0.9507	−2.25
Bus 1-5	0.8808	−26.93	0.9085	−2.24	0.8808	−26.93	0.9085	−2.24	0.922	−23.34	0.9505	−2.91
Bus 2	0.9994	−0.68	0.9994	−0.68	0.9994	−0.68	0.9994	−0.68	0.9994	−0.68	0.9994	−0.68
Bus 2-1	0.9845	−8.08	0.9846	−8.08	0.9845	−8.08	0.9846	−8.08	0.9845	−8.08	0.9845	−8.08
Bus 2-2	0.9676	−8.42	0.9676	−8.42	0.9676	−8.42	0.9676	−8.42	0.9676	−8.42	0.9676	−8.42
Bus 2-3	0.9617	−13.52	0.9617	−13.52	0.9617	−13.52	0.9617	−13.52	0.9617	−13.52	0.9617	−13.52
Bus 2-4	0.9613	−14.4	0.9613	−14.4	0.9613	−14.4	0.9613	−14.4	0.9613	−14.4	0.9613	−14.4
Bus 2-5	0.9612	−14.62	0.9612	−14.62	0.9612	−14.62	0.9612	−14.62	0.9612	−14.62	0.9612	−14.62
Bus 3	0.9995	-0.64	0.9995	−0.64	0.9995	−0.64	0.9995	−0.64	0.9995	−0.64	0.9995	−0.64

The data in the table are shown in the form of Figure 19 as follows:

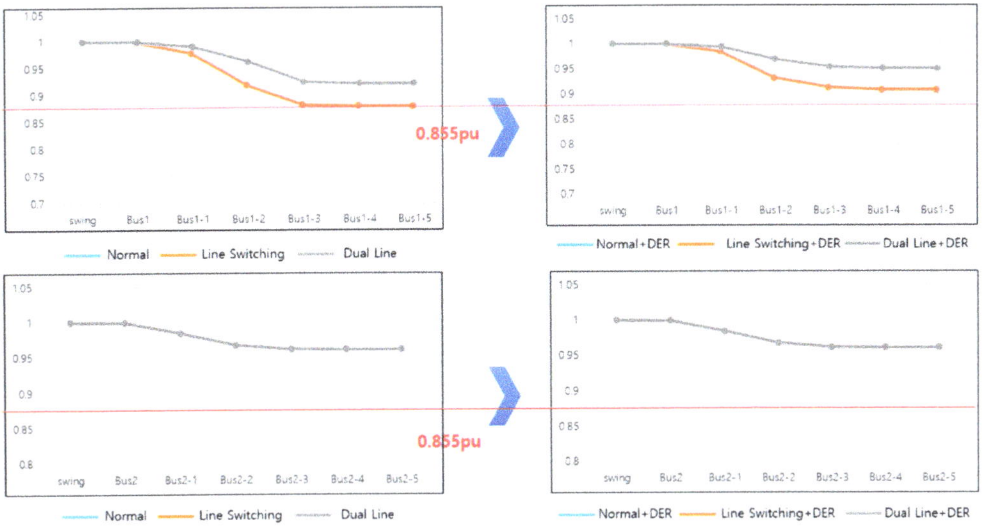

Figure 19. Power flow comparison graphs based on model B.

From the table and graphs, it is evident that the voltage and phase values are the same in the normal state and the line connection as a result of the PF calculation. The auxiliary transformer does not have other loads and it has the same capacity.

In the case of parallel operation of transformers, the load capacity is enough to supply one transformer, and as two transformers are connected, the load factor is lowered as a result; therefore, as a result of the simulation, it can be seen that the voltage is close to the reference voltage and the phase is fast, meaning that fast power supply is possible.

For line 2 (bottom), as a line that is irrelevant to the load switching and parallel operation of line 1, all three cases (normal, connected, parallel) show the same waveform.

As analyzed in Model A, buses 1 and 2 represent the HV terminal as the primary side of the transformer. Buses 1-1 and 2-1 are the transformer's secondary side, and the loads are connected from buses 1-2 and 2-2. As shown in the table, if one compares buses 1-2 and 2-2 with buses 1-3 and 2-3, it can be seen that the phase is largely left behind. This is because heavy LV loads are connected to buses 1-2 and 2-2, and the phase from the next bus is delayed; that is, it can be seen that when heavy loads are connected, the PF is delayed.

5.3. Model C

The last Model C is a system wherein only one S-substation is installed and connected to an existing transformer in the case of a transformer fault. To maintain the load at almost the same as that of other models and to match the capacity of the existing transformer to the 300 kVA that is typically used, only the heavy LV loads are partially adjusted. Figure 20 shows the equivalent circuit of Model C.

Similar to other system models, the PF results for each Model C situation are shown in the following Table 4.

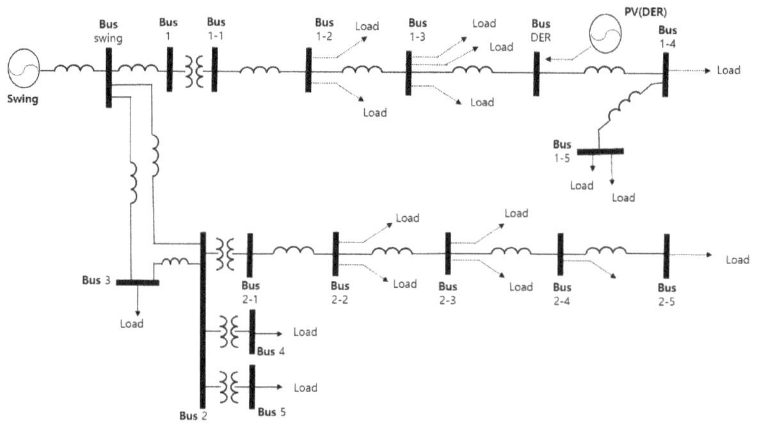

Figure 20. Underground distribution system based on model C.

Table 4. Comparison of power flow results by case.

BUS	Normal		Normal + DER		Line Switching		Line Switching + DER		Dual Line		Dual Line+ DER	
	V	Deg	V	Deg	V	Deg	V	Deg	V	Deg	V	Deg
swing	1	0	1	0	1	0	1	0	1	0	1	0
Bus 1	0.9995	−0.13	0.9996	−0.12	0.9999	0	1	0	0.9996	−0.13	0.9997	−0.12
Bus 1-1	0.9795	−7.87	0.9855	−7.03	0.9998	−0.01	0.9998	−0.01	0.9891	−7.91	0.9935	−7.27
Bus 1-2	0.9198	−9.01	0.9326	−8.12	0.738	−30.12	0.7804	−27.3	0.9393	−9.34	0.9483	−8.6
Bus 1-3	0.8835	−23.2	0.9137	−8.92	0.7089	−44.31	0.7539	−21.64	0.9022	−23.53	0.9301	−9.88
Bus 1-4	0.881	−26.27	0.9087	−1.58	0.7069	−47.38	0.7447	−9.33	0.8997	−26.59	0.9255	−2.91
Bus 1-5	0.8808	−26.93	0.9085	−2.24	0.7067	−48.03	0.7445	−9.99	0.8994	−27.25	0.9252	−3.57
Bus 2	0.9986	−0.97	0.9986	−0.97	0.9937	−1.31	0.9947	−1.28	0.9978	−0.97	0.998	−0.96
Bus 2-1	0.9837	−8.38	0.9838	−8.38	0.8463	−29.17	0.8768	−26.43	0.9488	−8.54	0.9549	−7.9
Bus 2-2	0.9668	−8.72	0.9668	−8.72	0.7857	−29.34	0.8216	−26.67	0.9328	−8.62	0.94	−8.01
Bus 2-3	0.9609	−13.82	0.9609	−13.82	0.7809	−34.44	0.8167	−31.77	0.9271	−13.72	0.9343	−13.11
Bus 2-4	0.9605	−14.7	0.9605	−14.7	0.7806	−35.32	0.8163	−32.65	0.9267	−14.6	0.9339	−13.98
Bus 2-5	0.9604	−14.92	0.9604	−14.92	0.7805	−35.54	0.8162	−32.87	0.9266	−14.82	0.9338	−14.2
Bus 3	0.9988	−0.87	0.9989	−0.87	0.9951	−1.13	0.9959	−1.1	0.9983	−0.86	0.9984	−0.86
Bus 4	0.9932	−5.54	0.9933	−5.54	0.9884	−5.88	0.9894	−5.86	0.9925	−5.54	0.9927	−5.53
Bus 5	0.9821	−10	0.9821	−10	0.9773	−10.34	0.9783	−10.32	0.9814	−10	0.9816	−9.99

Under normal operational state, there is no difference from other models, so the set standard voltage is maintained; however, the voltage drop appears to be larger when switching to another transformer due to a fault in the transformer installed in the S-substation. Although the voltage drop in the load connected to line 1 (top) is large, the voltage drop in the load connected to line 2 (bottom) is also large, so a normal power supply cannot be said to have been achieved. This is judged to be a problem that occurs because the transformer installed in the S-substation is operated at 1000 kVA, whereas the existing pad-mounted transformer is operated at 300 kVA.

As shown in the PF results (Table 4 and Figure 21), since there are more loads connected to the 1000 kVA transformer, it is not appropriate to only connect to the 300 kVA transformer; therefore, when three 300 kVA transformers in operation are connected or two 500 kVA transformers are connected, the loads must be switched to the connected transformer when there is a fault in the 1000 kVA transformer; furthermore, to maintain standard voltage, the load factor of the transformer should be limited to 50% or less. The PF results when three 300 kVA transformers are connected and the load factor is less than 50% are as following Figure 22.

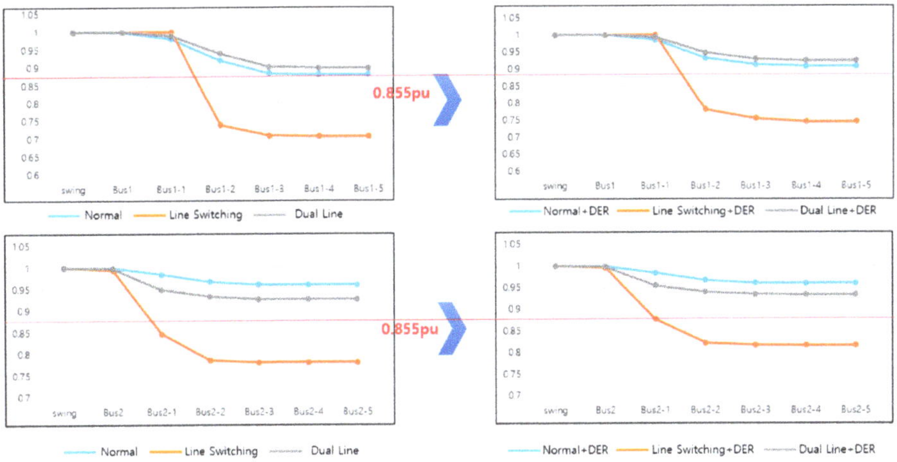

Figure 21. Power flow comparison graphs based on model C.

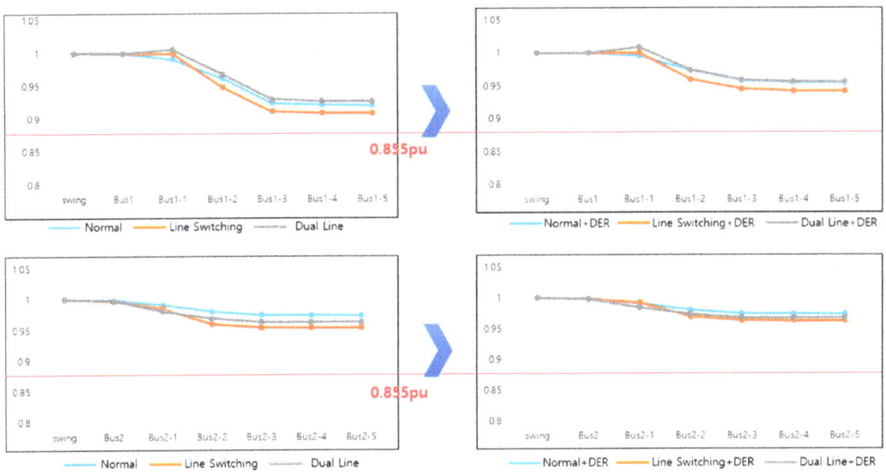

Figure 22. Power flow change graphs after 50% load factor.

When three 300 kVA transformers are connected and the load factor is kept below 50%, it can be confirmed that power is supplied stably.

6. Economic Analysis

6.1. Economic Analysis Issues

This paper also examines construction cost and VOLL for economic analysis [18].

6.2. Construction Cost

To calculate the construction cost, the quantity of equipment, cable length, management, etc., should be decided by models. The equipment output for Model A is as follows in Figure 23:

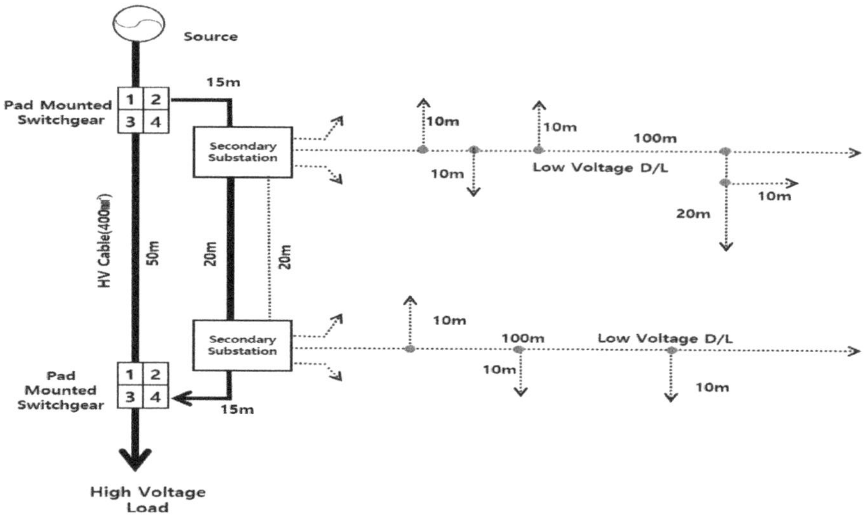

Figure 23. Required quantity of each piece of equipment for model A.

The equipment and materials for each model are as follows in Table 5:

Table 5. Equipment and materials required.

Equipment	Model A	Model B	Model C
Switchgear	2 EA	2 EA	2 EA
RMU	2 EA	1 EA	1 EA
Manhole	2 EA	1.5 EA	1 EA
Transformer (1000 kVA)	2 EA	2 EA	1 EA
Transformer (300 kVA)	-	1 EA	3 EA
Joint Box	8 EA	8 EA	8 EA
HV cable (400 mm^2)	100 cm	65 cm	65 cm
HV cable (95 mm^2)	-	10 cm	50 cm
LV cable (240 mm^2)	310 cm	290 cm	310 cm
HV cable tube (200 mm)	100 m	75 m	65 m
LV cable tube (100 mm)	310 m	290 m	310 m

Calculations for the construction costs are presented in the following Table 6:

Table 6. Calculations of construction costs.

[Unit: One Thousand Won]					
Assortment		Labor Cost	Material Cost	Other Expenses	SUM
Model A	HV D/L	23,191	103,007	269	126,467
	LV D/L	35,056	30,946	135	66,137
	SUM	58,247	133,953	404	192,604
Model B	HV D/L	21,675	93,058	244	114,977
	LV D/L	40,081	41,075	165	81,321
	SUM	61,755	134,134	409	196,298
Model C	HV D/L	23,806	97,463	263	121,532
	LV D/L	46,108	47,523	192	93,823
	SUM	69,915	144,986	455	215,356

The results indicate that Model C has the highest construction cost, which is attributable to the large number of transformers and the increase in cables caused by line connections. In contrast, Model A, which involves the installation of two nearby S-substations, has the lowest construction cost and is economical.

6.3. Value of Lost Load

The Value of Lost Load (VOLL) represents the monetary cost associated with all the consequences that arise when consumers pay for electricity but do not receive it. For example, in the event of a power outage, VOLL includes the costs of damage and repairs to products under production due to the interruption of the power supply, labor costs incurred from work stoppages, additional expenses related to business resumption, fuel costs from operating emergency generators, and losses such as the death of livestock due to interrupted oxygen supply or fish mortality. Essentially, VOLL encompasses any costs expected by customers due to power outage. The formula for each industry is as follows in Table 7:

Table 7. VOLL function cost.

Industry	VOLL Calculation Method
Industrial Load ($Load_I$)	$y = e^{-0.0000199x^2+0.0200631x+1.1214620}$
Educational Load ($Load_E$)	$\mathrm{y} = e^{-0.0000273x^2+0.0274188x-0.07049408}$
General Load ($Load_G$)	$\mathrm{y} = e^{-0.0000287x^2+0.0280840x+0.1159222}$
Agricultural Load ($Load_A$)	$\mathrm{y} = e^{-0.0000401x^2+0.0392395x-1.2240430}$
Residential Load ($Load_R$)	$\mathrm{y} = e^{-0.0000377x^2+0.0331652x+4.3880440}$

Where X is the power outage duration time; if it is less than 1 min, it is calculated as 1 min. The calculated results are as follows in Table 8:

Table 8. VOLL cost per hour and use.

[Unit: One Thousand Won]						
Industry	**1 min**	**20 min**	**1 h**	**2 h**	**4 h**	**8 h**
$Load_I$	3132	4548	9522	25,597	120,345	476,654
$Load_E$	508	846	2321	8955	73,928	476,355
$Load_G$	1155	1947	5461	21,601	181,810	1,078,970
$Load_A$	306	634	2680	18,307	359,127	4,323,744
$Load_R$	83	154	514	2502	26,270	111,442

The formula is converted into a function by examining the amount of damage that is expected by the consumer when a power outage occurs. It is observed that, for short power outages, the amount of damage is large for industrial and general loads, while it increases rapidly over a long period of time for agricultural loads; for example, if the oxygen supply in aquaculture farms is interrupted due to a power outage, it may lead to huge damages, such as mass death.

The basic analysis is set on the premise that the ATCB is put in immediately after the occurrence of a fault; however, the operation time of the ATCB should be taken into account—the manufacturer's performance is confirmed to be 15 ms.

Due to the phase difference for each bus, the supply time is different even when the power is turned on again; however, as can be seen in the Table 9, the convert to time for each bus is very small compared to the operation time of the ATCB, so it has little effect.

Table 9. VOLL cost of model A.

Bus ID	Classification of VOLL	V_LF (PU)	Vangle_LF (Deg)	Phase Difference (Deg)	Convert to Time	ATCB Operating Time	Sum	VOLL Cost
Bus 2-1	Reference Value	0.9619	−13.16	-	-	-	-	-
Bus 1-2	$Load_I$ 300 kW	0.8204	−14.08	−0.92	4.2677×10^{-7}	1.50×10^{-1}	1.50×10^{-1}	3132
Bus 1-2	$Load_E$ 500 kW	0.8204	−14.08	−0.92	4.2677×10^{-7}	1.50×10^{-1}	1.50×10^{-1}	508
Bus 1-3	$Load_A$ 50 kW	0.788	−28.27	−15.11	7.0092×10^{-6}	1.50×10^{-1}	1.50×10^{-1}	306
Bus 1-3	$Load_G$ 10 kW	0.788	−28.27	−15.11	7.0092×10^{-6}	1.50×10^{-1}	1.50×10^{-1}	1155
Bus 1-3	$Load_G$ 30 kW	0.788	−28.27	−15.11	7.0092×10^{-6}	1.50×10^{-1}	1.50×10^{-1}	1155
Bus 1-4	$Load_I$ 20 kW	0.7858	−31.34	−18.18	8.4333×10^{-6}	1.50×10^{-1}	1.50×10^{-1}	3132
Bus 1-5	$Load_A$ 10 kW	0.7856	−32	−18.84	8.7395×10^{-6}	1.50×10^{-1}	1.50×10^{-1}	306
Bus 1-5	$Load_R$ 5 kW	0.7856	−32	−18.84	8.7395×10^{-6}	1.50×10^{-1}	1.50×10^{-1}	83
				Sum (unit: 1000 won)				9777

7. Results and Discussion

7.1. Selecting the Optimal Distribution System

An optimal model is selected based on the results of the review of the PF by system model and economic analysis. The analysis results based on models are shown in Table 10.

Table 10. Analysis results based on models.

Model	Power Flow	Economics	Field Applicability
A	Voltage drop occurs when connecting lines when the transformer operates at over 50%	Installation of two underground structures for S-substation	Need to secure installation space for underground burial of two structures
B	Since there are no loads of auxiliary transformers, the voltage/phase is stable even when replaced with a peripheral voltage	Increased construction cost (due to the larger structure and auxiliary transformer)	Difficulty in construction due to additional installation of auxiliary transformer in the underground structure
C	Voltage drop occurs when connecting lines when the transformer operates at over 50%	Increased construction cost (due to multiple connections of other transformers for line conversion)	It is necessary to connect a number of other lines for alternative supply of large power transformers.

The detailed analysis results are as follows:

7.1.1. Distribution System

- The currents of Model A~C are stable in normal operation without fault in D/Ls.
- The PF analysis indicates that the parallel operation of transformers appears to be the optimal operation method.
- It is difficult to supply stable power due to voltage drop during load switching due to transformer faults. → Normal power supply is possible when the transformer load is operated at less than 50% capacity.
- Heavy LV loads (2~500 kW) need to be kept close to the transformer in consideration of the voltage drop.

7.1.2. Economics

- Model A has the lowest construction cost and Model C has the highest.
- Construction cost is affected by the number of structures/equipment and the length of the D/Ls.

7.1.3. Construction

- Since the S-substation is a structure installed underground, construction methods are also important.

- If an auxiliary transformer is installed, the area occupied increases, and if an ATCB is installed, additional area is required.
- In the case of Model C, the existing transformer is connected, but due to its heavy weight, many interconnected lines are required.

7.1.4. Optimal Model Selection

- As a result of the analysis, it is judged that Model A is the most suitable. Due to the characteristics of the S-substation, the transformer capacity is large, but they can be installed near each other to complement each other, and the construction efficiency can be improved by installing an underground structure nearby.
- In consideration of load arrangement, heavy LV loads should be placed close to the transformer, and the load factor should be kept below 50% in consideration of the transformer's fault. Moreover, S-substations should be installed in proximity to each other to reduce the load factor and voltage drop. The detailed criteria are as follows [4,19]:
  - The transformer should be kept at under 50% of the load factor.
  - S-substations should be installed near heavy LV loads.
  - S-substations should be installed adjacent to each other.
  - Load arrangement should consider voltage drop.
  - LV boards should be installed to secure LV system reliability.

7.2. Example of Application of Model A of Power Distribution System

The underground distribution system to which Model A is applied is shown in Figure 24. It is different from Figure 3. When two S-substations are installed, it can be confirmed that they are connected through ATCB to prevent power outages. In addition, considering the voltage drop, heavy LV loads are placed close to the S-substation.

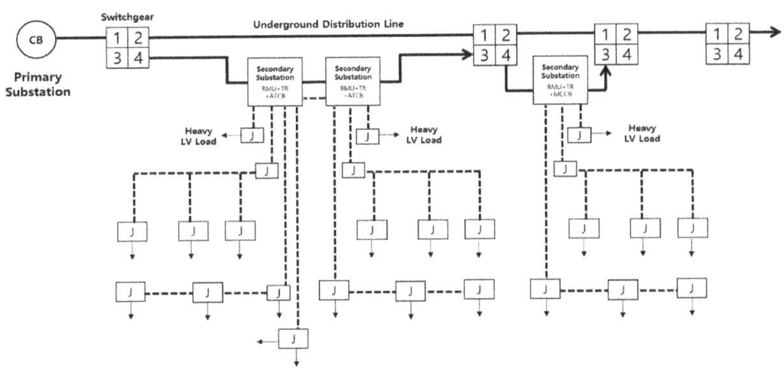

Figure 24. Underground LV distribution system for S-Substation based on Model A.

8. Conclusions

This paper addresses the optimal LV distribution system for applying the S-Substation to an underground distribution system. Three models were proposed, and both PF and economic analysis were conducted to determine the optimal distribution system model. PF analysis was conducted to determine the voltage and phase of each model; additionally, frequency variations were observed. MATLAB/SIMULINK was utilized for accurate analysis, and changes in power flow were also examined with the connection of DERs.

In LV distribution systems, voltage drop must be considered based on load placement, as the voltage drop can vary depending on the location of heavy loads, potentially falling outside the standard voltage range at the end of the D/Ls; therefore, load placement was also considered in this study. The voltage and phase values from the PF analysis are derived from the BUS. To support this analysis, we examined the standard voltage ranges in Korea, strategies for load deployment to prevent voltage drops, and how phase differences change over time. For underground areas in Korea, it is crucial to consider the placement of the PDE once potential load locations are confirmed. Unlike the transmission system, the distribution system connects to various loads, including motor, nonlinear, and unbalanced loads. The loads could fluctuate in capacities and positions, causing PF results to change occasionally; additionally, since the S-substation is an underground structure, factors such as installation methods and securing underground space must be considered. Therefore, it is necessary to analyze the field conditions and PF specific to each location where the S-substation is installed. This is because construction costs fluctuate greatly depending on the field conditions.

Construction costs and VOLL were compared across the models for economic feasibility. The optimal location and selected model can be determined by incorporating the PF analysis results and economic feasibility discussed in this paper. Ultimately, Model A was selected as the optimal model. An exemplary system was presented to demonstrate the application of this model to the underground distribution system.

Korea is more sensitive to power outages and faults than other countries; therefore, reliable equipment and system configurations are increasingly important. To build a more reliable LV distribution system, the following areas of research should be considered:

- Underground LV closed-loop or multi-LV distribution systems.
- Power quality management in LV distribution systems.
- Protection coordination methods for LV distribution systems.
- Fault detection methods in LV distribution systems.

In the future, we believe that the models presented in this paper as well as other models can be studied and applied in the field.

Author Contributions: Conceptualization, B.S., H.L. and S.C.; methodology, B.S.; software, B.S.; validation, B.S. and H.L.; formal analysis, B.S. and H.L.; investigation, B.S. and H.L.; resources, B.S. and H.L.; data curation, B.S. and H.L.; writing—original draft preparation, B.S.; writing—review and editing, B.S. and S.C.; visualization, B.S.; supervision, S.C.; project administration, B.S.; funding acquisition, B.S. and S.C. All authors have read and agreed to the published version of the manuscript.

Funding: This work was supported in part by the Korea Electric Power Research Institute (KEPRI) grant funded by the Korea Electric Power Corporation (KEPCO) (R23DA01) and in part by the Korea Institute of Energy Technology Evaluation and Planning (KETEP) grant funded by the Ministry of Trade, Industry & Energy (MOTIE) of the Republic of Korea (No. RS-2023-00234707).

Institutional Review Board Statement: Not applicable.

Informed Consent Statement: Not applicable.

Data Availability Statement: The data used in this study were obtained from the Korea Power Exchange. The period for data disclosure has expired.

Conflicts of Interest: The authors declare no conflicts of interest.

References

1. Shin, B.H.; Lee, H.S.; Shin, D.Y.; Hur, S.Y.; Kim, S.M. A study on the application of power electronics technology in secondary substation to improve power quality. In Proceedings of the 27th International Conference on Electricity Distribution (CIRED 2023), Rome, Italy, 12–15 June 2023.
2. Zhu, Y.; Tomsovic, K. Optimal distribution power flow for systems with distributed energy resources. *Int. J. Electr. Power Energy Syst.* **2007**, *29*, 260–267. [CrossRef]
3. Ibrahim, I.A.; Hossain, M.J. Low voltage distribution networks modeling and unbalanced (optimal) power flow: A comprehensive review. *IEEE Access* **2021**, *9*, 143026–143084. [CrossRef]
4. Yang, C.; Sun, Y.; Zou, Y.; Zheng, F.; Liu, S.; Zhao, B.; Wu, M.; Cui, H. Optimal power flow in distribution network: A review on problem formulation and optimization methods. *Energies* **2023**, *16*, 5974. [CrossRef]
5. Chamorro, H.; Diaz, N. Hierarchical power flow control in low voltage microgrids. In Proceedings of the 2013 North American Power Symposium (NAPS), Manhattan, KS, USA, 22–24 September 2013; IEEE: New York, NY, USA, 2013; pp. 1–5.

6. Stanelytė, D.; Radziukynas, V. Analysis of voltage and reactive power algorithms in low voltage networks. *Energies* **2022**, *15*, 1843. [CrossRef]
7. Babś, A.; Matusewicz, M.; Noske, S.; Falkowski, D. Smart Secondary Substation as the source of the flexibility services. In Proceedings of the CIRED 2021-The 26th International Conference and Exhibition on Electricity Distribution, Online Conference, 20–23 September 2021; IET: New York, NY, USA, 2021; pp. 477–480.
8. Apellaniz, I.; Arostegui, J.; Tejedo, J.R.; Sánchez, J.A. Smart Secondary Substation. A reality and a big opportunity for innovative solutions for predictive maintenance and life extension. In Proceedings of the 25th International Conference on Electricity Distribution, Madrid, Spain, 3–6 June2019.
9. Bjørtuft, T.R.; Granhaug, O.; Hagen, S.T.; Kuhlefelt, J.H.; Salge, G.; Skryten, P.K.; Stangherlin, S. Internal arc fault testing of gas insulated metal enclosed MV switchgear. In Proceedings of the CIRED 2005-18th International Conference and Exhibition on Electricity Distribution, Turin, Italy, 6–9 June 2005; IET: New York, NY, USA, 2005; pp. 1–5.
10. Cho, H.; Lee, U.; Park, Y. The temperature distribution and rise test of pole mount cast resin transformer for power distribution. In Proceedings of the IEEE/PES Transmission and Distribution Conference and Exhibition, Yokohama, Japan, 6–10 October 2002; IEEE: New York, NY, USA, 2002; pp. 1854–1857.
11. Liu, X.; Wang, J.; Sun, J. Temperature rise test and calculation method of liquid-immersed transformer. In Proceedings of the 2022 IEEE International Conference on High Voltage Engineering and Applications (ICHVE), Chongqing, China, 25–29 September 2022; IEEE: New York, NY, USA, 2022; pp. 1–4.
12. Shin, B.H.; Oh, G.D.; Kim, S.C.; Jung, K.H. Power Flow and Economics Analysis for RMU-based Low-voltage Distribution Networks Operation. In Proceedings of the CIRED 2021-The 26th International Conference and Exhibition on Electricity Distribution, Online Conference, 20–23 September 2021; IET: New York, NY, USA, 2021; pp. 2432–2436.
13. Marszal-Pomianowska, A.; Widén, J.; Le Dréau, J.; Heiselberg, P.; Bak-Jensen, B.; de Cerio Mendaza, I.D. Operation of power distribution networks with new and flexible loads: A case of existing residential low voltage network. *Energy* **2020**, *202*, 117715. [CrossRef]
14. Ring Main Unit in Electrical Distribution System. Available online: https://www.allumiax.com/blog/ring-main-unit-in-electrical-distribution-system (accessed on 17 August 2024.).
15. Etta, B.-O.; Idoniboyeobu, D. Improved Performance of Elelenwo 11kv Electric Power Distribution Network Using Newton Raphson Load Flow. *IRE J.* **2021**, *5*, 19–25.
16. Glover, J.D.; Mulukutla, T.J.O.; Sarma, S. *POWER SYSTEM Analysis & Design*; CENGAGE Learning: Boston, MA, USA, 2017; pp. 310–403.
17. Heidari-Akhijahani, A.; Safdarian, A.; Vrakopoulou, M. A linear ac power flow model for unbalanced multi-phase distribution networks based on current injection equations. *IEEE Trans. Power Syst.* **2021**, *36*, 3806–3809. [CrossRef]
18. Tsao, Y.-C.; Beyene, T.D.; Thanh, V.-V.; Gebeyehu, S.G.; Kuo, T.-C. Power distribution network design considering the distributed generations and differential and dynamic pricing. *Energy* **2022**, *241*, 122828. [CrossRef]
19. Seme, S.; Lukač, N.; Štumberger, B.; Hadžiselimović, M. Power quality experimental analysis of grid-connected photovoltaic systems in urban distribution networks. *Energy* **2017**, *139*, 1261–1266. [CrossRef]

Disclaimer/Publisher's Note: The statements, opinions and data contained in all publications are solely those of the individual author(s) and contributor(s) and not of MDPI and/or the editor(s). MDPI and/or the editor(s) disclaim responsibility for any injury to people or property resulting from any ideas, methods, instructions or products referred to in the content.

energies

MDPI

Article

Study on Electrical Characteristics Analysis and Electrical Circuit Model Design of Vanadium Redox Flow Battery Systems Based on Current and Flow Rate Conditions

Seongjun Lee [1], Hyeonhong Jung [2] and Yoon-Gyung Sung [1,*]

[1] Department of Mechanical Engineering, Chosun University, Gwangju 61452, Republic of Korea; lsj@chosun.ac.kr
[2] Mobase Electronics, Suwon 16648, Republic of Korea; gusghd0707@naver.com
* Correspondence: sungyg@chosun.ac.kr; Tel.: +82-62-230-7181

Abstract: Recent research has focused on vanadium redox flow batteries (VRFBs) to address the short lifetimes and fire risks associated with lithium battery systems. While VRFBs offer advantages in safety, they suffer from low energy density and efficiency compared with lithium batteries. To improve VRFB performance, studies are exploring improvements in materials such as anodes, cathodes, and separators and optimizing operations by controlling electrolyte flow rates. However, the impact of current magnitude on VRFB efficiency has been less studied, with few analyses addressing both current and flow rate effects. This research proposes an experimental procedure to evaluate charge/discharge efficiency, energy efficiency, and system efficiency across varying current magnitudes and electrolyte flow rates, using a 40 W VRFB stack composed of four 10 W cells in series. In addition, we introduce a design method for an electrical equivalent circuit model that simulates the VRFB stack, reflecting experimental findings. The model's accuracy was validated by comparing it with data from 11 full charge/full discharge cycle tests, which varied current and electrolyte amounts.

Keywords: vanadium redox flow battery (VRFB); energy storage system (ESS); efficiency; electrical circuit model (ECM); flow rate; electrolyte; current magnitude; charge/discharge cycle test

1. Introduction

Because the use of fossil fuels, such as oil and coal, causes global warming, environmental pollution, and resource depletion, industries that use eco-friendly renewable energy sources, such as solar and wind power, are attracting attention to solve these problems. However, renewable energy power generation systems have intermittent characteristics that depend on environmental conditions such as solar irradiance and wind speed. Energy storage systems (ESSs) that overcome these renewable energy problems and enable stable and efficient power and energy management are growing steadily [1,2]. The electrical energy sources used in ESSs include lead–acid batteries, lithium batteries, sodium–sulfur batteries, redox flow batteries, supercapacitors, and compressed air. Table 1 lists the energy densities, energy efficiencies and lifetimes of the electrical energy sources in an ESS [2]. Currently, lithium-ion batteries with high energy densities and efficiencies are widely used in ESSs to maximize energy efficiency [2,3].

However, energy storage devices that use lithium-ion batteries have several problems. As can be seen from the fire accidents of large-capacity batteries applied to electric vehicles, energy storage devices composed of lithium-ion batteries using flammable nonaqueous organic solvents face the problem of fire when thermal runaway occurs because of internal short circuits [4–6]. In particular, energy storage devices manufactured by combining a large number of lithium-ion battery cells in series and in parallel have a higher possibility of fire occurrence if the maximum and minimum voltages of the applied battery cells cannot be managed and if the failure status cannot be accurately predicted to maintain the cells or modules. Therefore, when lithium-ion batteries are used in large-capacity energy storage applications, considerable effort must be made to prevent fires and prevent them from spreading. Accordingly, to estimate the battery status, recent research on fault detection using gas sensors and force sensors, as well as voltage and temperature measurements, has been conducted [7]. In addition, research is being conducted on estimating SOC and temperature by training an artificial intelligence model using measurement signals using non-destructive techniques such as ultrasonic sensors [8]. In addition, lithium-ion batteries cause environmental pollution during disposal after their lifespan ends [9]. To solve these problems, research is being conducted to improve lithium-ion battery deterioration diagnosis and reuse. However, this increases the operational complexity and cost of the system [10,11].

Table 1. Energy and efficiency of energy storage system [2].

	Lead–Acid	Lithium-Ion	NaS	VRFB	Super Capacitors	CAES *
Energy Density (Wh/L)	50–80	200–500	150–250	16–33	2–10	3–6
Energy Efficiency (%)	70–90	85–95	80–90	70–85	90	45–60
Lifetime (cycles)	500–800	2000–3000	4000–40,000	1500–15,000	50,000	>10,000

* CAES: Compressed Air Energy Storage.

Therefore, research is being conducted on battery systems that can replace lithium-ion batteries in ESSs, and the Vanadium Redox Flow Battery (VRFB), which uses an aqueous solvent with a longer lifespan than lithium batteries, is being studied extensively [12–19]. In a VRFB system, positive and negative electrolytes are stored in two electrolyte tanks, and the flow is supplied in parallel to the series-connected cells inside the stack through a pump. The transferred electrolyte undergoes an oxidation/reduction reaction at the electrode, ions move to the ion exchange membrane of the Nafion material, and energy is stored in the electrolyte. The oxidation/reduction reaction of the VRFB is shown in Figure 1 and can be expressed by the anode, cathode, and overall electrochemical reaction formulas in Equations (1) to (3). The ionic reaction of the vanadium electrolyte is that when charging, the V^{3+} ions at the negative electrode are reduced to V^{2+} ions, and the V^{4+} ions at the positive electrode are oxidized to V^{5+} ions. During the discharging process, the V^{5+} ions at the positive electrode are oxidized to V^{4+} ions, and the V^{2+} ions at the negative electrode are oxidized to V^{3+} ions, thereby performing charging and discharging.

$$VO_2^+ + 2H^+ + e^- \leftrightarrow VO^{2+} + H_2O \text{ @Positive Electrode} \quad (1)$$

$$V^{2+} \leftrightarrow V^{3+} + e^- \text{ @Negative Electrode} \quad (2)$$

$$VO_2^+ + 2H^+ + V^{2+} \leftrightarrow VO^{2+} + H_2O + V^{3+} \text{ @Overall} \quad (3)$$

The output of the VRFB was determined by the area of the electrode, and the voltage of the system was determined by the number of cells connected in series. Unlike lithium-ion batteries, which increase capacity by connecting cells with the same characteristics in parallel, the VRFB is determined by the amount of electrolyte stored in the tank. Therefore, VRFBs have the advantages of increasing capacity and being more flexible in mechanical design than existing cylindrical, square, and pouch-shaped lithium-ion battery packaging designs [13]. However, VRFBs have a lower energy density and efficiency than lithium-ion batteries, and to achieve the same energy capacity as an ESS using lithium-ion batteries, the VRFB system requires a larger installation area and initial investment costs.

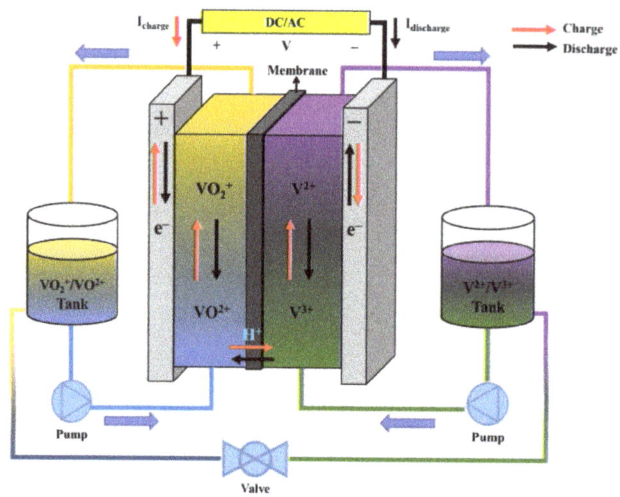

Figure 1. Diagram of VRFB energy storage system [19].

In order to overcome the disadvantages of low energy density and efficiency of VRFBs, research is being conducted on the materials of vanadium electrolytes and felt electrodes, as well as on membranes that affect crossover by ion selectivity [20,21]. To this end, VRFBs require an analysis of system efficiency, including charge/discharge Coulomb efficiency, energy efficiency, and pump loss through charge/discharge experiments according to current and flow conditions. In addition, to analyze the electrical characteristics of the VRFB system according to the flow rate and current conditions, electrical equivalent circuit modeling of the VRFB and parameter analysis are required. Unlike the electrical ECM model of a lithium battery, the VRFB must be modeled such that the shunt current loss that occurs as the electrolyte is connected in parallel to the cells inside the stack and a capacity change depending on the amount of electrolyte is possible.

Previous studies on improving the VRFB efficiency are as follows. Kim presented the discharge characteristics of a 1 kW/1 kWh VRFB using mixed-acid electrolytes and the advantages of using electrolytes [22]. The study presented the VRFB discharge characteristics and efficiency results according to the current and flow rate conditions but did not present the parameter analysis results of the electrical equivalent circuit model. Karrech analyzed charge and discharge experimental results according to variable flow rate conditions and presented a flow rate control method for optimal efficiency [23]. However, the shunt current loss was not considered because it was conducted on a single cell and its characteristics depended on the current size. Ma. X conducted a study on the optimal flow condition at the VRFB stack level but did not mention the influence of the current size for optimal efficiency control in detail [24]. Thus, to improve the efficiency of the VRFB battery system, it is necessary to comprehensively consider the influence of the shunt current and analyze the VRFB system characteristics on the charge/discharge current and flow conditions during operation.

Previous studies on the design of electrical equivalent circuit modeling are as follows: Fornaro presented the same electrical equivalent circuit model used in this study and estimated its parameters using recursive least squares (RLS) [25]. However, the study presented parameter estimation results from short-term simulations that did not consider the influence of the shunt current loss. However, in a series-connected stack system, the influence of shunt current should be considered. Yu Zhang also presented the same electrical equivalent circuit model considering the shunt resistance and showed that the stack could be well simulated with the estimated parameters using RLS [26]. This study considered the influence of the shunt current and flow rate; however, because the simulation results were for a short period, a method to simulate the change in electrolyte volume owing to crossover was not presented.

Therefore, in this study, we propose an experimental method that can analyze the charge/discharge Coulombic efficiency, energy efficiency, and system efficiency according to the current size and electrolyte flow rate conditions and present the results of the system efficiency analysis for a 40 W VRFB stack composed of four 10 W cells in series. In addition, we systematically present an electrical equivalent circuit model that can simulate the VRFB stack and an experimental procedure and analysis method for estimating its parameters.

2. Experimental Setup and Characteristic Experimental Profile Design of VRFB

2.1. VRFB Experimental Setup

The VRFB stack used in this study has the configuration shown in Figure 2. The stack was manufactured with a structure in which a pair of bipolar plates, flow frames, electrodes, and membranes formed a cell, and four pairs were connected in series. The VRFB cell/stack used in this study has four input/output ports on one endplate for electrolyte transport. The four ports comprise two inlet ports through which the electrolyte is transferred to the cells inside the stack and two outlet ports through which the electrolyte that has undergone oxidation/reduction reactions inside the cell is transferred to the outside. The felt and bipolar plate are made of graphite for electrical conductivity, and the ion separation membrane is NR212, with a thickness of 50 μm [27].

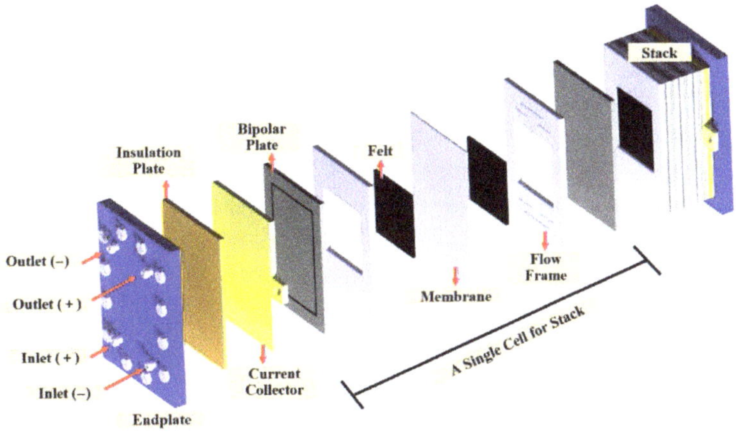

Figure 2. VRFB stack configuration diagram.

The charge/discharge experimental configuration of a 40 W VRFB stack with four 10 W cells connected in series used in this study is shown in Figure 3a, and a diagram of the actual constructed system is shown in Figure 3b. The VRFB stack system had two tanks for storing the electrolyte consisting of 1.6 M $V^{3.5+}$/2 M H_2SO_4, a diaphragm pump, and a flowmeter to measure the flow rate. Additionally, both electrolyte tanks were connected to a flow tube to balance the electrolyte in case of electrolyte imbalance due to crossover. The volume of the electrolyte was measured every 5 min during the experiment using cameras applied to positive and negative electrolyte tanks. The VRFB was charged and discharged using a battery cycler, and the measured stack voltage, current, temperature, flow rate, voltage, and current consumed by the pump were stored in a data logger at a sampling time of 100 ms. Table 2 lists the main specifications of the VRFB stack and the peripheral devices used in the experiment described above.

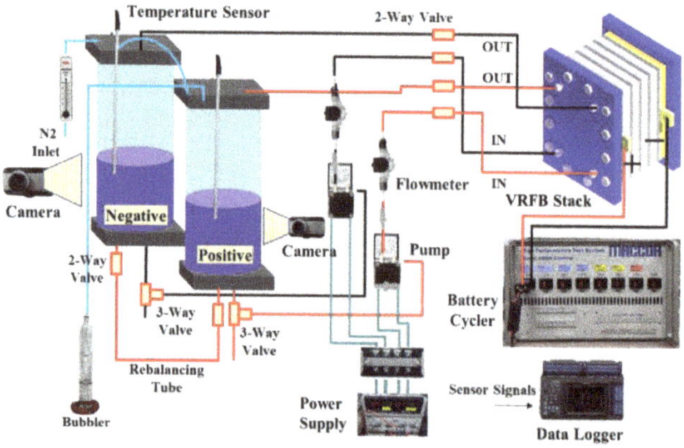

(a) Experiment configuration of VRFB system

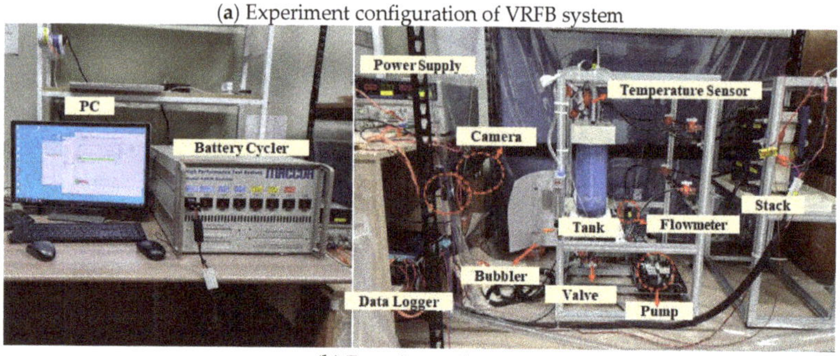

(b) Experimental setup

Figure 3. VRFB stack system configuration; (**a**) experiment configuration of VRFB system, (**b**) experimental setup.

Table 2. Specification for VRFB stack system.

40 W VRFB Stack System		
VRFB Stack	Electrode Area	100 cm^2 (10 × 10)
	Electrolyte	1.6 M $V^{3.5+}$/2 M H_2SO_4
	Membrane	Nafion-212
	Electric Collector	Brass
	Electrode	Carbon Felt
Battery Cycler	Maccor 4300 K	
Electrolyte Volume	210 mL (±5 mL)	
Pump	KNF, NF60	
Flowmeter	Omega, FTB-312	
Thermal Sensor	Teflon PT100/Thermocouple (T-type)	

2.2. Design of Experimental Procedure for Electrical Characteristics of VRFB

It is difficult to experiment with VRFBs under the same capacity (electrolyte volume) conditions because of electrolyte crossover when the experiments are conducted over a long period. This is because the VRFB operating point changes, making it challenging to analyze the characteristics under various experimental conditions. The crossover of the electrolyte can be reduced through improvements such as ion concentration control, changes in membrane thickness, and material modifications [28,29]. However, since this paper analyzes the electrical characteristics of an already-constructed VRFB system, the experiments were conducted following the procedure shown in Figure 4, minimizing the impact of crossover.

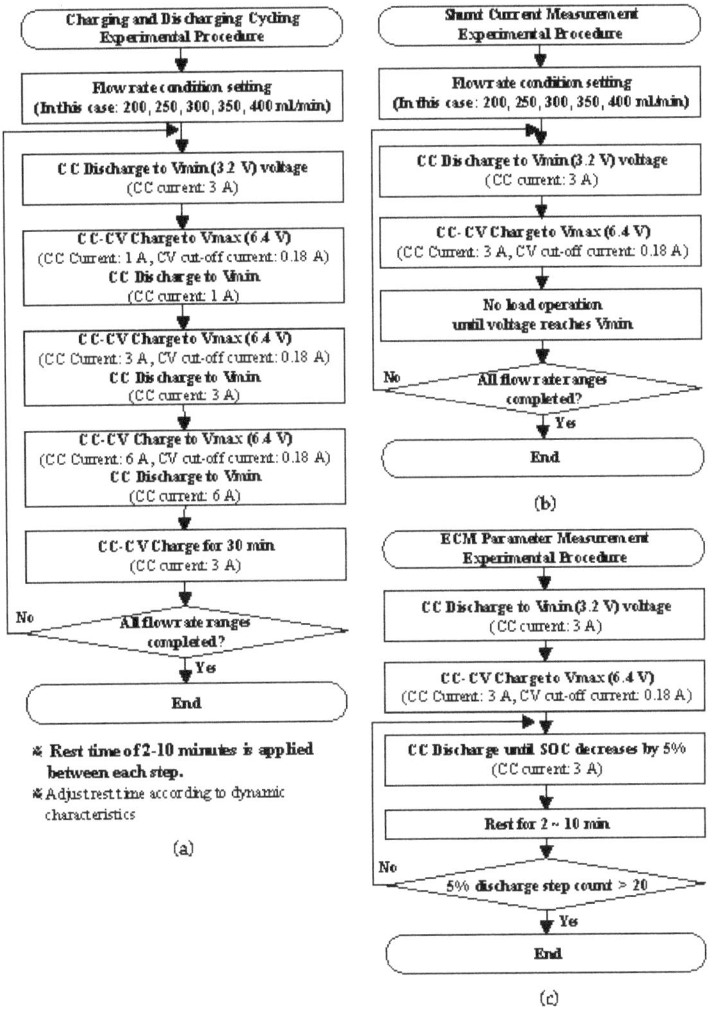

Figure 4. VRFB experimental profile; (**a**) charging and discharging experimental procedure, (**b**) shunt current measurement experimental procedure, and (**c**) ECM parameter measurement experimental procedure.

The experimental procedure for analyzing the Coulombic, energy, and system efficiencies of the VRFB stack is shown in Figure 4a. First, the charging and discharging cycles were performed continuously under current conditions of 1, 3, and 6 A (10, 30, and 60 mA/cm^2) with a fixed flow rate. The charging and discharging capacities were measured according to the current magnitude, and the Coulomb efficiency was analyzed. Subsequently, the same experiment was carried out by changing the flow rate condition so that the influence of each flow rate could be analyzed. In this study, tests were performed in the range of 200–400 mL/min at intervals of 50 mL/min intervals. The VRFB stack used in this experiment was charged by the Constant Charge–Constant Voltage (CC-CV) method with a full charge voltage of 6.4 V (1.6 V per cell) and a cut-off current of 0.18 A for terminating CV charge, and the full discharge voltage is 3.2 V (0.8 V per cell) and discharged with CC current.

The VRFB stack has a shunt current loss owing to internal self-discharge because the electrolyte is connected in parallel to each cell. The experimental procedure for extracting the shunt current and resistance parameters is shown in Figure 4b. First, the pump was driven under conditions of no load current while the battery stack was fully charged at a fixed flow rate. In this case, the voltage of the battery gradually decreased owing to the shunt current, and the experiment was conducted until the battery was fully discharged (cut-off voltage of 3.2 V). The shunt current was obtained by dividing the capacity measured when fully charged by the time required for complete discharge. Subsequently, the same experiment was conducted by changing the flow rate conditions, and the shunt current loss was analyzed according to the flow rate.

The experimental profile for measuring the open-circuit voltage of the VRFB and parameters of the equivalent electrical circuit are shown in Figure 4c. After the battery was fully charged, a pulse current was applied to discharge the SOC to 5%, followed by a rest period. This pulse current, including the rest time, was repeatedly applied until the stack voltage reached the full discharge voltage. Through this experiment, the parameters R_i, R_d, and C_d of the electrical equivalent circuit were obtained from the voltage response characteristics when a pulsed current was applied, and the open-circuit voltage (OCV) at a specific SOC was obtained for each rest period.

3. Efficiency Analysis and ECM Modeling of VRFB

In this section, we present the energy efficiency analysis results of the VRFB system according to the current magnitude and flow rate changes using the voltage/current and pump power consumption data measured during the charge/discharge test of the VRFB, using the test procedure shown in Figure 4. The Coulomb, voltage, energy, and system efficiency of the VRFB were calculated using Equations (4)–(7). Equation (4) represents the Coulomb efficiency, which is defined as the ratio of the charge capacity to the discharge capacity. Voltage efficiency is an index for analyzing the influence of the polarization voltage or overvoltage on the battery voltage and is defined in Equation (5). The energy efficiency is expressed as the ratio of the discharge energy to the charge energy, as shown in Equation (6). The system efficiency was calculated using Equation (7), including the energy consumed by the pump used to drive the VRFB stack and the energy of the VRFB stack itself during battery charging and discharging.

$$Coulombic\ Efficiency(\%) = \frac{Discharge\ Capacity(Ah)}{Charge\ Capacity(Ah)} = \frac{\int I_{Dchg}(t)dt}{\int I_{Chg}(t)dt} \times 100 \tag{4}$$

$$Voltage\ Efficiency(\%) = \frac{\int V_{Dchg}(t)dt}{\int V_{Chg}(t)dt} \times 100 \tag{5}$$

$$Energy\ Effifciency(\%) = \frac{\int V_{Dchg}(t) \times I_{Dchg}(t)dt}{\int V_{Chg}(t) \times I_{Chg}(t)dt} \times 100 \tag{6}$$

$$System\ Efficiency(\%) = \frac{\int \left(V_{Dchg}(t) \times \left|I_{Dchg}(t)\right| - P_{pump}\right)dt}{\int \left(V_{Chg}(t) \times \left|I_{Chg}(t)\right| + P_{pump}\right)dt} \times 100 \tag{7}$$

3.1. *Efficiency Analysis of VRFB According to Current and Flow Rate*

Figure 5 shows the voltage and current waveforms, electrolyte volume changes, and pump power consumption results when the VRFB stack was charged and discharged under 100% depth of discharge (DoD) conditions; the current profile is shown in Figure 4a when the electrolyte flow rate was 250 mL/min. As can be seen in (a) voltage and (b) current in Figure 5, as the charge and discharge currents increase, the voltage decrease size increases, and the CV charge arrival time becomes faster, showing the characteristic of a longer charging time. Figure 5c shows the volume of the negative electrolyte measured during the test. The volume of the negative electrolyte decreases when the state of charge is low (as the voltage decreases) and increases when the state of charge increases (as the voltage increases). And the average volume of the electrolyte during the experimental profile was 174 mL. The pump power required to transport the electrolyte in the VRFB during the experiment is shown in Figure 5d. The pump consumed an average of 3.27 W of power when operated at a flow rate of 250 mL/min flow rate.

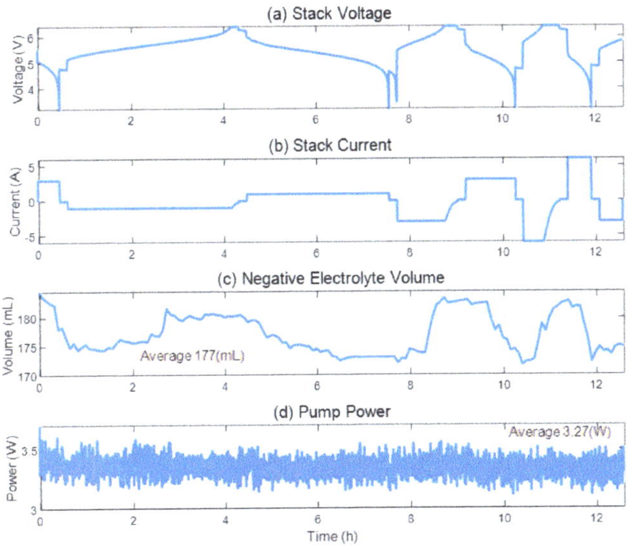

Figure 5. Experimental results of VRFB system according to current magnitude at 250 mL/min flow rate: (**a**) stack voltage; (**b**) stack current; (**c**) negative electrolyte volume; (**d**) pump consumed power.

To analyze the effect of the flow rate conditions on the charge and discharge capacities, experiments were conducted by changing the flow rate conditions under the same conditions as the current profile in Figure 5. The experimental results at the flow rates of 200 mL/min are shown in Figure 6a, 300 mL/min in Figure 6b, 350 mL/min in Figure 6c, and 400 mL/min in Figure 6d, respectively. At this time, as can be seen in the third waveform of the experimental waveforms in Figures 5 and 6, for the negative electrolyte volume, a total of five experiments for characteristic comparison were conducted with similar electrolyte volumes, averaging 174–179 mL. If the electrolyte volume changes significantly as the experiment progresses, it should be maintained by rebalancing before the experiment is conducted. The pump consumes an average power of 3.01, 3.27, 4.58, 6.03, and 7.85 W when controlling the flow rates of 200, 250, 300, 350, and 400 mL/min, respectively.

Figure 7 shows the changes in the charge/discharge capacity, Coulomb efficiency, and voltage efficiency according to the flow rate and current magnitude in the previous cycling experiment. Figure 7a shows that the charge capacity is the largest under a current of 1 A, and the charge capacity decreases as the current increases. It can also be observed that the battery charge capacity decreased as the flow rate increased, even when the current was the same. However, as confirmed from the discharge capacity results in Figure 7b, at currents above 3 A, the same characteristic of decreasing discharge capacity was observed as the current increased. However, when discharging the battery with a constant current of 1 A, the result was smaller than the discharge capacity measured during the 3 A test. This indicates that the VRFB has poor low-current discharge characteristics, owing to the shunt current loss that occurs when the pump is driven, as mentioned in Section 3.2. When a small current is used from the battery, energy transfer takes a long time. Therefore, as the time at which the pump is operated increases, the energy loss owing to the shunt current also increases, resulting in poor efficiency.

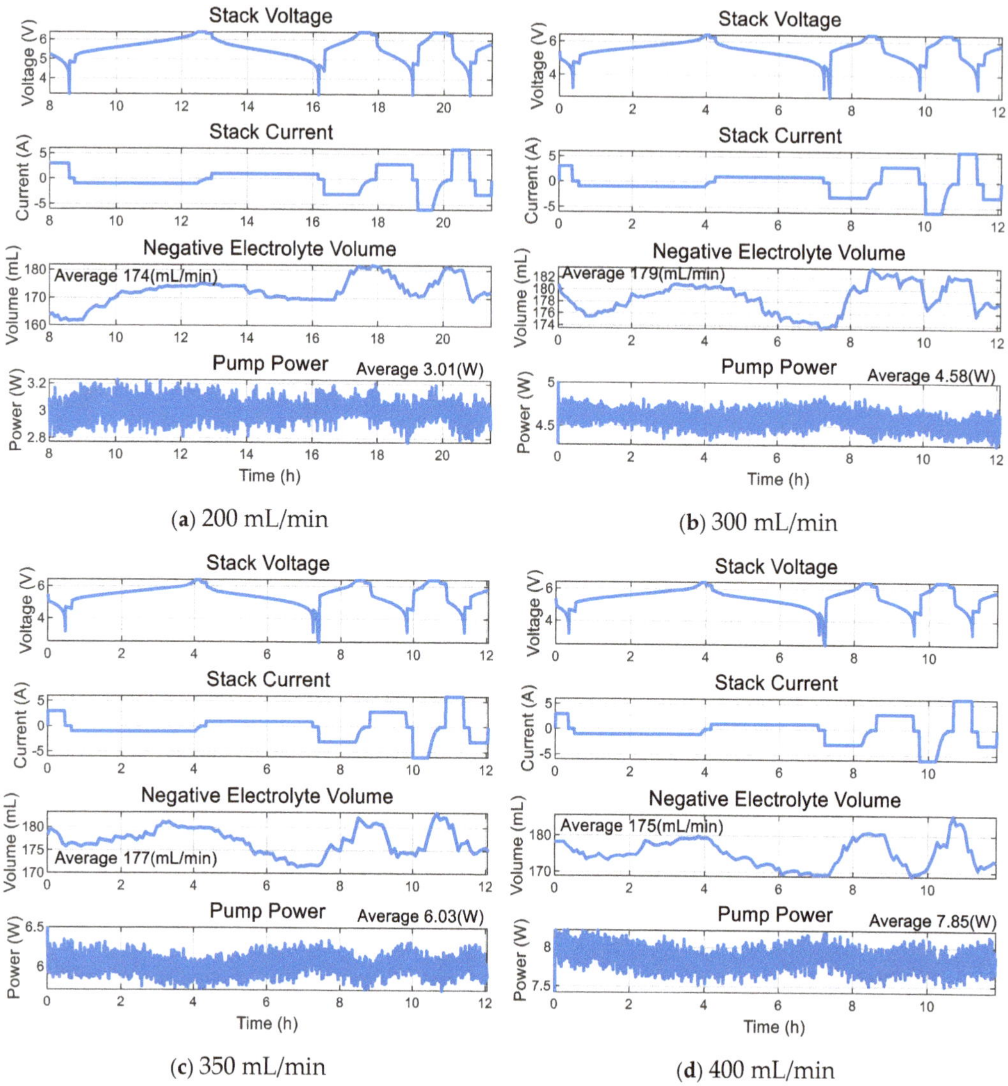

Figure 6. Stack voltage, current, cathode electrolyte amount, and pump power consumption of VRFB according to flow conditions: (**a**) 200 mL/min; (**b**) 300 mL/min; (**c**) 350 mL/min; (**d**) 400 mL/min.

Figure 7c shows the Coulomb efficiency calculated using Equation (4). As mentioned above, when charging/discharging at 1 A under all flow rate conditions, the Coulomb efficiency is 84.9%, which is not good, whereas the results at 3 A and 6 A confirm that the Coulomb efficiency improves to approximately 92.5% as the current increases. Therefore, to increase the usable discharge capacity and improve the Coulomb efficiency, it can be seen that the VRFB system tested in this study should be operated at high current under low-flow rate conditions. In this study, results below 200 mL/min could not be tested due to the limited speed control range of the pump used in the experimental setup. If the flow rate continues to decrease, it is expected that the charge/discharge capacity will decrease due to insufficient electrolyte supply, and it is judged that appropriate low-flow conditions can be determined based on this. In addition, it should be noted that the results above are from the experimental set environment of this study. In a situation where the VRFB is operated in conjunction with a renewable energy generation system, the generated energy should be transferred to the VRFB and the utility grid as much as possible, so further research on maximum efficiency operation control is required.

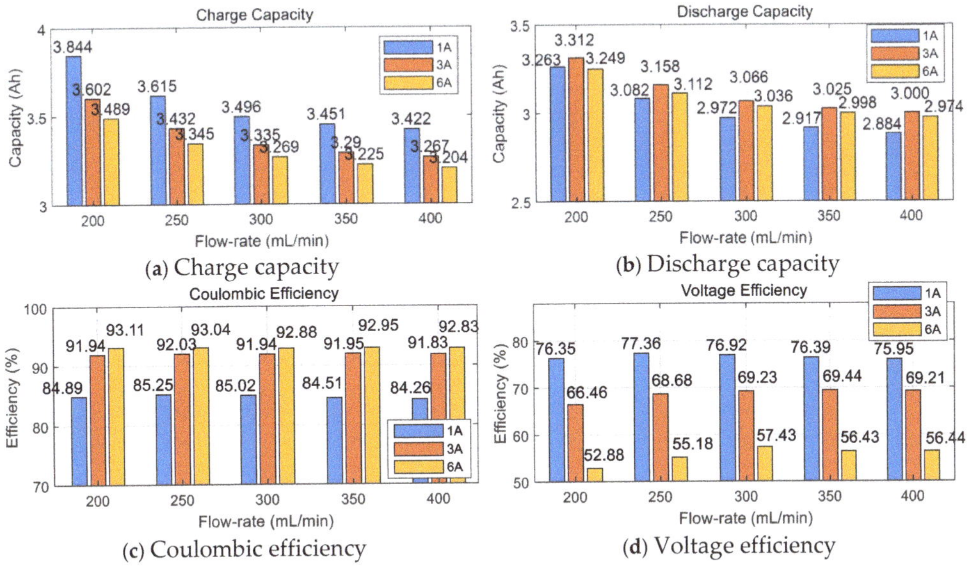

Figure 7. VRFB charge/discharge characteristic analysis under different current and flow conditions: (**a**) charge capacity; (**b**) discharge capacity; (**c**) Coulombic efficiency; (**d**) voltage efficiency.

Figure 7d shows the voltage efficiency calculated by Equation (5). As the current increases, the voltage efficiency deteriorates owing to battery overvoltage. The voltage efficiency does not show a large difference depending on the flow rate change, but it is expected that the efficiency will be somewhat higher overall regardless of the current when operating at a flow rate of 250 mL/min to 300 mL/min. However, the voltage efficiency mentioned above is specific to the context of this study, and the efficiency characteristics may vary when the electrode material is altered or an electrode using a catalyst is used [30].

Figure 8 shows the charge and discharge energies of the characteristic analysis experiment based on the current, flow rate, and energy efficiency calculated using Equation (6). As shown in Figure 8a, the charge energy is the largest when the current is 1 A at a flow rate of 200 mL/min. However, the charged energy can only be transferred to the load side at 17.65 Wh due to a low efficiency of 79.6%. When the battery is charged and discharged at 3 A at 200 mL/min, 79.3% of the charged energy can be transferred, similar to when it is charged and discharged at 1 A.

As shown in Figure 8c, because the energy efficiency decreased as the flow rate increased, it was expected that the energy efficiency could be improved by operating under low-flow conditions. In addition, the VRFB shows that the energy efficiency of the stack itself, which does not consider the loss of peripheral devices such as pumps, is not good, as the voltage efficiency increases as the current increases; thus, it shows that the results can be obtained by operating under low-current conditions to obtain high efficiency. However, because the VRFB has pump loss, the system efficiency, including this, should be analyzed to determine an efficient operation method.

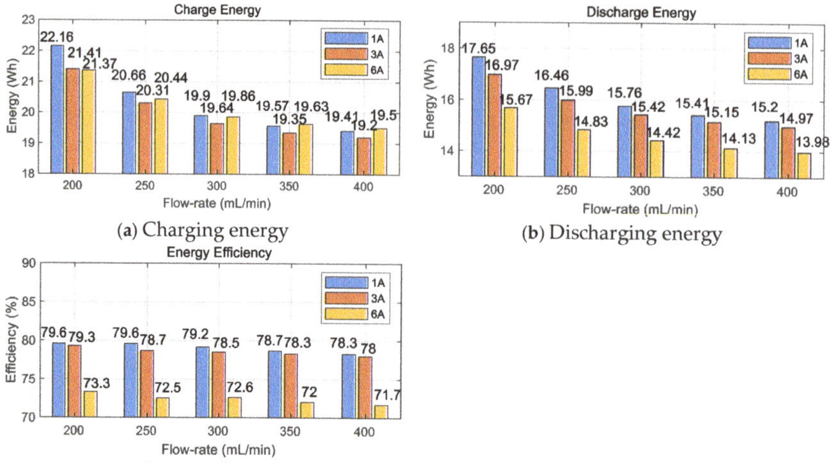

(**a**) Charging energy

(**b**) Discharging energy

(**c**) Energy efficiency

Figure 8. VRFB charge/discharge energy and energy efficiency under different current and flow conditions: (**a**) charge energy; (**b**) discharge energy; (**c**) energy efficiency.

Figure 9 shows the system efficiency calculated using Equation (7). This is inefficient because the energy consumed by the pump increases with the flow rate. Therefore, the system energy required for charging the VRFB stack in Figure 9a can be seen from the fact that more energy is required when the current is small and the flow rate is high. In addition, as can be observed from the discharge system energy in Figure 9b, driving with a high current under low-flow conditions is efficient. This is the effect of the increased pump loss owing to the longer usage time when charging and discharging the battery with a low current, even under low-flow conditions. In the case of the discharge system energy, more energy is required to drive the pump than to supply energy to the load side when used under flow conditions of 350 mL/min or higher (the discharge system energy value is represented as 0). Therefore, the system efficiency in Figure 9c was found to be the highest when the battery was used with a high current (6A) of 200 mL/min.

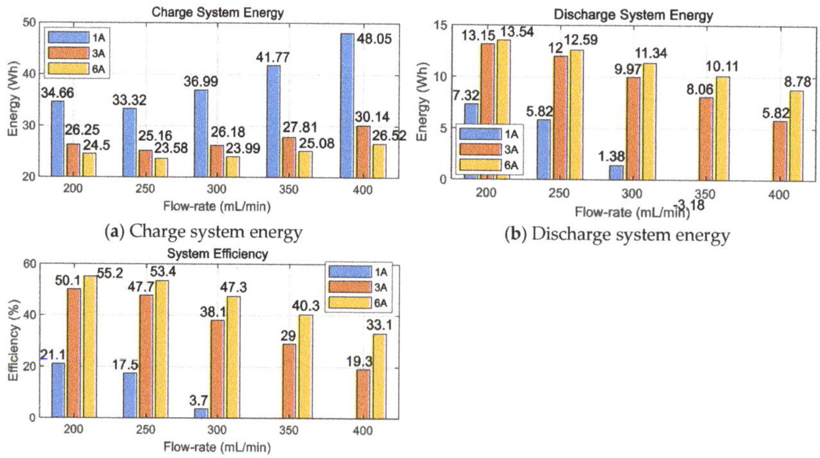

(**a**) Charge system energy

(**b**) Discharge system energy

(**c**) System efficiency

Figure 9. VRFB system efficiency under different current and flow conditions: (**a**) system energy during charging; (**b**) system energy during discharging; (**c**) system efficiency.

The VRFB stack system, with cells connected in series, requires optimal operating point analysis to improve efficiency owing to shunt current loss and pump-driving loss. The electrical characteristic test procedure proposed in this study is expected to facilitate rapid and accurate characteristic analysis, aiding in the derivation of an optimal operating method.

3.2. Design of Electrical Circuit Model of VRFB

This section presents the electrical circuit model and the parameter design method for the VRFB. In the electrical circuit model (ECM) of the VRFB in Figure 10, V_{ocv} represents the open-circuit voltage according to the state of charge (SOC), which is the state of charge. The series resistance R_i and parallel connection circuit of R_d/C_d in the ECM model were used to simulate the dynamic characteristics of the voltage according to the current. In addition, VRFBs have an internal leakage current called a shunt current, owing to the parallel connection configuration of the electrolyte applied to the series-connected cells. To simulate the shunt current, the ECM has a resistor connected in parallel with the open-circuit voltage [25,26].

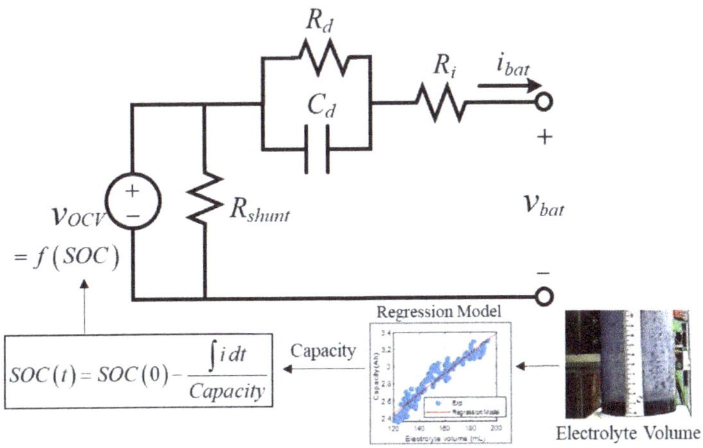

Figure 10. Electrical circuit model of VRFB.

The parameters of the open-circuit voltage V_{ocv} and series resistance R_i, R_d/C_d of the parallel circuit can be obtained by an experiment in which a pulse current is applied to reduce the SOC by approximately 5% from the fully charged state to the fully discharged voltage, as shown in Figure 11, tested using the procedure shown in Figure 4c. The waveform of the period in which the discharge pulse current was applied is shown in Figure 12. The series resistance R_i is calculated from the voltage drop ΔV when a pulse current with a magnitude of ΔI is applied using Ohm's law. The parameters of R_d and C_d are selected as values that can simulate the voltage relationship of Equation (8), which is obtained by subtracting the magnitude of ΔV owing to the series resistance from the battery terminal voltage during times t1 and t2 when the current is continuously applied and the change in V_{ocv} according to the current applied, which is shown in the red waveform in Figure 12. In this study, the R_d and C_d parameters were estimated by the least squares method using Matlab software. Figure 13 shows the electrical circuit parameters of the VRFB stack used in this study.

$$R_d i_{bat}\left(1 - e^{-\frac{t}{R_d C_d}}\right) = v_{bat} - R_i i_{bat} - v_{OCV} \tag{8}$$

To obtain the shunt resistance (R_{shunt}) of the electrical circuit model, the pump was driven under no-load conditions and the time to discharge from a fully charged state to a fully discharged voltage was measured for each flow rate. Figure 14 shows the results obtained when the VRFB stack was used under no-load conditions at a flow rate of 250 mL/min. The measured charge capacity of the battery before the no-load test was 2.89 Ah, and the battery was fully discharged after 39.5 h. Therefore, the VRFB stack can be modeled as having an R_{shunt} resistance of 76.6 Ω when operated at 250 mL/min.

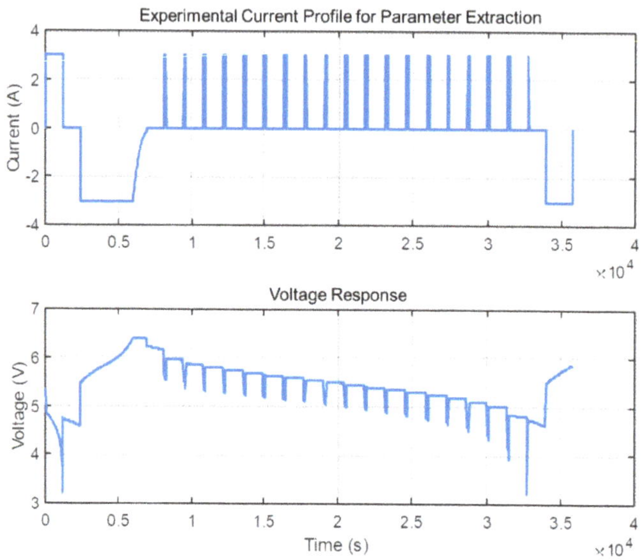

Figure 11. Current and voltage waveforms of experiment for parameter extraction of VRFB.

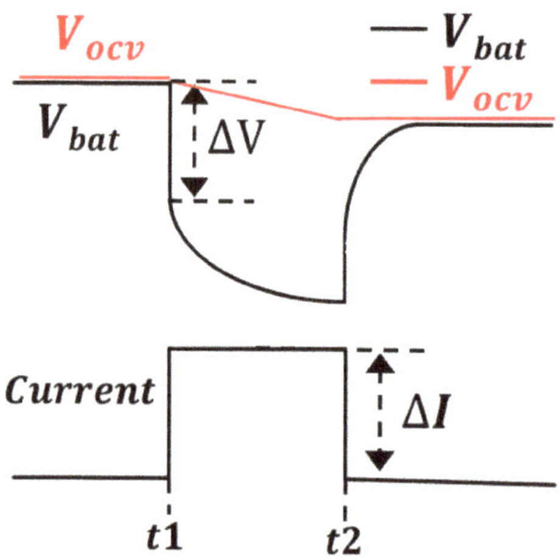

Figure 12. Battery voltage response during pulse current.

The shunt current/resistance values measured under flow rate conditions of 200–400 mL/min using the same test method are listed in Table 3. From Table 3, it can be confirmed that the shunt current increases slightly as the flow rate of the VRFB stack increases. Therefore, in this study, the shunt resistance was modeled as a power series function as a function of the flow rate, as shown in Equation (9).

$$R_{shunt} = -288.6 \times flow\_rate^{4.547} + 76.96 \ (\Omega) \tag{9}$$

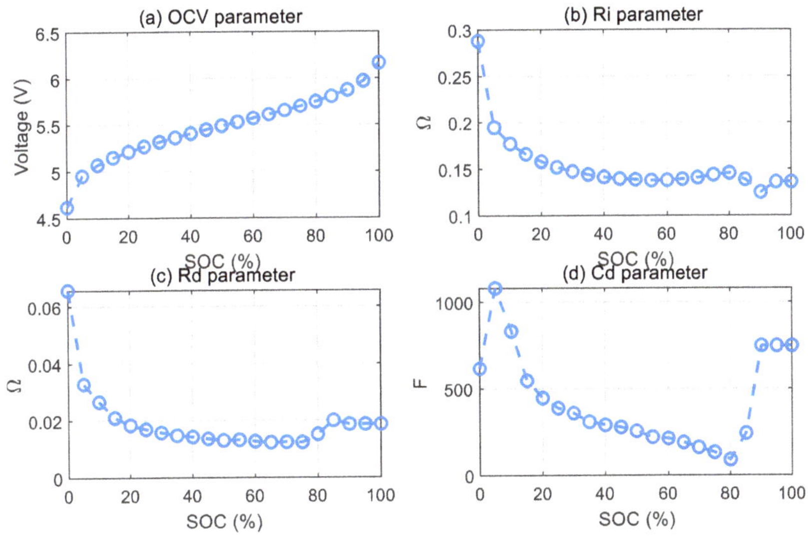

Figure 13. Model parameters of VRFB.

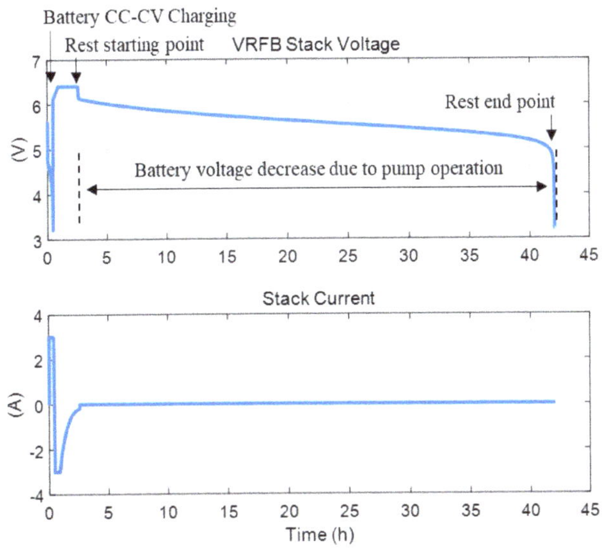

Figure 14. VRFB voltage/current experiment results when the pump is driven at a flow rate of 250 mL/min under no load.

Table 3. Shunt current and resistance by flow rate.

	200 mL/min	250 mL/min	300 mL/min	350 mL/min	400 mL/min
I_{shunt}	73.4 mA	73 mA	74.3 mA	75 mA	77.3 mA
R_{shunt}	76.7 Ω	76.6 Ω	75.6 Ω	74.6 Ω	72.5 Ω

Figure 15 shows the experimental results of the VRFB cycled ten times under 100% DoD conditions. As shown in the third and fourth waveforms, the volume of the electrolyte changes during the charging and discharging processes, and the capacity decreases in proportion to the change in the electrolyte. Therefore, in this study, a capacity estimation model for the electrolyte volume was designed and modeled to enable the capacity change by reflecting it in the simulation model.

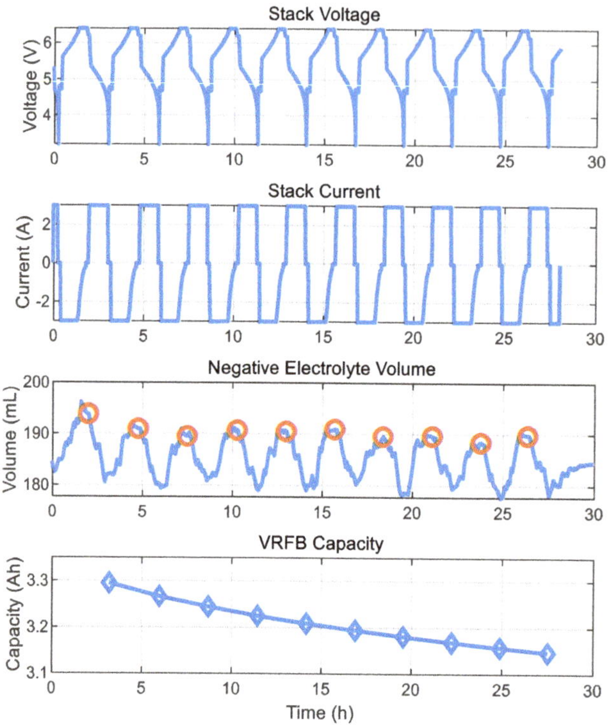

Figure 15. Experimental results of electrolyte volume and capacity change during cycling of VRFB.

The capacity estimation model was designed by analyzing the correlation between the average charge/discharge capacity and negative electrolyte volume in the fully charged state of the battery, which is indicated by the red circle in the third waveform in Figure 15. Figure 16 shows the capacity measurement and estimation results of the linear regression model for the electrolyte volume at 175 fully charged points measured during the long-term cycle of the VRFB battery. For this VRFB system, a capacity estimation model was designed using Equation (10).

$$Capacity = 0.01173 \times electrolyte\ volume\ (mL) + 1.04\ \ (Ah) \tag{10}$$

The validity of the electrically equivalent circuit model of the designed VRFB stack was verified through two charge/discharge experiments. First, as shown in Figure 17a, when the VRFB was used at a flow rate of 250 mL/min, a discharge pulse current of 3 A was applied at approximately every 5% SOC section, and the error between the battery voltage and the simulation model voltage was analyzed. Figure 17b shows the stack voltage measured in the experiment and the simulation model voltage, and the difference between these two voltages is shown in Figure 17c. Except for the fully discharged voltage point with high nonlinearity, the maximum absolute value error of the simulation model is within 0.17 V, and it can be seen that it accurately simulates the actual battery voltage with a root mean square error (RMSE) of 0.025 V.

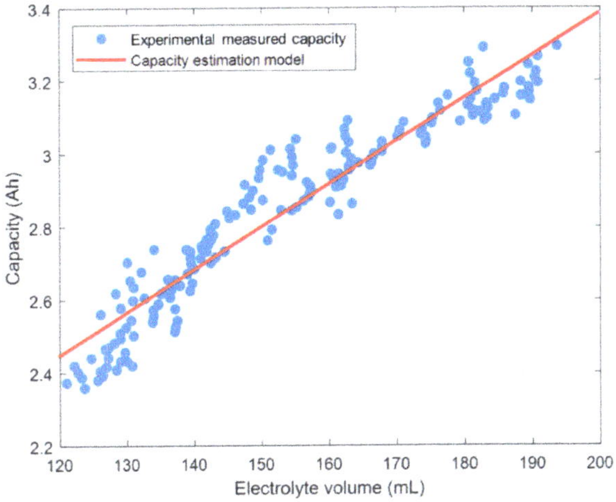

Figure 16. Charge/discharge average capacity experimental data and capacity estimation model for 175 electrolyte volumes.

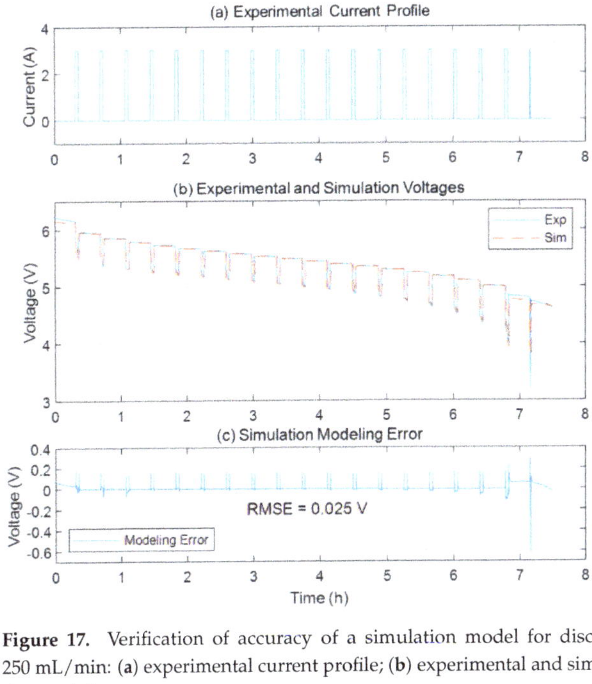

Figure 17. Verification of accuracy of a simulation model for discharge pulse current under 250 mL/min: (**a**) experimental current profile; (**b**) experimental and simulation voltages; (**c**) simulation modeling error.

The experimental and modeling simulation results for 11 full charge/discharge cycles with varying currents and electrolyte volumes are shown in Figure 18. The VRFB was fully charged/discharged for the first five cycles with a current of 3 A and then cycled twice, each with currents of 1, 3, and 6 A. Except for a somewhat large voltage error at the full discharge point, it was confirmed that the model could simulate the actual battery voltage within an RMSE of 0.096 V, even when the electrolyte volume and current magnitude were varied.

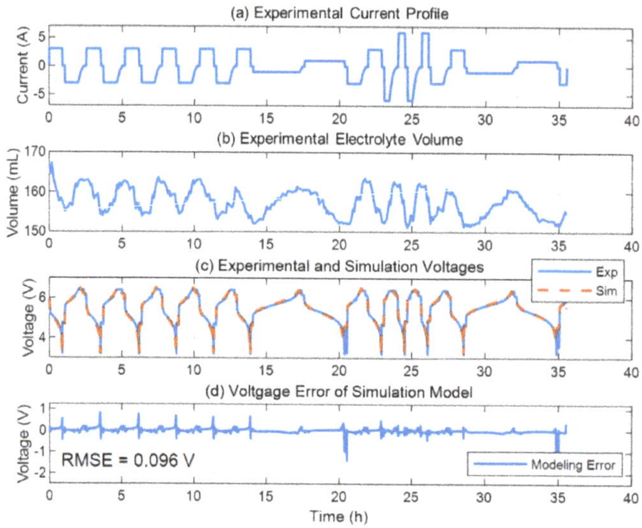

Figure 18. Verification of simulation model accuracy when current magnitude and electrolyte volume change under 250 mL/min.

4. Conclusions

In this study, we propose an experimental procedure and analysis method for analyzing the system efficiency of VRFBs according to the current and flow rate conditions and an electrical equivalent circuit modeling design method for simulating VRFBs.

According to the results of the system efficiency analysis, it was confirmed that the VRFB system has a significant impact on the system efficiency owing to the shunt current loss and pump loss for electrolyte transport. For the VRFB system used in this study, it was found that it is efficient to operate a VRFB with a high current under low-flow conditions. In addition, even if the VRFB system is changed, its efficiency characteristics can be quickly analyzed by applying the test and analysis procedures presented in this study.

In addition, this paper presents an electrical circuit modeling and parameter design method that can simulate the electrical output characteristics of VRFBs when the electrolyte volume and flow rate conditions are changed. The proposed simulation model can be used to investigate efficient operational control methods for VRFB systems. The accuracy of the proposed simulation model was verified through comparison with the full charge/discharge cycle experimental results of a 40 W VRFB stack comprising four 10 W cells in series.

Author Contributions: Conceptualization, S.L.; Methodology, S.L.; Software, S.L. and H.J.; Validation, Y.-G.S.; Formal Analysis, Y.-G.S.; Investigation, H.J.; Resources, S.L.; Data Curation, H.J.; Writing—Original Draft Preparation, S.L.; Writing—Review and Editing, Y.-G.S.; Visualization, S.L. and H.J.; Supervision, Y.-G.S.; Project Administration, S.L.; Funding Acquisition, S.L. All authors have read and agreed to the published version of the manuscript.

Funding: This research was supported by "Regional Innovation Strategy (RIS)" through the National Research Foundation of Korea (NRF) funded by the Ministry of Education (MOE) (2021RIS-002) and a National Research Foundation of Korea (NRF) grant funded by the Korean Government (MSIT) (No. NRF-2021R1F1A1063150).

Data Availability Statement: The original contributions presented in the study are included in the article, further inquiries can be directed to the corresponding author.

Conflicts of Interest: The authors declare no conflicts of interest.

References

1. Guney, M.S.; Tepe, Y. Classification and Assessment of Energy Storage Systems. *Renew. Sustain. Energy Rev.* **2017**, *75*, 1187–1197. [CrossRef]
2. Castillo, A.; Gayme, D.F. Grid-Scale Energy Storage Applications in Renewable Energy Integration: A Survey. *Energy Convers. Manag.* **2014**, *87*, 885–894. [CrossRef]
3. Shao, Y.; Zhang, H.; Gao, Y.; Jin, B. Dynamic power distribution strategy using multi-objective collaborative optimization for hybrid energy storage systems. *J. Power Electron.* **2023**, *23*, 1517–1528. [CrossRef]
4. Diaz, L.B.; He, X.; Hu, Z.; Restuccia, F.; Marinescu, M.; Barreras, J.V.; Patel, Y.; Offer, G.; Rein, G. Meta-Review of Fire Safety of Lithium-Ion Batteries: Industry Challenges and Research Contributions. *J. Electrochem. Soc.* **2020**, *167*, 090559. [CrossRef]
5. Kong, L.; Li, C.; Jiang, J.; Pecht, M.G. Li-Ion Battery Fire Hazards and Safety Strategies. *Energies* **2018**, *11*, 2191. [CrossRef]
6. Chen, Y.; Kang, Y.; Zhao, Y.; Wang, L.; Liu, J.; Li, Y.; Liang, Z.; He, X.; Li, X.; Tavajohi, N.; et al. A Review of Lithium-Ion Battery Safety Concerns: The Issues, Strategies, and Testing Standards. *J. Energy Chem.* **2021**, *59*, 83–99. [CrossRef]
7. Cai, T.; Mohtat, P.; Stefanopoulou, A.G.; Siegel, J.B. Li-ion Battery Fault Detection in Large Packs Using Force and Gas Sensors. *IFAC-Pap.* **2020**, *53*, 12491–12496. [CrossRef]
8. Zhang, R.; Li, X.; Sun, C.; Yang, S.; Tian, Y.; Tian, J. State of Charge and Temperature Joint Estimation Based on Ultrasonic Reflection Waves for Lithium-Ion Battery Applications. *Batteries* **2023**, *9*, 335. [CrossRef]
9. Mrozik, W.; Rajaeifar, M.A.; Heidrich, O.; Christensen, P. Environmental Impacts, Pollution Sources and Pathways of Spent Lithium-Ion Batteries. *Energy Environ. Sci.* **2021**, *14*, 6099–6121. [CrossRef]
10. Birkl, C.R.; Roberts, M.R.; McTurk, E.; Bruce, P.G.; Howey, D.A. Degradation Diagnostics for Lithium Ion Cells. *J. Power Sources* **2017**, *341*, 373–386. [CrossRef]
11. Velázquez-Martínez, O.; Valio, J.; Santasalo-Aarnio, A.; Reuter, M.; Serna-Guerrero, R. A Critical Review of Lithium-Ion Battery Recycling Processes from a Circular Economy Perspective. *Batteries* **2019**, *5*, 68. [CrossRef]
12. Alotto, P.; Guarnieri, M.; Moro, F. Redox Flow Batteries for the Storage of Renewable Energy: A Review. *Renew. Sustain. Energy Rev.* **2014**, *29*, 325–335. [CrossRef]
13. Weber, A.Z.; Mench, M.M.; Meyers, J.P.; Ross, P.N.; Gostick, J.T.; Liu, Q. Redox Flow Batteries: A Review. *J. Appl. Electrochem.* **2011**, *41*, 1137–1164. [CrossRef]
14. Jeong, D.; Jung, S. Numerical Analysis of Cycling Performance of Vanadium Redox Flow Battery. *Int. J. Energy Res.* **2020**, *44*, 5209–5222. [CrossRef]
15. Bhattarai, A.; Ghimire, P.C.; Whitehead, A.; Schweiss, R.; Scherer, G.G.; Wai, N.; Hng, H.H. Novel Approaches for Solving the Capacity Fade Problem during Operation of a Vanadium Redox Flow Battery. *Batteries* **2018**, *4*, 48. [CrossRef]
16. Xiong, B.; Zhao, J.; Wei, Z.; Skyllas-Kazacos, M. Extended Kalman Filter Method for State of Charge Estimation of Vanadium Redox Flow Battery Using Thermal-Dependent Electrical Model. *J. Power Sources* **2014**, *262*, 50–61. [CrossRef]
17. Kim, J.; Park, H. Experimental Analysis of Discharge Characteristics in Vanadium Redox Flow Battery. *Appl. Energy* **2017**, *206*, 451–457. [CrossRef]
18. Cunha, Á.; Martins, J.; Rodrigues, N.; Brito, F.P. Vanadium Redox Flow Batteries: A Technology Review. *Int. J. Energy Res.* **2015**, *39*, 889–918. [CrossRef]
19. Jung, H.; Lee, S. A Study on Capacity and State of Charge Estimation of VRFB Systems Using Cumulated Charge and Electrolyte Volume under Rebalancing Conditions. *Energies* **2023**, *16*, 2478. [CrossRef]
20. Iwakiri, I.; Antunes, T.; Almeida, H.; Sousa, J.P.; Figueira, R.B.; Mendes, A. Redox Flow Batteries: Materials, Design and Prospects. *Energies* **2021**, *14*, 5643. [CrossRef]
21. Sun, C.; Negro, E.; Nale, A.; Pagot, G.; Vezzù, K.; Zawodzinski, T.A.; Meda, L.; Gambaro, C.; Di Noto, V. An efficient barrier toward vanadium crossover in redox flow batteries: The bilayer [Nafion/$(WO_3)x$] hybrid inorganic-organic membrane. *Electrochim. Acta* **2021**, *378*, 138133. [CrossRef]
22. Kim, S.; Thomsen, E.; Xia, G.; Nie, Z.; Bao, J.; Recknagle, K.; Wang, W.; Viswanathan, V.; Luo, Q.; Wei, X.; et al. 1 Kw/1 Kwh Advanced Vanadium Redox Flow Battery Utilizing Mixed Acid Electrolytes. *J. Power Sources* **2013**, *237*, 300–309. [CrossRef]
23. Karrech, A.; Regenauer-Lieb, K.; Abbassi, F. Vanadium Flow Batteries at Variable Flow Rates. *J. Energy Storage* **2022**, *45*, 103623. [CrossRef]
24. Ma, X.; Zhang, H.; Sun, C.; Zou, Y.; Zhang, T. An Optimal Strategy of Electrolyte Flow Rate for Vanadium Redox Flow Battery. *J. Power Sources* **2012**, *203*, 153–158. [CrossRef]
25. Fornaro, P.; Puleston, T.; Puleston, P.; Serra-Prat, M.; Costa-Castelló, R.; Battaiotto, P. Redox flow battery time-varying parameter estimation based on high-order sliding mode differentiators. *Int. J. Energy Res.* **2022**, *46*, 16576–16592. [CrossRef]
26. Zhang, Y.; Zhao, J.; Wang, P.; Skyllas-Kazacos, M.; Xiong, B.; Badrinarayanan, R. A comprehensive equivalent circuit model of all-vanadium redox flow battery for power system analysis. *J. Power Sources* **2015**, *290*, 14–24. [CrossRef]
27. Schwenzer, B.; Zhang, J.; Kim, S.; Li, L.; Liu, J.; Yang, Z. Membrane Development for Vanadium Redox Flow Batteries. *ChemSusChem* **2011**, *4*, 1388–1406. [CrossRef]
28. Bae, D.; Kanellos, G.; Wedege, K.; Dražević, E.; Bentien, A.; Smith, W.A. Tailored energy level alignment at MoOX/GaP interface for solar-driven redox flow battery application. *J. Chem. Phys.* **2020**, *152*, 124710. [CrossRef]

29. Jirabovornwisut, T.; Arpornwichanop, A. A review on the electrolyte imbalance in vanadium redox flow batteries. *Int. J. Hydrogen Energy* **2019**, *44*, 24485–24509. [CrossRef]
30. Chen, S.; Sun, C.; Zhang, H.; Yu, H.; Wang, W. Electrochemical Deposition of Bismuth on Graphite Felt Electrodes: Influence on Negative Half-Cell Reactions in Vanadium Redox Flow Batteries. *Appl. Sci.* **2024**, *14*, 3316. [CrossRef]

Disclaimer/Publisher's Note: The statements, opinions and data contained in all publications are solely those of the individual author(s) and contributor(s) and not of MDPI and/or the editor(s). MDPI and/or the editor(s) disclaim responsibility for any injury to people or property resulting from any ideas, methods, instructions or products referred to in the content.

energies

MDPI

Article

Forecast-Aided Converter-Based Control for Optimal Microgrid Operation in Industrial Energy Management System (EMS): A Case Study in Vietnam

Yeong-Nam Jeon [1] and Jae-ha Ko [2,*]

[1] Department of AI Mechanical Convergence Engineering, Donggang University, 50 Dongmun-daero, Buk-gu, Gwangju 61200, Republic of Korea; snower24@naver.com
[2] Department of Electrical Engineering, Honam University, 417 Eodeung-daero, Gwangsan-gu, Gwangju 62399, Republic of Korea
* Correspondence: jhko@honam.ac.kr; Tel.: +82-062-940-3630

Abstract: This study proposes a forecast-aided energy management strategy tailored for industrial microgrids operating in Vietnam's tropical climate. The core novelty lies in the implementation of a converter-based EMS that enables bidirectional DC power exchange between multiple subsystems. To improve forecast accuracy, an artificial neural network (ANN) is used to model the relationship between electric load and localized meteorological features, including temperature, dew point, humidity, and wind speed. The forecasted load data is then used to optimize charge/discharge schedules for energy storage systems (ESS) using a Particle Swarm Optimization (PSO) algorithm. The strategy is validated using real-site data from a Vietnamese industrial complex, where the proposed method demonstrates enhanced load prediction accuracy, cost-effective ESS operation, and multi-microgrid flexibility under weather variability. This integrated forecasting and control approach offers a scalable and climate-adaptive solution for EMS in emerging industrial regions.

Keywords: load forecasting; artificial neural networks; Vietnam; meteorological data; MAPE; time series; energy prediction; industrial forecasting; energy efficiency; PSO; EMS; converter-based control; microgrid; DC grid; energy trading; battery scheduling; forecast-aided EMS

1. Introduction

In recent years, power systems have been undergoing significant transformation, transitioning towards more advanced and intelligent configurations such as microgrids and smart grids. These modern systems integrate renewable energy sources directly with energy storage systems (ESS), incorporating technologies like high-efficiency photovoltaic (PV), solar thermal, combined heat and power (CHP), wind power generation, and ESS. Such integrated systems enable proactive electricity supply and demand planning, allowing for optimal economic operation and smart autonomous management, operation, and maintenance (O&M) through the latest IT technologies and intelligent controls [1–4]. Given the variability and intermittent nature of renewable energy sources, effective power generation forecasting and demand prediction technologies have become increasingly important to ensure stability and efficiency in these evolving systems.

Accurate electric load forecasting plays a vital role in modern power systems by enabling better alignment between energy supply and demand, improving generation scheduling, and reducing operational risk [5–7]. With the increasing deployment of renewable energy sources and distributed energy storage systems (ESS), microgrids—especially those implemented in industrial contexts—require more intelligent forecasting and control mechanisms [8,9]. In particular, load prediction helps improve the efficiency of microgrid operation not only at the grid scale but also at the level of small-scale, site-specific microgrid clusters [9,10].

Electric load forecasting methodologies are traditionally classified into statistical and artificial-intelligence-based techniques. Statistical approaches such as regression analysis and time-series models, including autoregressive integrated moving average (ARIMA), have long been used for their interpretability and effectiveness under linear assumptions. These models capture temporal trends and seasonal components but may lack robustness when handling non-linear interactions among influencing variables [11–13]. Unlike in temperate regions, where seasonal heating and cooling demands predominantly influence load patterns, tropical climates such as Vietnam exhibit less pronounced seasonal variation and more frequent short-term fluctuations due to humidity, solar intensity, and unpredictable weather. These conditions result in non-cyclical, high-frequency load–weather interactions that require more localized and adaptive forecasting approaches.

Artificial intelligence techniques, particularly artificial neural networks (ANNs), have gained popularity due to their ability to model complex, non-linear relationships between input variables such as temperature, humidity, and past load data. ANNs can learn these relationships directly from data without predefined assumptions, making them suitable for environments with fluctuating patterns and irregular behaviors. This flexibility is especially beneficial for load forecasting in microgrid systems, where energy usage patterns are heavily influenced by local conditions and building-specific characteristics [12–14].

This study focuses on a demonstration site in Vietnam where a microgrid system has been deployed across three distinct buildings with unique energy consumption characteristics. These buildings are interconnected through a direct current (DC) grid and enabled with 144 kW class bidirectional DC/DC converters to allow mutual power exchange. To ensure reliable and efficient operation of the microgrid, a combined forecasting and control approach is proposed.

The primary contribution of this research lies in the integration of weather-based ANN forecasting models with a converter-based EMS strategy. The EMS uses ANN predictions along with real-time state of charge (SoC) information from each building's energy storage system to determine optimal power dispatch priorities. This strategy allows for proactive control of power flows, enabling balanced energy sharing and enhancing microgrid efficiency [15,16].

The remainder of this paper is organized as follows. Section 2 presents the meteorological and load data analysis, followed by the ANN forecasting methodology in Section 3. Section 4 evaluates forecasting accuracy across buildings. Section 5 introduces the converter-based EMS design and simulation scenarios. Section 6 summarizes key findings and proposes directions for future research.

2. Data Analysis

2.1. Overview of the Vietnam Demonstration Site

The demonstration site, located at the Taekwang Vina industrial complex in Dong Nai Province, Vietnam, consists of three distinct facilities: VT2, Mold, and a kindergarten. VT2 is a footwear manufacturing plant that operates many machines and equipment, resulting in high and fluctuating electricity demand. Mold serves as a mold fabrication facility for shoe outer structures, exhibiting a moderately lower and more stable load profile compared to VT2. The kindergarten, designed to accommodate the children of factory employees, has minimal power usage and operates independently from the energy-intensive production sites. Understanding the functional role and load profile of each building provides essential context for the design and operation of the EMS. These contextual differences informed the selection of target buildings for forecasting and ESS scheduling in this study.

As shown in Figure 1, the system operates as a microgrid system integrating various distributed energy resources, including photovoltaic (PV) systems, a combined solar and heat power (CSP/CHP) unit, an energy storage system (ESS), and a centralized Total Operation Control (TOC) center. PV capacity includes 900 kWp on VT2 and 400 kWp on the Mold building. The ESS consists of inverters (1.29 MW), converters (144 kW), and a battery system (720 kWh), supported by MPPT and redundancy inverters. Through the 144 kW rated DC/DC converter, the system allows for a maximum bidirectional power exchange of up to 144 kW between microgrid subsystems, enabling both energy import and export depending on demand and operational strategy. These distributed resources are managed in real time via the TOC to optimize power flow, ensure reliability, and reduce energy costs within the site.

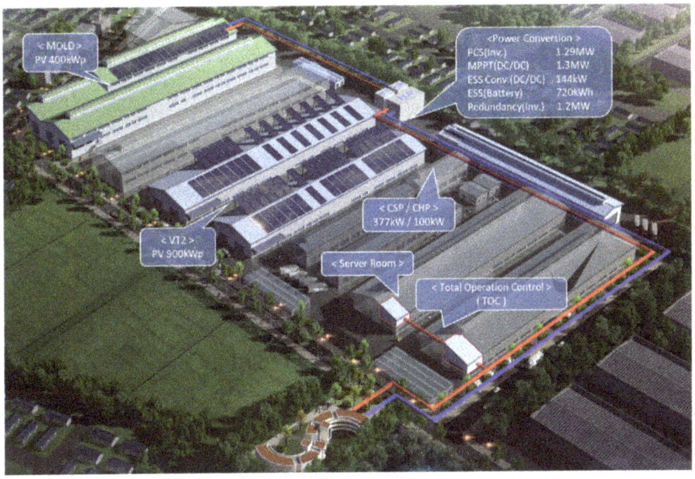

Figure 1. Overview of microgrid construction in Ho Chi Minh Industrial Complex.

This demonstration site follows a network of microgrids architecture, enabling bidirectional energy exchange between local subsystems. Depending on operational strategies, power can be imported from or exported to other buildings within the industrial complex. This configuration enhances resilience, supports dynamic load balancing, and enables coordinated energy management across distributed assets. Such networked microgrid structures have been shown to improve overall system flexibility and operational efficiency in renewable-integrated environments [17–19]. This makes the site an ideal testbed for advanced forecasting and energy optimization techniques in real-world settings.

Additionally, the microgrid architecture includes distributed subsystems deployed at both the VT2 and Mold buildings. As illustrated in Figure 2, each building operates an independent PV-ESS-inverter system interconnected by a DC grid. The VT2 building is equipped with MPPT, ESS converters, and dual PCS inverters feeding into ACB #1 and #2, while the Mold building operates similarly with PCS #3 and ACB #3. Both sites interface with the local Taekwang Vina transformers, enabling grid-tied and isolated microgrid functionality depending on operational strategy.

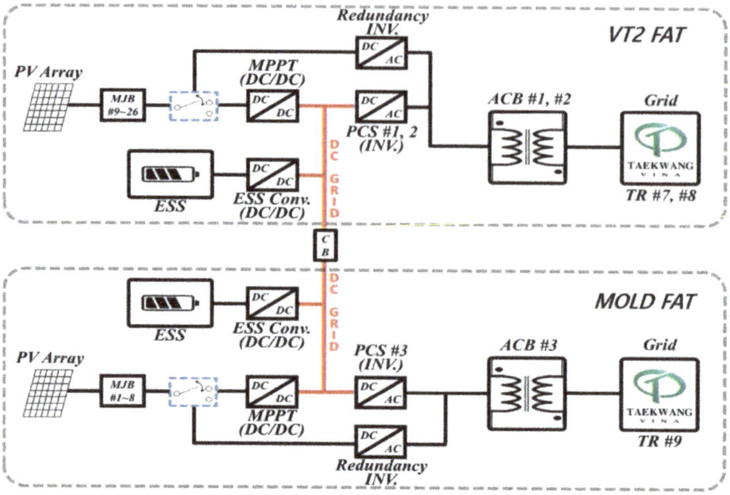

Figure 2. Configuration diagram of multiple microsystems based on DC grid.

2.2. *Analysis of Meteorological Data in Vietnam*

Meteorological variables are generally considered to be among the key factors influencing electric load demand. For this study, temperature, dew point, humidity, and wind speed were selected as input variables for load forecasting [12–14].

To determine which weather parameters are most closely correlated with electric load, Pearson correlation analysis was employed. This method quantifies the linear relationship between each meteorological variable and power consumption.

Figure 3 presents the meteorological data collected from Ho Chi Minh City, Vietnam, which consists of hourly observational records. These historical data serve as the foundation for training and validating the forecasting model, providing key input variables such as temperature, dew point, humidity, and wind speed.

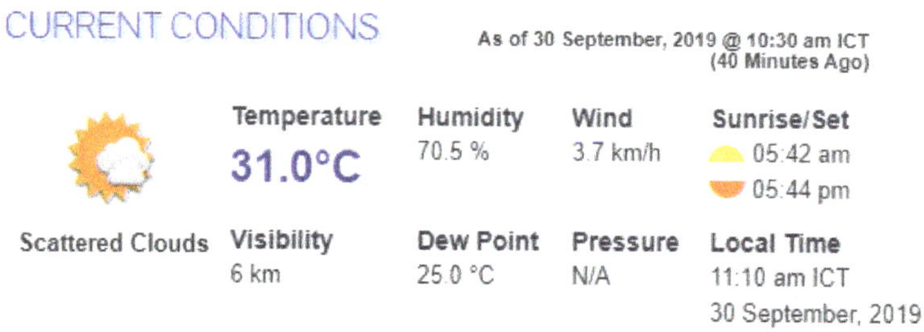

Figure 3. Example of Ho Chi Minh meteorological data.

The available meteorological dataset includes various features, from which temperature, dew point, humidity, and wind speed were selected for modeling. Figures 4–7 visualize the trends and characteristics of these variables—temperature, dew point, humidity, and wind speed—over the data collection period from 10 March to 27 November 2018.

Unlike temperate climate regions in Northeast Asia, which exhibit distinct seasonal variations across spring, summer, fall, and winter, tropical climates in Southeast Asia, including Vietnam, do not show such pronounced seasonal temperature changes. Likewise, other weather features such as humidity and wind speed in tropical regions tend to display weaker seasonal patterns compared to those in temperate zones.

This observation suggests that, unlike in temperate regions, seasonality need not be a major consideration in the model structure for Vietnam. However, since power load is generally known to respond sensitively to heating and cooling needs driven by temperature, a strong correlation between load and temperature is expected.

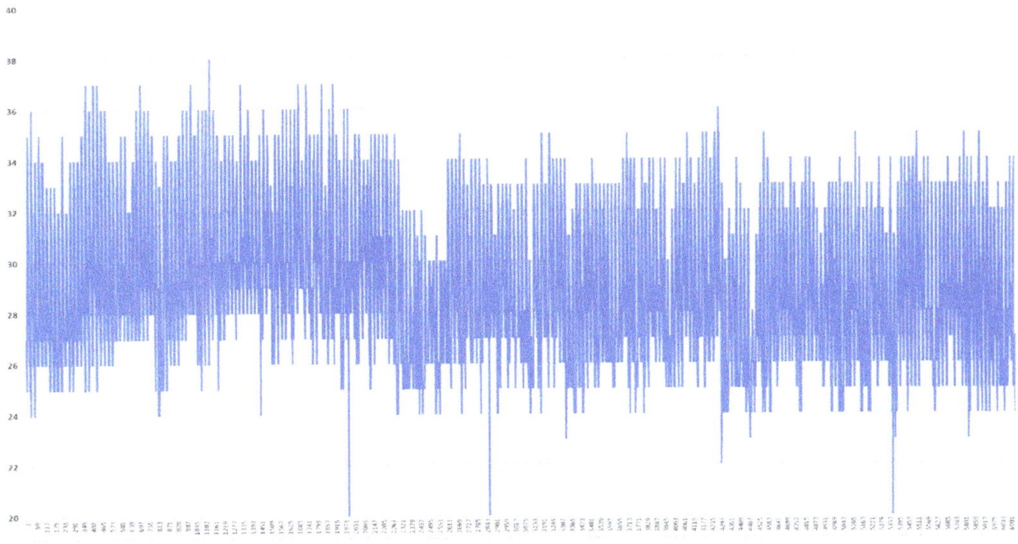

Figure 4. Temperature data.

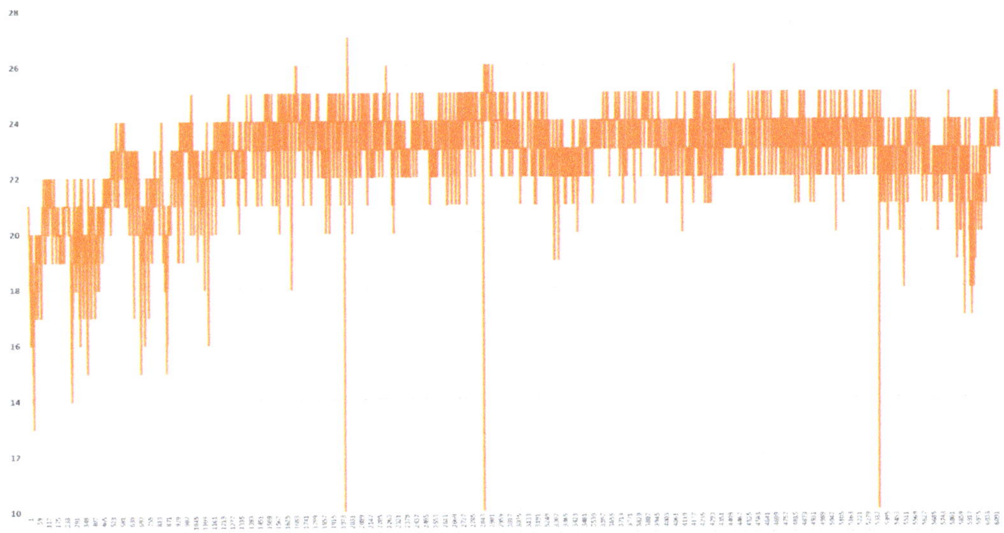

Figure 5. Dew point data.

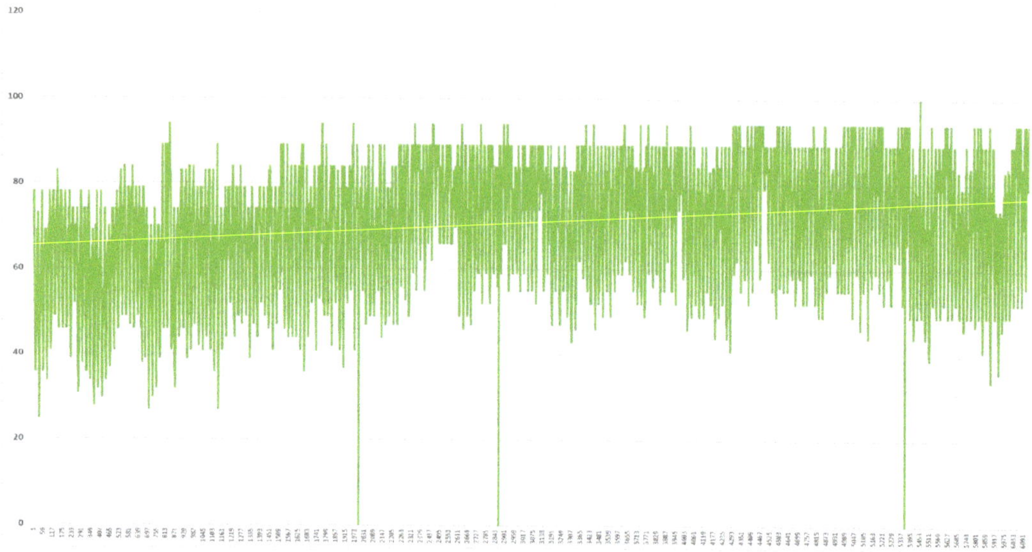

Figure 6. Humidity data.

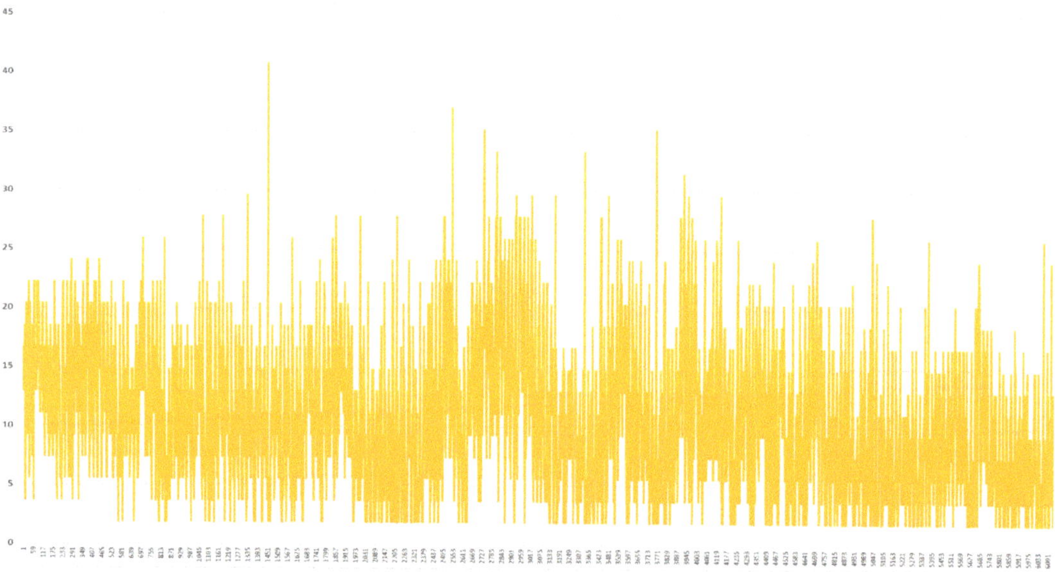

Figure 7. Wind speed data.

2.3. *Analysis of Load Data in the Vietnam Demonstration Site*

To further illustrate the load behavior at each site, representative load profiles were visualized. Figures 8–10 illustrate the hourly load patterns of these buildings—VT2, Mold, and Kindergarten—for the period between 10 March and 27 November 2018. While some data points show abrupt spikes or drops in load, overall, the load tends to be higher during the winter season (December to March) and the summer season (July to September) compared to other periods.

Additionally, a broader review of historical load patterns reveals a general increasing trend in power consumption from 2015 to 2017, suggesting long-term load growth at the site.

The electric load data collected from the Vietnam demonstration factory site was obtained from three key buildings within the site—VT2, Mold, and Kindergarten—each representing different usage patterns and operational characteristics. By analyzing historical load trends, we were able to identify temporal patterns and assess variations between weekdays and weekends. This insight formed the basis for training individual ANN models per building to optimize forecast accuracy under localized conditions [20].

While some data points show abrupt spikes or drops in load, overall, the load tends to be higher during the winter season (December to March) and the summer season (July to September) compared to other periods.

Additionally, a broader review of historical load patterns reveals a general increasing trend in power consumption from 2015 to 2017, suggesting long-term load growth at the site. The electric load data were collected from the Vietnam demonstration factory site. Load data were obtained from three key buildings within the site—VT2, Mold, and Kindergarten—each representing different usage patterns and operational characteristics. By analyzing historical load trends, we were able to identify temporal patterns and assess variations between weekdays and weekends. This insight formed the basis for training individual ANN models per building to optimize forecast accuracy under localized conditions.

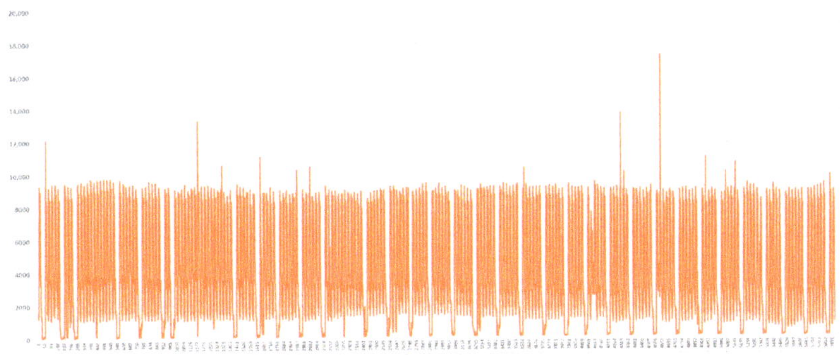

Figure 8. VT2 load pattern.

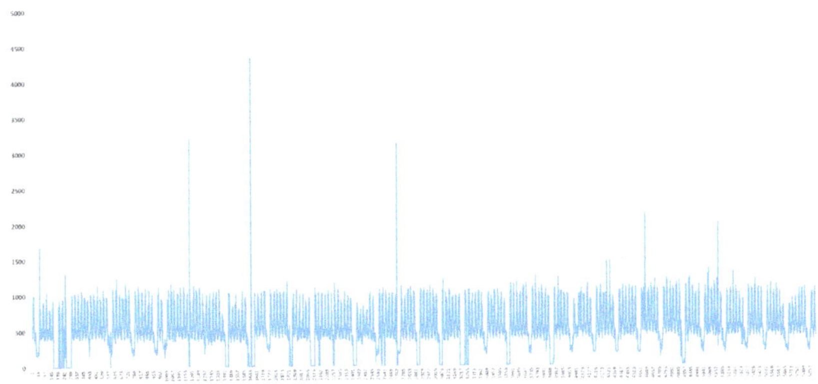

Figure 9. Mold load pattern.

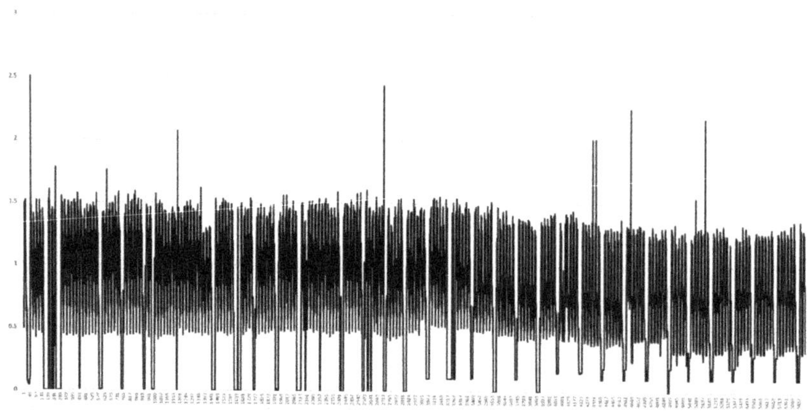

Figure 10. Kindergarten load pattern.

2.4. Correlation Analysis

Correlation analysis in this study is supported by the use of the Pearson correlation coefficient, which quantitatively evaluates the degree of linear relationship between a meteorological variable and electric load. The formula used is as follows:

$$R_m = \frac{n\sum X_m Y - (\sum X_m)(\sum Y)}{\sqrt{\left(n\sum X_m^2 - (\sum X_m)^2\right)\left(n\sum Y^2 - (\sum Y)^2\right)}} \quad (1)$$

where

R_m: Pearson correlation coefficient between meteorological variable X_m and power load Y;

n: number of data points;

X_m: meteorological data (e.g., temperature and humidity);

Y: power consumption data.

Interpretation:

$R_m = 1$: perfect positive linear relationship;

$R_m = -1$: perfect negative linear relationship;

$R_m = 0$: no linear relationship.

This formula is a standard statistical tool that helps in determining the strength and direction of a linear relationship between two continuous variables. It forms the basis of variable selection in our forecasting model.

Correlation analysis is a statistical method used to assess the linear relationship between two variables. The correlation coefficient quantifies this relationship, where a value close to 1 or −1 indicates strong positive or negative correlation, respectively, and values near 0 suggest little to no linear correlation.

In this study, Pearson correlation coefficient was used to evaluate the relationship between meteorological variables (temperature, dew point, humidity, and wind speed) and power consumption. This method was applied over the entire period of data collection (10 March to 27 November 2018) to identify which variables have the strongest association with load patterns.

Visualizations using scatter plots were also generated to provide an intuitive understanding of these relationships. To enhance this analysis, we visualized the relationship between VT2 building load and meteorological variables over the entire data collection period (10 March to 27 November 2018), as shown in Figures 11–14. The scatter plots below illustrate these correlations clearly, allowing us to visually interpret the strength and direction of each variable's influence on power demand. The results show that, while temperature and dew point exhibit weak linear correlations with load, humidity and wind speed have even weaker or negligible associations, as quantified in the following figures.

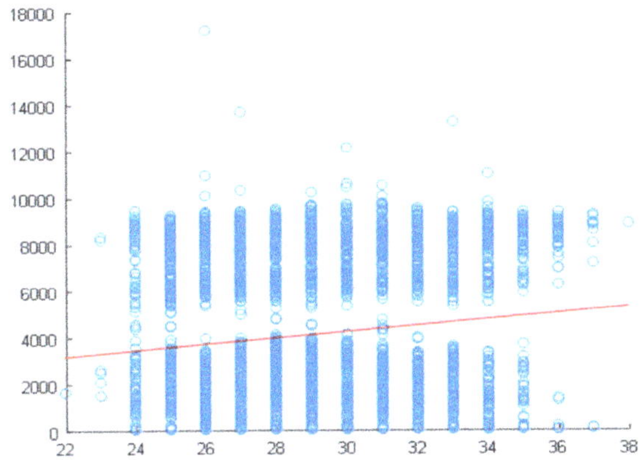

Figure 11. VT2 load vs. temperature.

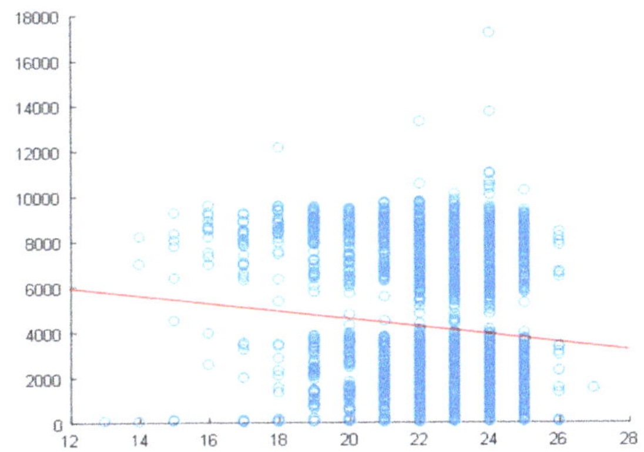

Figure 12. VT2 load vs. dew point.

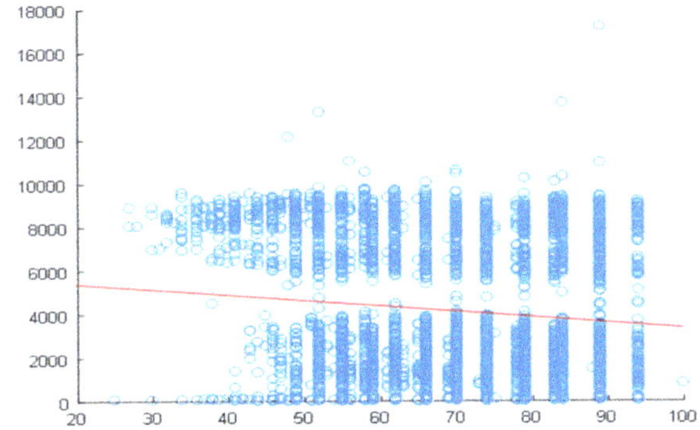

Figure 13. VT2 load vs. humidity.

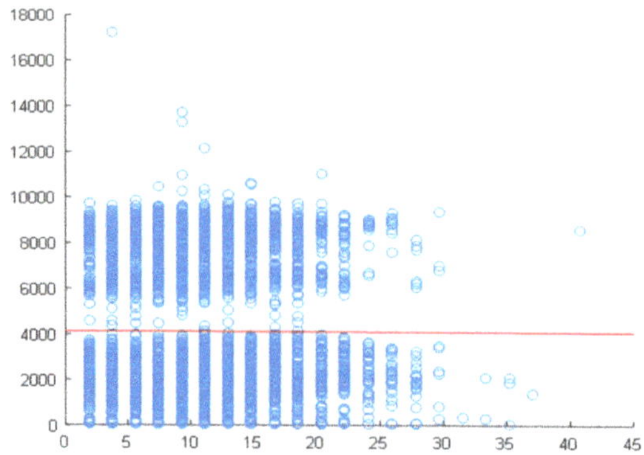

Figure 14. VT2 load vs. wind speed.

Temperature exhibits a weak positive linear correlation with power load, indicated by a Pearson correlation coefficient of 0.1194. Dew point also shows a very weak negative linear relationship with load, with a coefficient of −0.0828. Similarly, humidity demonstrates a weak negative correlation of −0.1077 with load. In contrast, wind speed displays virtually no linear association with load, as shown by a near-zero correlation coefficient of −0.0020.

Temperature and power load for the Mold building exhibit a weak linear correlation, with a Pearson correlation coefficient of 0.1073. Dew point and load show a very weak negative linear correlation, with a coefficient of −0.0598. Humidity shows a similarly weak negative relationship with load, with a coefficient of −0.0906. Wind speed demonstrates virtually no linear correlation with Mold load, as shown by a correlation coefficient of −0.0279. These relationships are illustrated in Figures 15–18.

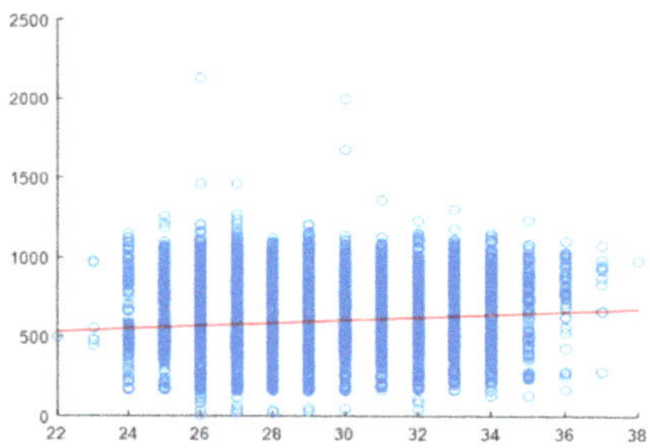

Figure 15. Mold load vs. temperature.

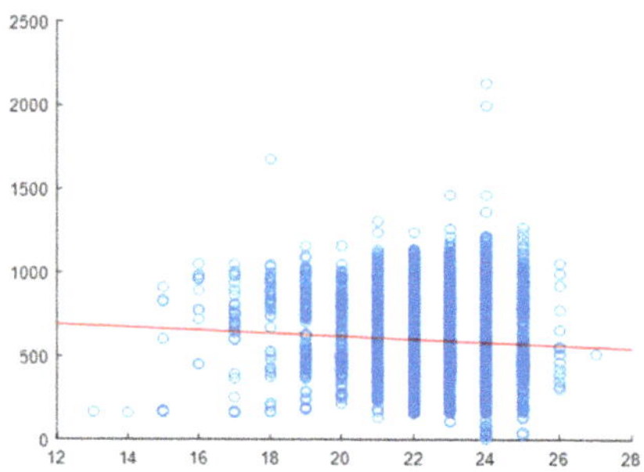

Figure 16. Mold load vs. dew point.

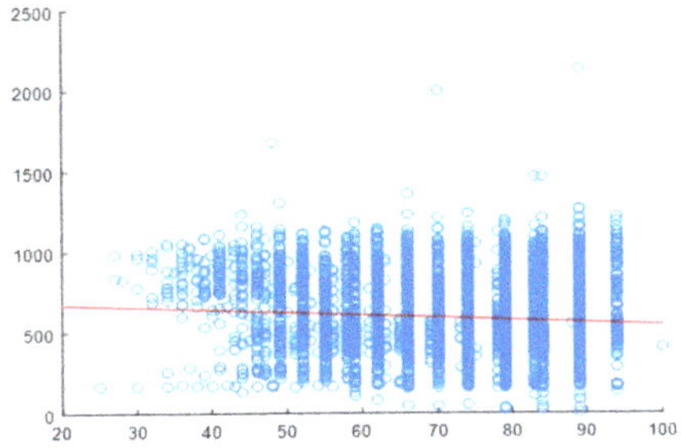

Figure 17. Mold load vs. humidity.

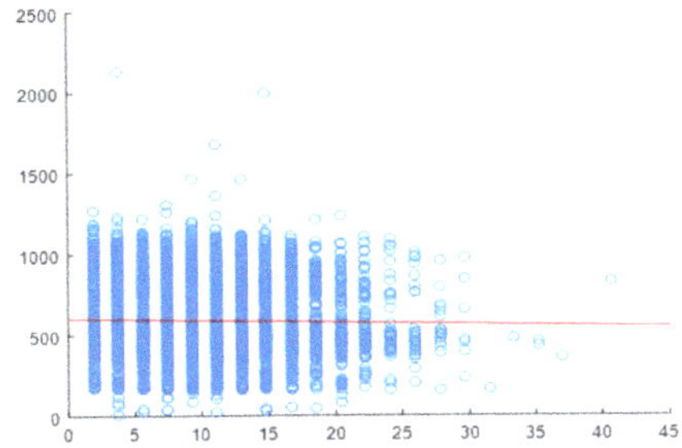

Figure 18. Mold load vs. wind speed.

Kindergarten building load data show similar patterns. Temperature shows a weak positive linear relationship with load, with a Pearson coefficient of 0.1149. Dew point and load have a very weak negative correlation of −0.0729, while humidity correlates negatively with a coefficient of −0.1104. Wind speed shows a very weak positive correlation with kindergarten load, with a coefficient of 0.0548. These relationships are illustrated in Figures 19–22.

The results above differ somewhat from the expectations formed during the meteorological data analysis in Section 2.1. It is commonly understood that electric load is highly sensitive to temperature due to heating and cooling demands, which would suggest a strong correlation between temperature and power load. However, the observed Pearson correlation coefficients indicate that the relationship is not significantly strong. Similar weak or negligible correlations were found for other meteorological variables as well.

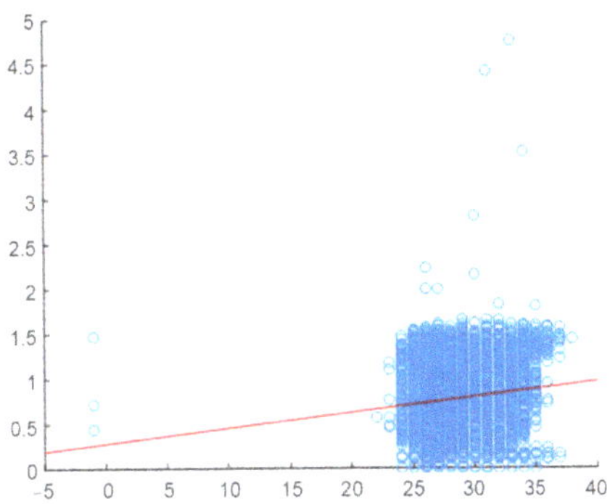

Figure 19. Kindergarten load vs. temperature.

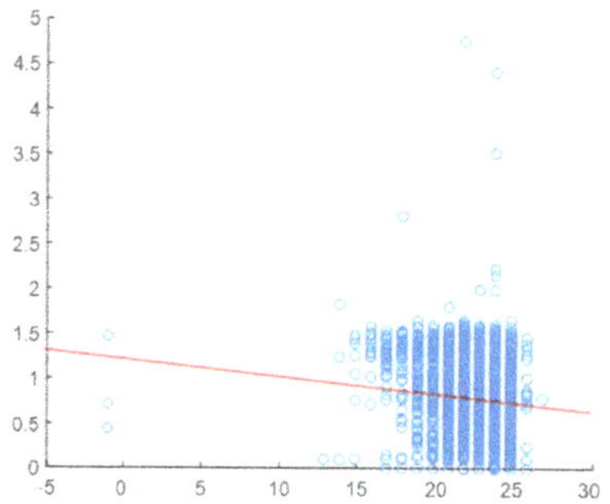

Figure 20. Kindergarten load vs. dew point.

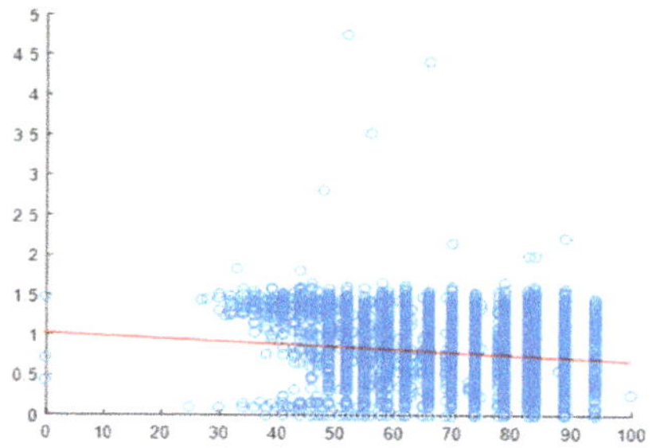

Figure 21. Kindergarten load vs. humidity.

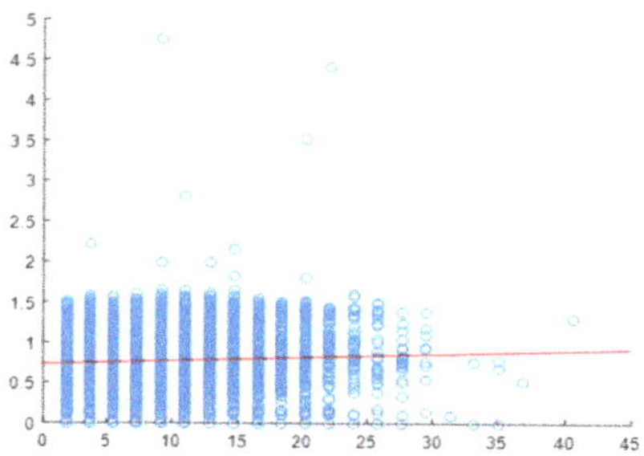

Figure 22. Kindergarten load vs. wind speed.

The weak overall correlation between temperature and electric load observed in this study can primarily be attributed to the non-linear and intermittent nature of weather influences under Vietnam's tropical climate. In such environments, sudden changes in humidity, solar radiation, or wind may affect energy usage in brief, irregular patterns that are not consistently captured over long-term aggregations. Additionally, the load characteristics of the studied buildings reflect the operational realities of industrial footwear manufacturing for globally recognized consumer brands. As original equipment manufacturing (OEM) sites, their production schedules tend to follow market-driven demand and fashion cycles rather than seasonal climate changes, further decoupling electric load from meteorological conditions.

2.5. Daily Correlation Trend Analysis

While the previous section analyzed the overall correlation between meteorological variables and power load across the entire dataset, this section presents the distribution of Pearson correlation coefficients calculated on a daily basis. Given that power forecasting is often performed over 24 h periods, understanding the day-to-day variation in correlation can provide more granular insight into the strength of relationships over time.

Histograms were generated to visualize the daily correlation coefficient distributions for each building (VT2, Mold, and Kindergarten) across the four meteorological variables: temperature, dew point, humidity, and wind speed. These distributions offer a more nuanced view of the dynamic relationships between weather conditions and power demand, which can vary significantly based on operational schedules, weather fluctuations, and building use patterns.

For the VT2 building, Figures 23–26 present the histograms showing the distribution of daily Pearson correlation coefficients between power load and each meteorological variable.

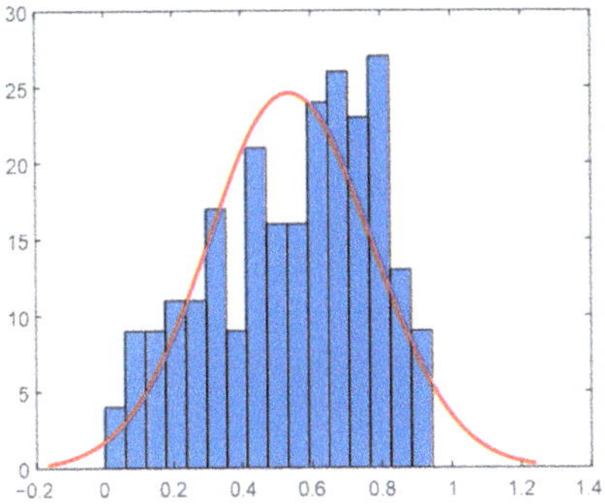

Figure 23. Histogram of daily Pearson correlation coefficients between VT2 load and temperature.

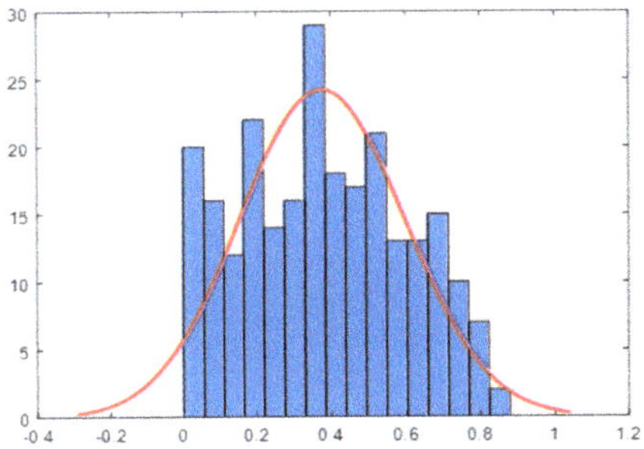

Figure 24. Histogram of daily Pearson correlation coefficients between VT2 load and dew point.

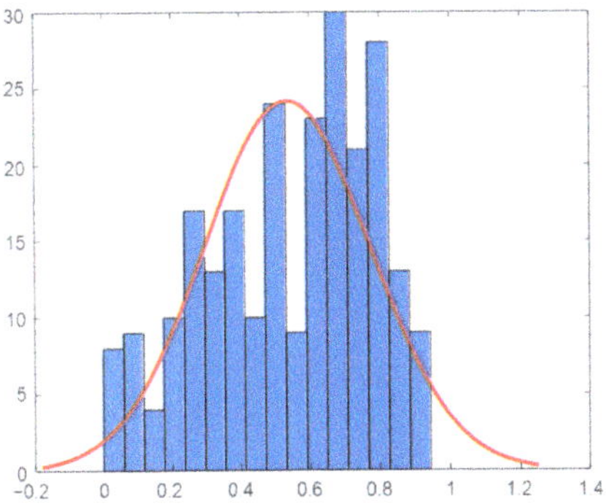

Figure 25. Histogram of daily Pearson correlation coefficients between VT2 load and humidity.

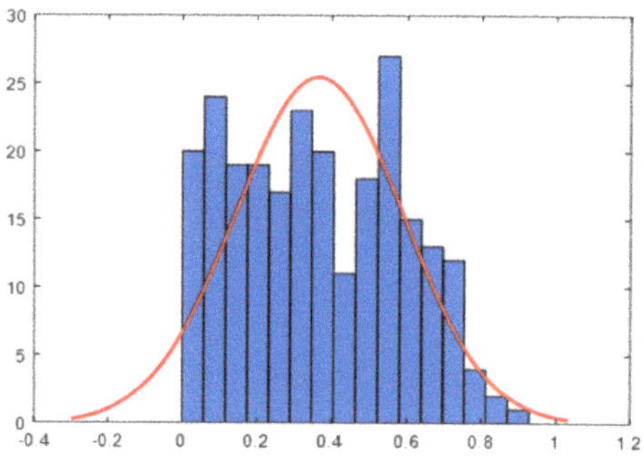

Figure 26. Histogram of daily Pearson correlation coefficients between VT2 load and wind speed.

Temperature and power load demonstrate a strong linear relationship, with most daily Pearson correlation values ranging between 0.6 and 0.8. This suggests a consistent and strong dependency between temperature and load throughout each day. Dew point shows a moderate relationship with daily load values, where the majority of correlation coefficients fall between 0.3 and 0.5, indicating a weaker but present linear trend.

Humidity also reflects a strong correlation pattern, with values similarly clustered within the 0.6 to 0.8 range. Wind speed, on the other hand, exhibits a notably weaker correlation, with most values distributed between 0 and 0.4, suggesting limited influence on the daily power load.

For the Mold building, Figures 27–30 illustrate the daily Pearson correlation coefficient distributions for each meteorological factor.

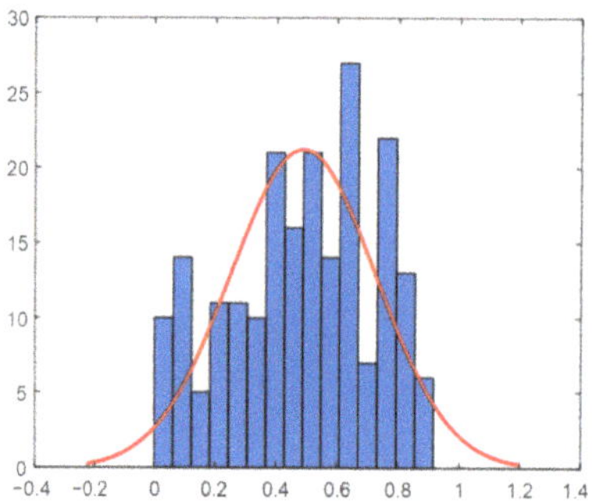

Figure 27. Histogram of daily Pearson correlation coefficients between Mold load and temperature.

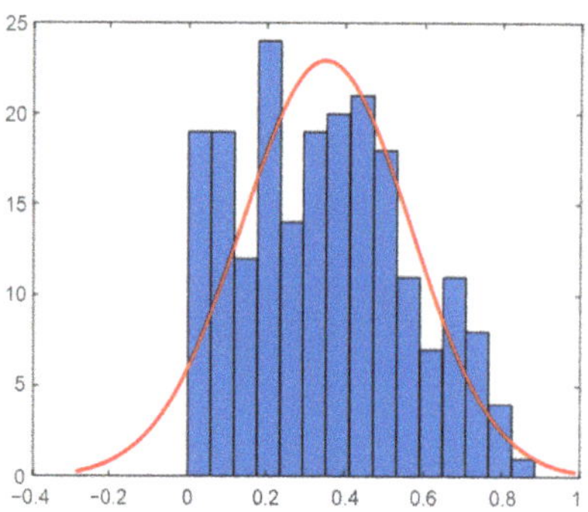

Figure 28. Histogram of daily Pearson correlation coefficients between Mold load and dew point.

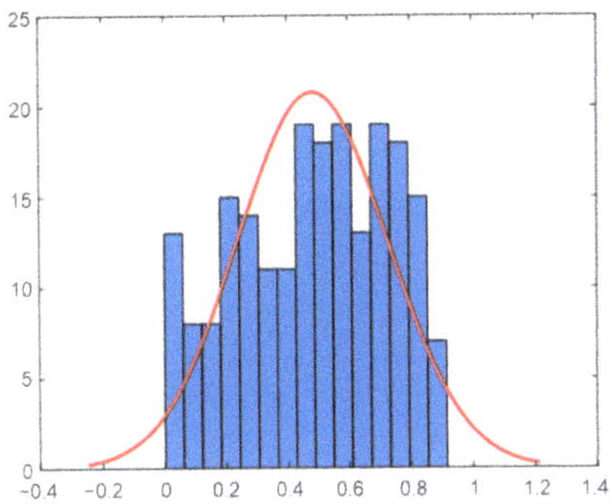

Figure 29. Histogram of daily Pearson correlation coefficients between Mold load and humidity.

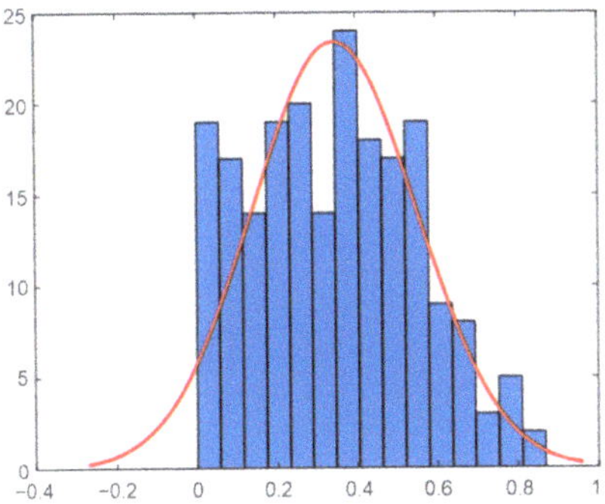

Figure 30. Histogram of daily Pearson correlation coefficients between Mold load and wind speed.

Temperature and power load show a moderately strong linear relationship, with most daily correlation values falling within the 0.4 to 0.6 range. Dew point also displays a moderate correlation, where daily coefficients typically range from 0.3 to 0.5, indicating a consistent linear association across days.

Humidity demonstrates a stronger linear relationship with load, with correlation values predominantly between 0.6 and 0.8. In contrast, wind speed correlations are generally weaker, with daily coefficients mostly distributed between 0 and 0.4, suggesting minimal impact on daily load patterns.

For the kindergarten building, Figures 31–34 illustrate the histograms representing the distribution of daily Pearson correlation coefficients with respect to the selected meteorological factors.

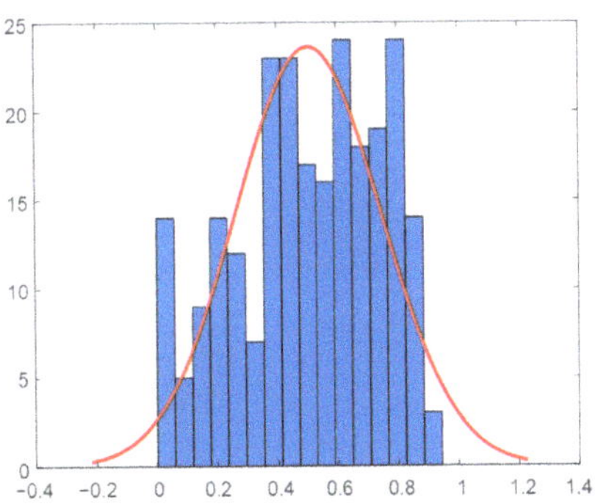

Figure 31. Histogram of daily Pearson correlation coefficients between kindergarten load and temperature.

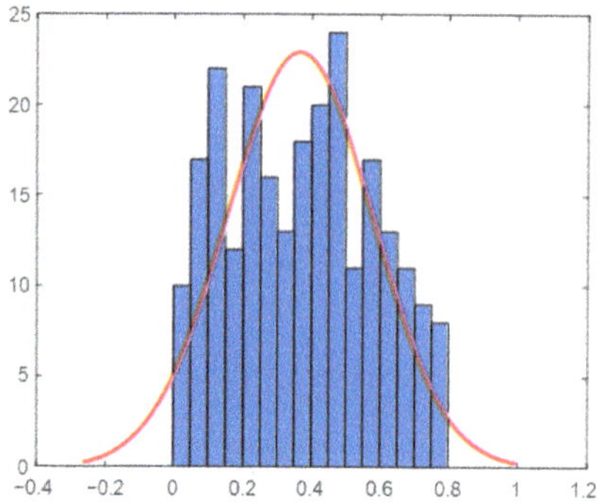

Figure 32. Histogram of daily Pearson correlation coefficients between kindergarten load and dew point.

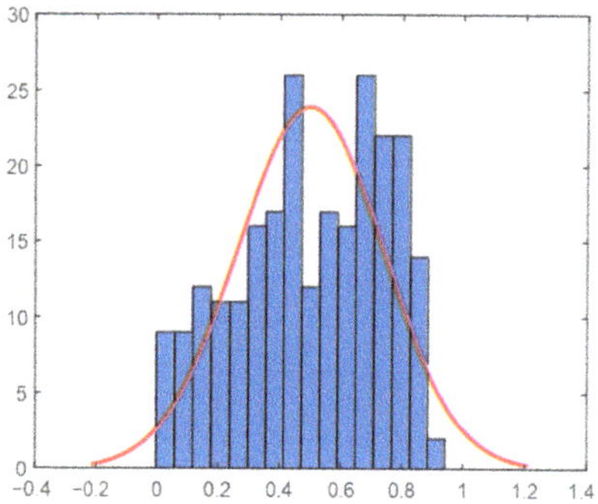

Figure 33. Histogram of daily Pearson correlation coefficients between kindergarten load and humidity.

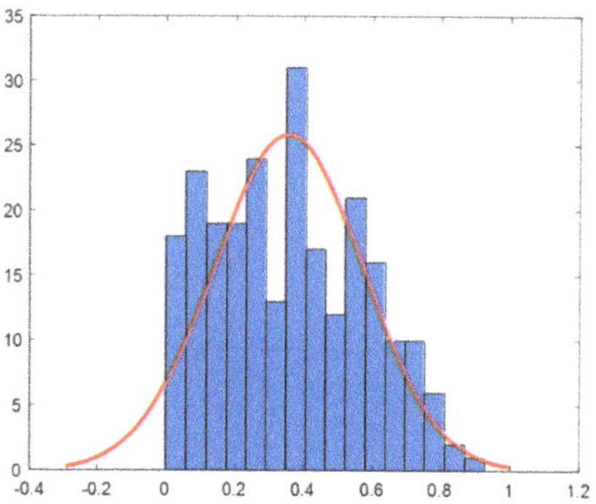

Figure 34. Histogram of daily Pearson correlation coefficients between kindergarten load and wind speed.

Temperature and load correlations are consistently strong, with values mainly distributed between 0.4 and 0.8. This suggests a reliable and moderately strong relationship throughout the day. Dew point shows a weaker but still notable linear association, with daily correlation coefficients ranging between 0.2 and 0.5.

Humidity demonstrates a strong linear correlation with load, as coefficients are primarily within the 0.6 to 0.8 range. Wind speed again shows only a weak relationship, with most correlation values falling between 0 and 0.4.

The daily correlation histograms in Figures 23–34 reveal a different trend from the aggregated analysis. For example, in the VT2 building, daily correlations between temperature and load frequently range from 0.6 to 0.8, suggesting a strong and consistent relationship at the intra-day level. This indicates that HVAC usage may spike during specific temperature conditions or time windows (e.g., mid-day peak hours), even in the absence of pronounced seasonal changes. Such transient behaviors are masked in the overall analysis due to temporal averaging. This reinforces the importance of day-level analysis when forecasting load in climates with minimal seasonality but high daily variability.

The following Table 1 summarizes the key findings from Sections 2.4 and 2.5, highlighting both overall and daily Pearson correlation coefficient trends between meteorological variables and electric load for each of the three buildings.

Table 1. Summary of correlation analysis results.

Building	Variable	Overall Correlation	Daily Correlation Range	Strength of Relationship
VT2	Temperature	0.1194	0.6–0.8	Strong
	Dew Point	−0.0828	0.3–0.5	Moderate
	Humidity	−0.1077	0.6–0.8	Strong
	Wind Speed	−0.002	0–0.4	Weak
Mold	Temperature	0.1073	0.4–0.6	Moderate
	Dew Point	−0.0598	0.3–0.5	Moderate
	Humidity	−0.0906	0.6–0.8	Strong
	Wind Speed	−0.0279	0–0.4	Weak
Kindergarten	Temperature	0.1149	0.4–0.8	Moderate to Strong
	Dew Point	−0.0729	0.2–0.5	Weak to Moderate
	Humidity	−0.1104	0.6–0.8	Strong
	Wind Speed	0.0548	0–0.4	Weak

The observed discrepancy between overall and daily correlation coefficients can be attributed to several interacting factors. Daily operational schedules, such as work shifts and equipment usage in VT2 and Mold buildings, vary significantly across weekdays and weekends, causing fluctuations in hourly load patterns. In addition, transient weather conditions—such as sudden humidity surges or temperature drops—may influence short-term cooling or ventilation demands. Each building's unique usage profile, including the kindergarten's non-production-based occupancy, also contributes to the varying sensitivity of load to meteorological changes. These factors, when combined, lead to non-linear and temporally dynamic relationships between weather and load, which are better captured in daily correlation analyses than in aggregated values.

3. Methodology

This section outlines the structure and implementation of the load forecasting model developed in this study. The model is based on artificial neural networks (ANNs), which are capable of capturing complex and non-linear relationships between weather variables and electric load. By employing building-specific models and incorporating multiple weather variables, the proposed methodology aims to enhance short-term forecasting accuracy in industrial microgrid environments.

3.1. ANN Model Design

The output of the Softplus activation function used in the hidden layers is defined as:

$$f(x) = \ln(1 + e^x) \quad (2)$$

This activation function is continuous and differentiable, which makes it well-suited for gradient-based optimization. Unlike the ReLU function that outputs zero for negative inputs, the Softplus function maintains a smooth gradient for all input values, enabling more stable learning during training. It serves as a smooth approximation to the ReLU function while avoiding issues related to gradient vanishing in deeper layers.

$$\begin{aligned} h^{(1)} &= \ln\left(1 + e^{W^{(1)}x + b^{(1)}}\right) \\ h^{(2)} &= \ln\left(1 + e^{W^{(2)}h^{(1)} + b^{(2)}}\right) \\ \hat{y} &= W^{(3)}h^{(2)} + b^{(3)} \end{aligned} \tag{3}$$

This equation describes the sequence of transformations applied to the input data through each layer of the neural network. The first hidden layer computes a non-linear mapping of the input weather variables, enabling the model to capture direct influences on load. The second hidden layer processes the output of the first to uncover higher-order dependencies and interactions among the features. The output layer then generates the final predicted load value. This architecture allows the model to approximate highly complex, non-linear relationships in the data that traditional models may fail to capture, making it particularly suitable for energy forecasting in diverse environmental conditions.

This equation illustrates the full forward propagation process of the ANN used in this study. It includes two hidden layers, each applying the Softplus activation function, and a linear output layer that produces the final power load prediction.

This function allows the network to handle non-linearity while avoiding issues related to vanishing gradients commonly seen in sigmoid or tanh functions.

The proposed forecasting model utilizes a multi-layer perceptron (MLP) architecture. The input layer receives four normalized weather variables: temperature, dew point, humidity, and wind speed. The model is structured to produce hourly electric load forecasts for a 24 h horizon, enabling day-ahead prediction of power demand.

Each ANN consists of one input layer, two hidden layers, and one output layer. The hidden layers use the Softplus activation function, which was selected for its ability to handle non-linearities while maintaining smooth gradients for optimization. The output layer generates the predicted hourly load values. These two hidden layers enable the network to capture more complex non-linear relationships between inputs and outputs. The first hidden layer identifies basic interaction patterns among weather variables, while the second layer builds upon these to learn higher-level abstractions. This multi-layered structure increases the model's capacity to approximate intricate patterns in load behavior, improving forecast accuracy over single-layer designs.

This model is particularly suitable for capturing non-linear dependencies between weather variables and load patterns. Given the frequent short-term fluctuations in tropical climates like Vietnam, the ANN structure helps maintain prediction robustness even under abrupt weather shifts or irregular factory operation schedules, which are difficult to handle with traditional linear approaches.

3.2. Model Training and Evaluation

The loss function minimized during training is the Mean Squared Error (MSE), expressed as:

$$\mathrm{MSE} = \left(\frac{1}{\mathrm{n}}\right) * \Sigma(\mathrm{y_i} - \hat{\mathrm{y}}_\mathrm{i})^2 \tag{4}$$

where $\mathrm{y_i}$ is the actual load and $\hat{\mathrm{y}}_\mathrm{i}$ is the predicted load for instance i.

The forecasting performance was evaluated using the Mean Absolute Percentage Error (MAPE), calculated as:

$$\mathrm{MAPE} = \left(\frac{100}{\mathrm{n}}\right) * \Sigma\left(\frac{|\mathrm{y_i} - \hat{\mathrm{y}}_\mathrm{i}|}{\mathrm{y_i}}\right) \tag{5}$$

This metric quantifies the accuracy of the forecasts as a percentage of the actual load.

The model was trained using historical weather data and corresponding load measurements for the VT2, Mold, and Kindergarten buildings. Data was split into training and validation sets with a typical 80:20 ratio. Mean Squared Error (MSE) was used as the loss function, and model performance was evaluated using the Mean Absolute Percentage Error (MAPE) metric.

The input data was standardized to have zero mean and unit variance to enhance learning efficiency and convergence. Model training was conducted independently for each building to account for their different load characteristics and operational profiles.

The ANN architecture consists of an input layer, two hidden layers with 16 and 8 neurons, respectively, and an output layer. The Softplus activation function is applied to the hidden layers, and a linear activation is used in the output. The network was trained using the Adam optimizer with a learning rate of 0.001. The batch size was set to 32, and training was conducted over 100 epochs. These parameters were selected based on preliminary experiments to balance convergence speed and prediction accuracy. Model performance was monitored using validation loss under K-fold cross-validation.

3.3. Forecasting Approach

Separate ANN models were trained for each building using the respective historical data. Forecasts were performed on a day-ahead basis, with the model generating 24 hourly load predictions at once. Figure 35 illustrates the end-to-end architecture and flow of the ANN-based forecasting model developed in this study. The model receives multiple meteorological variables (temperature, dew point, humidity, and wind speed) along with load history as inputs. The forecasting engine consists of a multi-layer perceptron with two hidden layers using the Softplus activation function. To ensure generalization, the model undergoes K-fold cross-validation, where each fold is used once for validation while the remaining folds are used for training. Hyperparameters, such as the number of hidden nodes and regularization coefficient, are tuned based on the average validation error. This approach enables the selection of an ANN configuration that minimizes generalization error and enhances prediction reliability. By tailoring the ANN model to each building's data, the method improves localized forecasting accuracy, capturing site-specific behaviors and weather sensitivities.

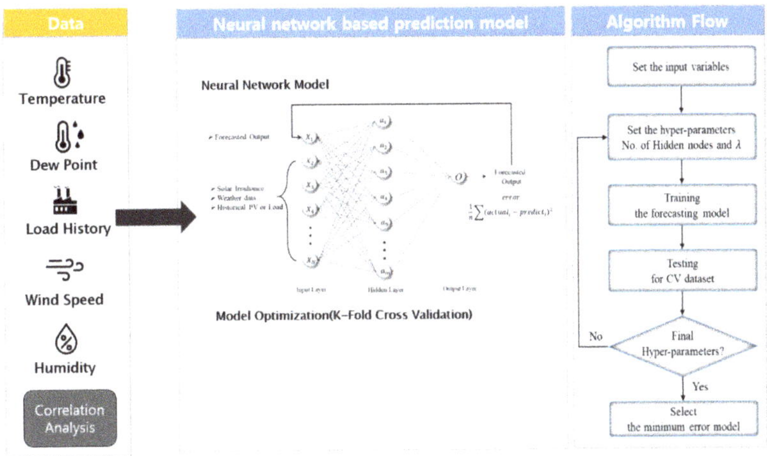

Figure 35. Overall structure and algorithmic flow of the ANN-based forecasting model, including input data sources, neural network architecture, and K-fold cross-validation for hyperparameter optimization.

4. Results and Performance Evaluation

This section presents the results of the electric load forecasting using the proposed ANN-based models for the VT2, Mold, and Kindergarten buildings. Each building's model was evaluated based on its ability to accurately predict 24 h ahead load values using meteorological inputs.

The models demonstrated consistent and reliable prediction performance across all three buildings. For each case, the ANN model captured the daily load pattern with sufficient accuracy, despite differing operational schedules and sensitivities to meteorological inputs. As shown in Table 2, MAPE scores ranged from 8.8% to 10.6%, suggesting effective learning and generalization by the ANN model.

Table 2. Forecasting accuracy for each building.

Building	MAPE (%)
VT2	10.2
Mold	8.8
Kindergarten	10.6

To further validate prediction performance, Figures 36–38 compare actual and predicted load values over a representative day for each building. In the figures, the actual load is shown in blue, while the predicted load is illustrated in red. Among the three buildings, both VT2 and Mold are factory buildings, while the kindergarten is an educational facility. VT2 has the highest average power consumption, whereas Mold represents one of the lower-load industrial sites. Interestingly, despite its lower total load, the Mold building shows better prediction alignment. This suggests that stable and predictable load behavior, rather than total consumption, may play a more critical role in achieving accurate forecasts. The results show that the ANN model effectively captures both peak and off-peak load trends, even during times of abrupt changes. This is especially evident in the Mold building, where the ANN demonstrates higher precision, likely due to consistent usage and more predictable weather–load interaction.

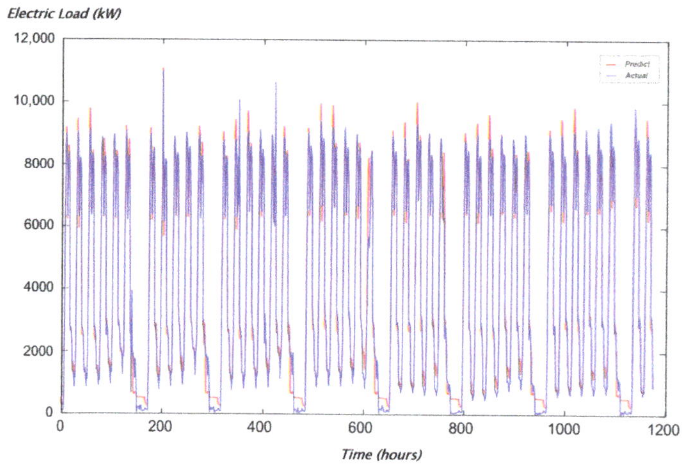

Figure 36. Actual vs. predicted load—VT2.

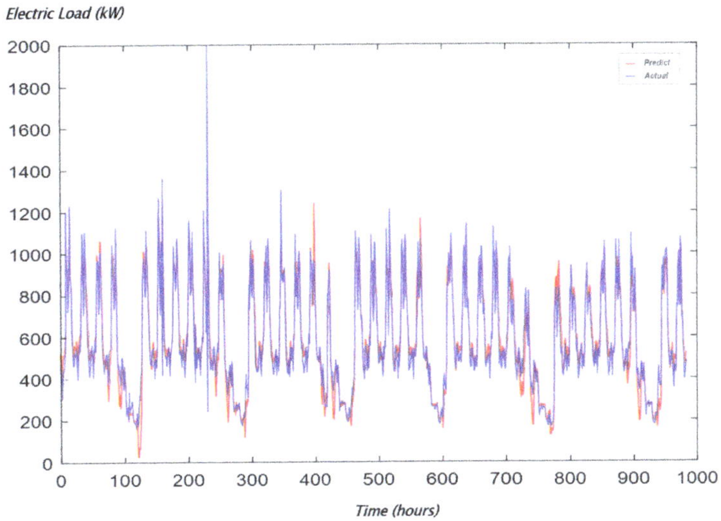

Figure 37. Actual vs. predicted load—Mold.

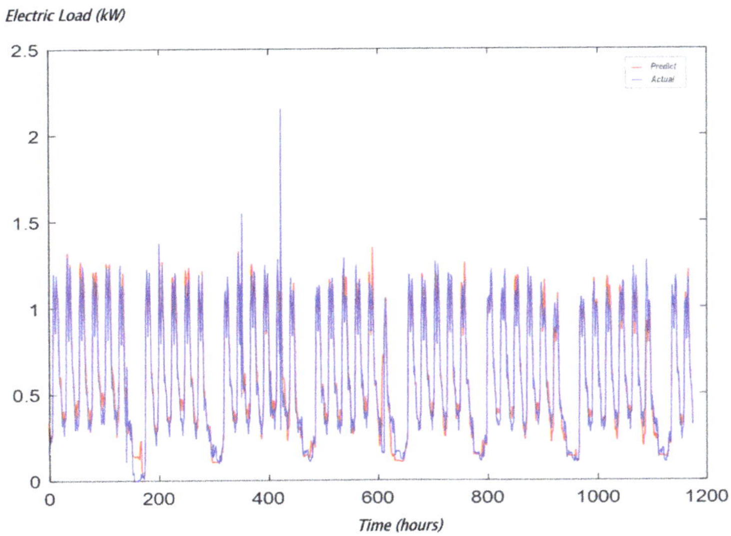

Figure 38. Actual vs. predicted load—Kindergarten.

In contrast, VT2 and Kindergarten buildings showed slightly higher MAPE values, possibly due to variable schedules or complex usage behaviors not directly explained by the four weather variables used. Nevertheless, the visual alignment of predicted and observed values in these figures supports the model's robustness.

Overall, these findings suggest that the proposed ANN model is well-suited for site-specific energy demand forecasting and has the potential to support microgrid optimization and operational planning.

5. EMS Operation Strategy Based on Load Forecasting

5.1. Forecast-Aided Load Scheduling

Although load forecasting was conducted for three sites—VT2, Mold, and a kindergarten—the EMS operation and ESS scheduling in this study focused only on the VT2 and Mold buildings. This is because the kindergarten's electric load was significantly smaller and less variable compared to the industrial facilities, rendering it unnecessary to include in the economic scheduling and control framework. However, its load patterns were still considered as part of the overall demand profile for monitoring purposes.

Figures 39 and 40 illustrates the forecasting system, which is built upon the integration of local climate data, historical load patterns, and facility schedule information. As shown in the diagram, key meteorological parameters such as temperature, dew point, humidity, and wind speed, along with factory-specific load history, are processed through Pearson correlation analysis to identify significant relationships. This data is then fed into a multi-layer artificial neural network (ANN) for prediction, the integration of local climate data, historical load patterns, and facility schedule information.

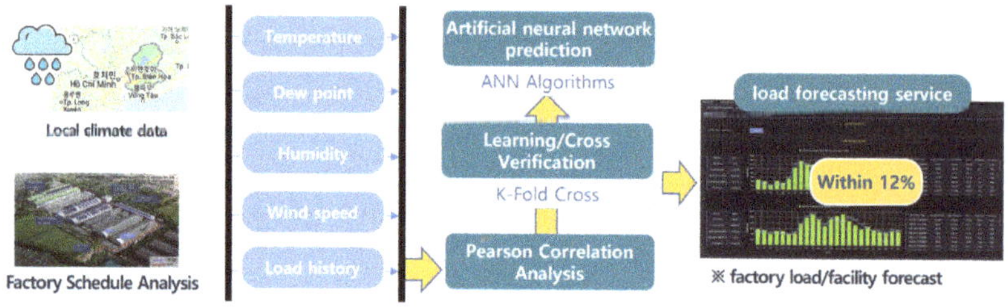

Figure 39. Overview of the load forecasting system using local weather, historical load data, and ANN model.

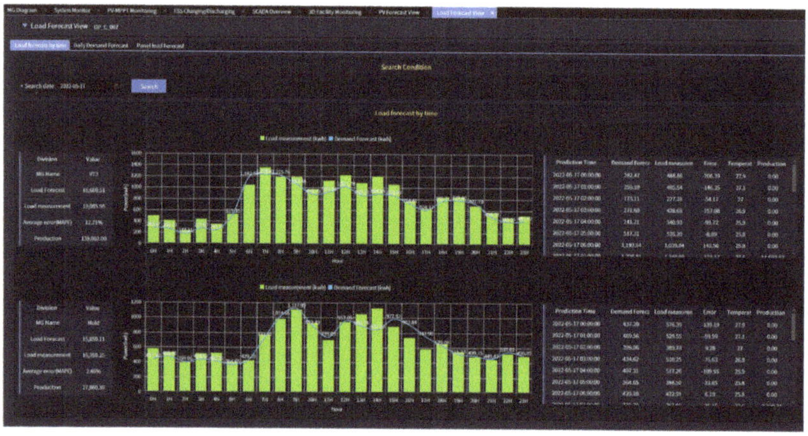

Figure 40. Comparison of actual measured load and ANN-based forecast results for VT2 and Mold on 17 May 2022.

To ensure reliability, a K-fold cross-validation approach is employed during model training, verifying performance across multiple data splits. The resulting ANN model can then deliver a 24 h ahead load forecast for each building.

This forecast supports real-time scheduling for industrial microgrids. It enables the EMS to proactively plan charging and discharging of energy storage, schedule surplus power exchanges, and maintain operational efficiency even under fluctuating load conditions. The model achieves a mean absolute percentage error (MAPE) within 12%, as shown in the following illustration.

This image shows actual versus predicted electric load values for the VT2 and Mold buildings on 17 May 2022. The upper chart corresponds to VT2, where an MAPE of 12.71% was observed, and the lower chart shows Mold with an MAPE of just 2.46%. Each bar represents measured load (green), while the blue line denotes the forecasted demand. The error values and temperature inputs are tabulated on the right side. The higher accuracy in Mold likely reflects a more consistent load pattern compared to VT2's operational fluctuations.

5.2. Economic Operation with PSO Algorithm

The proposed Energy Management System (EMS) integrates a Particle Swarm Optimization (PSO) algorithm to generate optimal charge and discharge schedules for each building's Energy Storage System (ESS). This optimization minimizes operational cost while satisfying system constraints and responding to forecasted load conditions.

5.2.1. PSO Algorithm for ESS Scheduling

The PSO algorithm emulates the social behavior of particles in a swarm to find the global optimum. In this context, each particle represents a candidate ESS schedule vector that minimizes the total energy cost.

Objective: minimize total electricity cost over time:

$$\min_{P_{ch}, P_{dch}} \sum_{t=1}^{T} \left(C_t \cdot \left(P_{grid,t} + P_{dch,t} - P_{ch,t} \right) \right) \quad (6)$$

Subject to:

$$SOC_{min} \leq SOC_t \leq SOC_{max}$$

$$0 \leq P_{ch,t}, P_{dch,t} \leq P_{max}$$

$$SOC_{t+1} = SOC_t + \eta_{ch} \cdot P_{ch,t} \cdot \Delta t - \frac{1}{\eta_{dch}} \cdot P_{dch,t} \cdot \Delta t$$

where

C_t: electricity price at time t;

$P_{ch,t}, P_{dch,t}$: charging and discharging power;

η_{ch}, η_{dch}: charging/discharging efficiency;

SOC_t: battery state of charge at time t.

To provide a clearer understanding of the PSO implementation, Figure 41 illustrates the detailed logic of the ESS scheduling algorithm and the swarm-based optimization process. In this study, the PSO algorithm utilizes a swarm size of 100, where each particle represents a candidate charge/discharge schedule for the ESS over a 24 h horizon. The particles are initialized randomly within feasible operational ranges, and their velocities and positions are updated iteratively based on inertia and acceleration parameters. The objective function is evaluated at each step, and the process converges within approximately 50 iterations. The global best solution obtained defines the optimal ESS schedule.

The DC-based multi-microgrid system allows energy sharing between VT2 and Mold. In the event of equipment failure or scheduling deviation, the system dynamically reallocates excess power where needed. The structure supports independent operation and cooperative balancing, improving the robustness of the overall EMS.

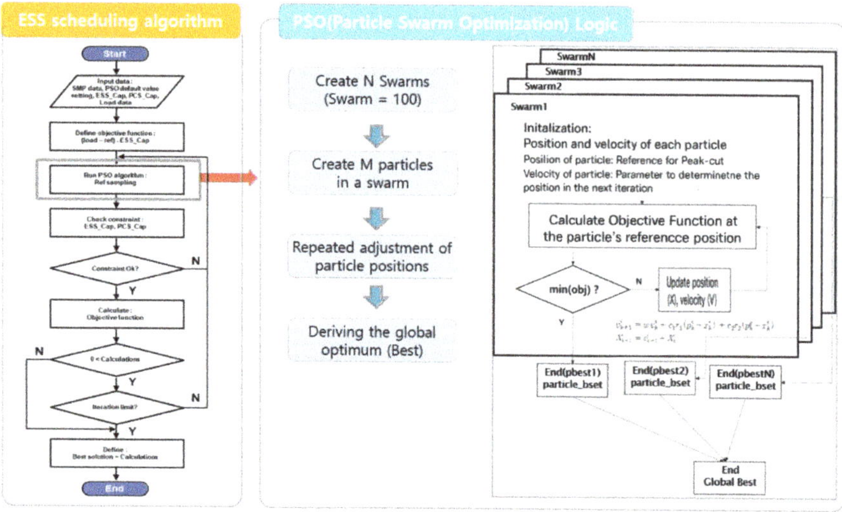

Figure 41. ESS scheduling algorithm and PSO optimization logic illustrating particle initialization, evaluation, and convergence.

5.2.2. VT2 ESS Operation Results

Figure 42 presents the PSO-based operation schedule for the VT2 ESS with 30 min resolution. During the off-peak hours from midnight to early morning (00:00–05:00), the system performs aggressive charging to elevate the state of charge (SOC). Later, the battery discharges during peak pricing windows (09:00–11:00 and 17:00–19:00), corresponding to the predicted high-demand periods.

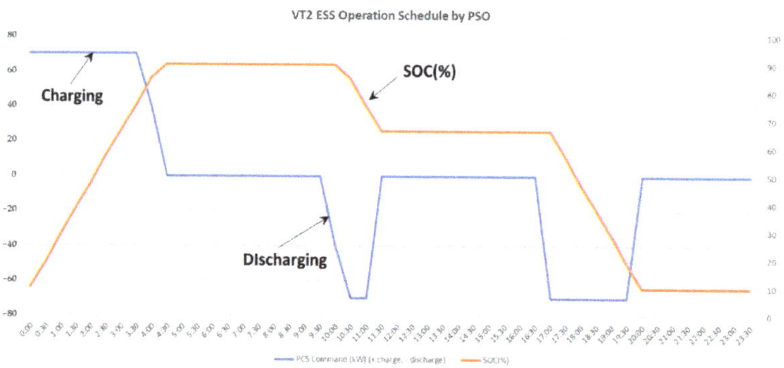

Figure 42. PSO-based ESS command and SOC profile for VT2 on a 30 min interval.

The SOC rises from 10 percent to 90 percent early in the day through continuous 70 kW charging, discharges at minus 70 kW during load peaks, and then stabilizes above 65 percent post-peak to maintain redundancy and supply backup.

This pattern demonstrates that the PSO-based control effectively anticipates future load using 24 h forecasts, prevents inefficient overlapping charge/discharge actions, and maintains battery longevity through shallow discharge cycles.

5.2.3. Mold ESS Operation Results

Figure 43 shows the PSO-generated schedule for the Mold ESS, which operates under the same control framework but with a more consistent load profile. The SOC increases steadily from 10 percent to 90 percent by 06:30 due to continuous overnight charging, then follows a two-stage discharge, first from 10:00 to 12:00 and again during the evening peak between 17:30 and 20:00.

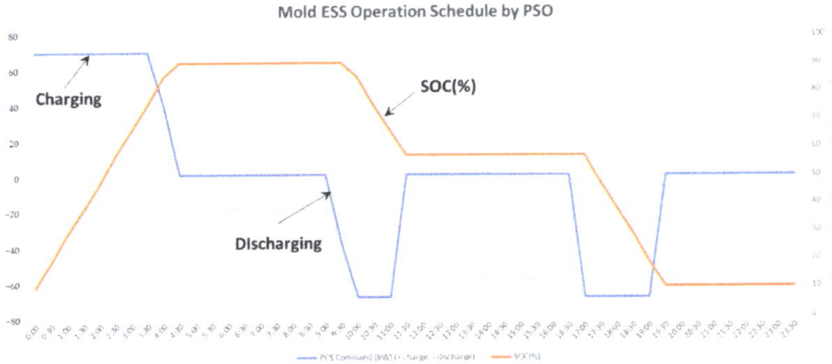

Figure 43. PSO-based ESS command and SOC profile for Mold on a 30 min interval.

Throughout the day, the SOC profile remains stable with minimal fluctuation, which is indicative of the steady-state operation of the Mold facility. This stable response enables Mold to engage in power trading activities efficiently while preserving its operational reliability.

While both VT2 and Mold ESS systems follow the same PSO-based scheduling approach, their operation patterns differ significantly: VT2 shows sharper SOC transitions due to more variable and peak-heavy load profiles, whereas Mold exhibits a smoother and more balanced charging–discharging behavior. This difference stems from the operational characteristics of each site—VT2's production-based shifts lead to higher fluctuations, while Mold maintains a more constant load. As a result, Mold is better positioned to consistently participate in energy trading, while VT2 focuses on peak shaving and internal efficiency.

5.3. Multi-Microgrid Flexibility and Fault Resilience

In addition to energy forecasting and ESS scheduling, maintaining power quality under variable load and generation conditions is critical in industrial microgrid settings [21]. Figure 44 illustrates the real-time multi-microgrid (MG) operational structure deployed at the VT2 and Mold buildings, where distributed energy resources (DERs) such as photovoltaic (PV) systems and energy storage systems (ESS) are coordinated through an integrated Energy Management System (EMS).

Within the EMS, a Power Management System (PMS) supports key functions such as inverter parallel operation, fault-tolerant DER coordination, and real-time energy flow optimization based on load and generation forecasts, electricity tariffs, and ESS state of charge (SOC). It further contributes to system stability through features like inverter margin control, automatic restart, and AC voltage monitoring.

The PMS incorporates a PSO-based ESS scheduling algorithm, which considers building-specific load profiles, time-of-use pricing, and battery constraints to generate cost-optimized operational schedules. This approach enhances both economic performance and system flexibility.

As shown in Figure 45, the EMS autonomously generates charging and discharging schedules for each building. Both VT2 and Mold systems perform off-peak charging and peak-time discharging, but the patterns vary depending on site-specific load behavior. This demonstrates the EMS's ability to implement adaptive, site-aware strategies using a shared optimization engine while preserving local autonomy and resilience.

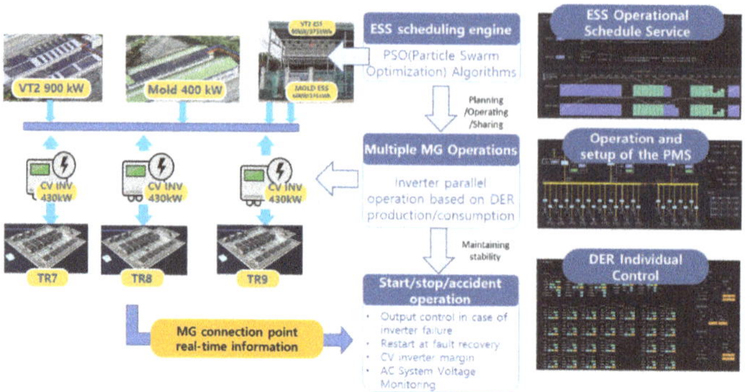

Figure 44. DER-integrated multi-microgrid control system with PMS and PSO-based scheduling algorithm.

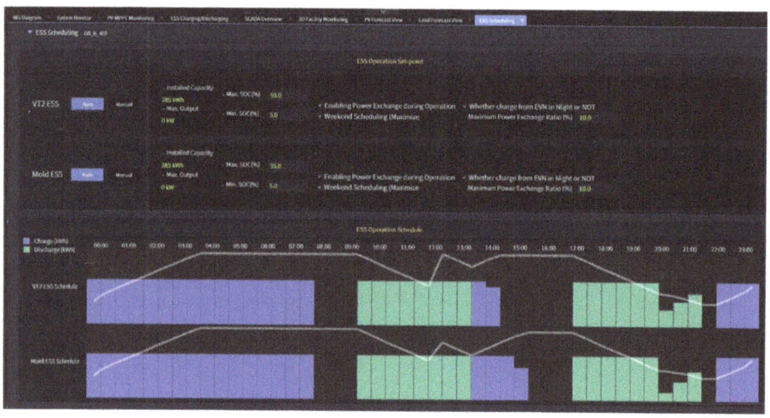

Figure 45. Automatically generated ESS operation schedule and SOC profile for VT2 and Mold.

5.4. ESS Power Transaction and Settlement

The EMS facilitates power trading between VT2 and Mold based on predictive analytics. By using forecasted load demand, time-of-use electricity prices, and current SOC levels, the system generates a power trade plan that optimizes cost and balances surplus and deficit.

When VT2 accumulates excess energy during low-price charging periods, it supplies Mold during high-price intervals. This inter-facility transaction is automatically scheduled and managed within the EMS, as illustrated in Figure 46.

The platform also performs real-time execution and settlement of these trades. On 18 May, for example, Mold sold 240 kWh of power to VT2, generating a total revenue of 610.08 kVND. Detailed transaction records, including timestamps, exchanged energy, and corresponding settlement values, are continuously logged for audit and operational transparency, as illustrated in Figures 47 and 48.

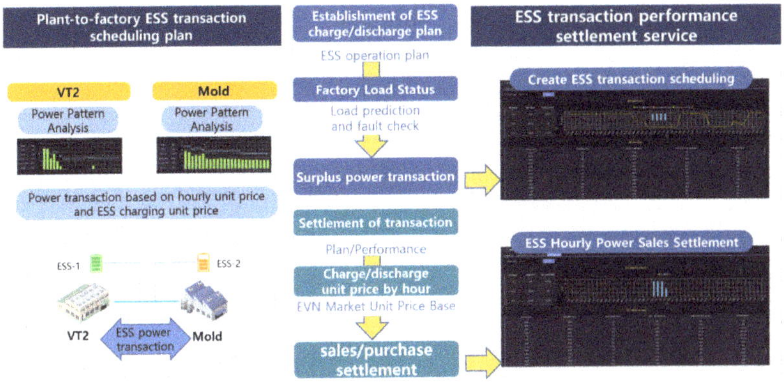

Figure 46. Overall ESS transaction scheduling system and workflow from forecasting to settlement.

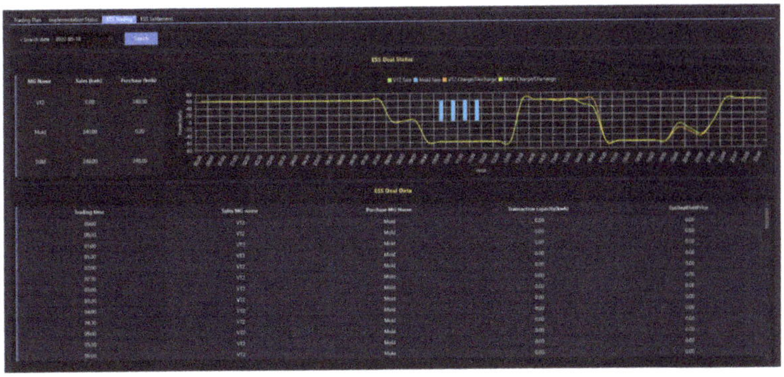

Figure 47. ESS transaction scheduling between VT2 and Mold based on forecasted surplus power and time-of-use pricing.

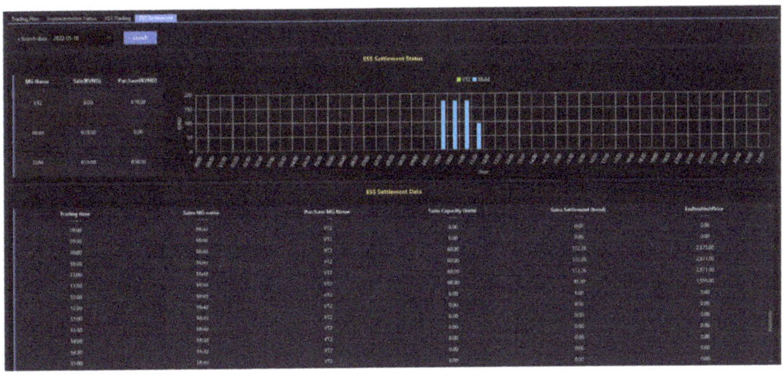

Figure 48. ESS deal status and financial settlement records for VT2 and Mold on 18 May.

6. Conclusions

This study presented an integrated energy management strategy that combines load forecasting, economic operation scheduling, and inter-facility energy trading for industrial microgrids. By leveraging meteorological data and historical load information, an artificial neural network (ANN) model was developed to achieve a 24 h ahead forecast with an error margin within 12%. This predictive capability enabled the implementation of Particle Swarm Optimization (PSO)-based ESS scheduling tailored to site-specific consumption patterns. The PSO algorithm successfully minimized operational costs by generating optimal charge/discharge schedules under time-of-use pricing and SOC constraints. Distinct operational behaviors were observed between VT2 and Mold sites due to differences in load variability, which the EMS accommodated effectively through dynamic schedule customization.

Furthermore, the deployment of a Power Management System (PMS) facilitated real-time control of DERs and multi-microgrid coordination. Through this system, surplus energy transactions between facilities were executed automatically, with corresponding financial settlements logged and verified.

While the current study focused on load forecasting, the integration of photovoltaic (PV) generation forecasting could further enhance the efficiency and economy of microgrid operations. Accurate supply-side forecasting would allow the EMS to better align generation and storage strategies, reduce curtailment, and improve renewable energy utilization.

In addition, future pricing models must evolve beyond simple time-of-use rates. Factors such as ESS state of health (SOH), the number of charge–discharge cycles, site-specific peak periods, and weekday-dependent tariff variations should be incorporated to enable truly optimized economic dispatch strategies.

Future research will expand on this by incorporating generation forecasts and developing coordinated scheduling algorithms for each component of the microgrid, including ESS, PV, and real-time pricing models. This modular optimization approach will further refine the system's ability to deliver cost-effective and reliable energy management across diverse industrial environments.

These results demonstrate the feasibility and scalability of a forecast-driven EMS that enhances flexibility, reliability, and cost-efficiency in industrial microgrid environments. Future work may explore extensions to multi-site coordination involving external grids or market-driven dispatch frameworks. Additionally, from a system architecture perspective, future research should also address the development and deployment of a robust DC grid infrastructure. Such work is essential to support stable, bidirectional energy exchange and coordinated converter-based control strategies across interconnected microgrid units.

One limitation of this study is the absence of explicit photovoltaic (PV) power generation forecasting in the EMS framework. Although PV output profiles were considered in the simulation environment, real-time or day-ahead PV prediction was not integrated into the control strategy. Incorporating accurate PV forecasting—using machine learning or hybrid models—would significantly enhance the effectiveness of PSO-based scheduling and enable more proactive coordination between renewable generation and energy storage systems.

In addition, the current EMS utilizes a simplified time-of-use (TOU) pricing model, which does not fully reflect practical economic dispatch conditions. Future research will incorporate advanced pricing mechanisms that account for the state of health (SOH) of ESS, degradation from charge/discharge cycles, dynamic price fluctuations, and site-specific peak periods. These enhancements will allow for more realistic and cost-optimized energy management, enabling the EMS to operate effectively in more complex and variable industrial microgrid environments. Future work will thus aim to jointly optimize both load and PV forecasts while integrating dynamic, cost-aware dispatch models to establish a more robust, adaptive, and economically efficient EMS.

Author Contributions: Concept, design, writing, J.-h.K.; Analysis, writing, J.-h.K. Writing—review and editing, J.-h.K. and Y.-N.J. All authors have read and agreed to the published version of the manuscript.

Funding: This work was supported by Korea Institute of Energy Technology Evaluation and Planning (KETEP) grant funded by the Korea government (MOTIE) (20218530050090, Development and demonstration of energy IoT platform and service based on multi channel AMI infrastructure for Vietnam).

Data Availability Statement: The raw data supporting the conclusions of this article will be made available by the authors on request.

Acknowledgments: This study was supported by a research fund from Honam University, 2022.

Conflicts of Interest: The authors declare no conflicts of interest.

References

1. Heldeweg, M.A.; Séverine, S. Renewable energy communities as 'socio-legal institutions': A normative frame for energy decentralization? *Renew. Sustain. Energy Rev.* **2020**, *119*, 109518. [CrossRef]
2. Urishev, B. Decentralized Energy Systems, Based on Renewable Energy Sources. *Appl. Sol. Energy* **2019**, *55*, 207–212. [CrossRef]
3. Yaqoot, M.; Diwan, P.; Kandpal, T.C. Review of barriers to the dissemination of decentralized renewable energy systems. *Renew. Sustain. Energy Rev.* **2016**, *58*, 477–490. [CrossRef]
4. Kim, J.S.; So, S.M.; Kim, J.-T.; Cho, J.-W.; Park, H.-J.; Jufri, F.H.; Jung, J. Microgrids platform: A design and implementation ofcommon platform for seamless microgrids operation. *Electr. Power Syst. Res.* **2019**, *167*, 21–38. [CrossRef]
5. Zheng, X.; Yang, M.; Yu, Y.; Wang, C. Short-Term Net Load Forecasting for Regions with Distributed Photovoltaic Systems Based on Feature Reconstruction. *Appl. Sci.* **2023**, *13*, 9064. [CrossRef]
6. Kong, W.; Dong, Z.Y.; Jia, Y.; Hill, D.J.; Zhang, Y. A Review of Deep Learning Methods for Short-Term Load Forecasting. In Proceedings of the 2021 International Conference on Smart Grid and Electrical Automation (ICSGEA 2021), Kunming, China, 29–30 May 2021; Springer: Singapore, 2021; pp. 485–494. [CrossRef]
7. Wang, J.; Liu, H.; Zheng, G.; Li, Y.; Yin, S. Short-Term Load Forecasting Based on Outlier Correction, Decomposition, and Ensemble Reinforcement Learning. *Energies* **2023**, *16*, 4401. [CrossRef]
8. Hamidi, M.; Raihani, A.; Bouattane, O. Sustainable Intelligent Energy Management System for Microgrid Using Multi-Agent Systems: A Case Study. *Sustainability* **2023**, *15*, 12546. [CrossRef]
9. Gutiérrez-Oliva, D.; Colmenar-Santos, A.; Rosales-Asensio, E. A Review of the State of the Art of Industrial Microgrids Based on Renewable Energy. *Electronics* **2022**, *11*, 1002. [CrossRef]
10. Ginzburg-Ganz, E.; Segev, I.; Balabanov, A.; Segev, E.; Kaully Naveh, S.; Machlev, R.; Belikov, J.; Katzir, L.; Keren, S.; Levron, Y. Reinforcement Learning Model-Based and Model-Free Paradigms for Optimal Control Problems in Power Systems: Comprehensive Review and Future Directions. *Energies* **2024**, *17*, 5307. [CrossRef]
11. Lv, L.; Wu, Z.; Zhang, L.; Gupta, B.B.; Tian, Z. An Edge-AI Based Forecasting Approach for Improving Smart Microgrid Efficiency. *IEEE Trans. Ind. Inform* **2022**, *18*, 7946–7954. [CrossRef]
12. Taylor, J.W.; McSharry, P.E.; de Menezes, L.M. A Comparison of Univariate Methods for Forecasting Electricity Demand up to a Day Ahead. *Int. J. Forecast.* **2006**, *22*, 1–16. [CrossRef]
13. Yildiz, H.B. Modeling and Forecasting Electricity Consumption of Residential Consumers by Using the ARIMA and Hybrid Models. *Energy* **2017**, *124*, 117–127. [CrossRef]
14. Zhang, Y.; Wang, X.; Li, Y.; Liu, Y. Load Forecasting with Machine Learning and Deep Learning Methods. *Appl. Sci.* **2022**, *13*, 7933. [CrossRef]
15. Li, X.; Peng, J.; Zhang, Y.; Lu, S. Short-Term Load Forecasting for Microgrids Based on LSTM Recurrent Neural Network. *Sustain. Cities Soc.* **2023**, 104775. [CrossRef]
16. Li, Q.; Sun, H.; Zhang, L.; Wen, Y.; Ma, C. Short-Term Load Forecasting Using a Deep Neural Network. *Energies* **2020**, *13*, 3870. [CrossRef]
17. Toure, I.; Payman, A.; Camara, M.-B.; Dakyo, B. Energy Management in a Renewable-Based Microgrid Using a Model Predictive Control Method for Electrical Energy Storage Devices. *Electronics* **2024**, *13*, 4651. [CrossRef]
18. Shi, L.; Cen, Z.; Li, Y.; Wu, F.; Lin, K.; Yang, D. Distributed Optimization of Multi-Microgrid Integrated Energy System with Coordinated Control of Energy Storage and Carbon Emissions. *Sustainability* **2024**, *16*, 3225. [CrossRef]
19. Merabet, A.; Al-Durra, A.; El Fouly, T.; El-Saadany, E.F. Multifunctional energy management system for optimized network of microgrids considering battery degradation and load adjustment. *J. Energy Storage* **2024**, *100*, 113709. [CrossRef]
20. Deb, C.; Zhang, F.; Yang, J.; Lee, S.E.; Shah, K.K. A Review on Time Series Forecasting Techniques for Building Energy Consumption. *Renew. Sustain. Energy Rev.* **2017**, *74*, 902–924. [CrossRef]
21. Ko, J.-h. High-Definition Dynamic Voltage Restorer Systems Using Equivalent Time Sampling Techniques and Circular Structural Memory Filters. *Appl. Sci.* **2024**, *14*, 6896. [CrossRef]

Disclaimer/Publisher's Note: The statements, opinions and data contained in all publications are solely those of the individual author(s) and contributor(s) and not of MDPI and/or the editor(s). MDPI and/or the editor(s) disclaim responsibility for any injury to people or property resulting from any ideas, methods, instructions or products referred to in the content.

energies

Article

A Switched-Capacitor-Based Quasi-H7 Inverter for Common-Mode Voltage Reduction

Thi-Thanh Nga Nguyen [1], Tan-Tai Tran [2], Minh-Duc Ngo [1,*] and Seon-Ju Ahn [3,*]

1 Department of Automation, Thai Nguyen University of Technology (TNUT), 666, 3/2 Street, Tich Luong Ward, Thai Nguyen City 24132, Vietnam; nguyenthithanhnga-tdh@tnut.edu.vn
2 Faculty of Electrical Engineering Technology, Industrial University of Ho Chi Minh City, Ho Chi Minh City 71700, Vietnam; trantantaikdd@gmail.com
3 Department of Electrical Engineering, Chonnam National University, Buk-gu, Gwangju 61186, Republic of Korea
* Correspondence: ngoduc198-tdh@tnut.edu.vn (M.-D.N.); sjahn@chonnam.ac.kr (S.-J.A.); Tel.: +84-978658291 (M.-D.N.); +82-62-530-1738 (S.-J.A.)

Abstract: This paper proposes a novel three-phase two-level DC-AC inverter with significantly reduced common-mode voltage. The proposed inverter combines a conventional three-phase H7 configuration with a voltage multiplier network, effectively doubling the DC-link voltage relative to the input. Compared to existing solutions, the topology achieves a remarkably low common-mode voltage, limited to only 16.6% of the DC-link voltage. Additionally, the voltage stress across the additional switches remains at half of the DC-link voltage. The paper details the operating principles, mathematical formulation, and circuit-level analysis of the proposed inverter. Simulation results are provided to validate its performance. Furthermore, a hardware prototype has been implemented using a DSP TMS320F28379D microcontroller manufactured by Texas Instruments, headquartered in Dallas, TX, USA in conjunction with an Altera Cyclone® IV EP4CE22F17C6N FPGA-based digital control platform manufactured by Intel Corporation, headquarters in Santa Clara, CA, USA. Experimental results are presented to confirm the effectiveness and feasibility of the proposed design.

Keywords: switched-capacitor network; common-mode voltage; H7 inverter; DC-AC inverter

1. Introduction

Common-mode voltage (CMV) plays a crucial role in various applications, particularly in transformerless photovoltaic (PV) and electric motor drives [1–5]. In conventional pulse-width modulation (PWM) inverters, CMV with high frequency and amplitude is known to cause several undesirable effects, including electromagnetic interference (EMI), ground leakage currents, and bearing current generation [6,7]. In a typical three-phase system, the CMV is calculated as the average of the voltages between nodes A-O, B-O, and C-O and is expressed as:

$$V_{CM} = \frac{(V_{AO} + V_{BO} + V_{CO})}{3}. \quad (1)$$

where the point "O" represents the negative terminal of the input DC source, and V_{AO}, V_{BO}, and V_{CO} denote the phase voltages measured with respect to this point, as illustrated in Figure 1.

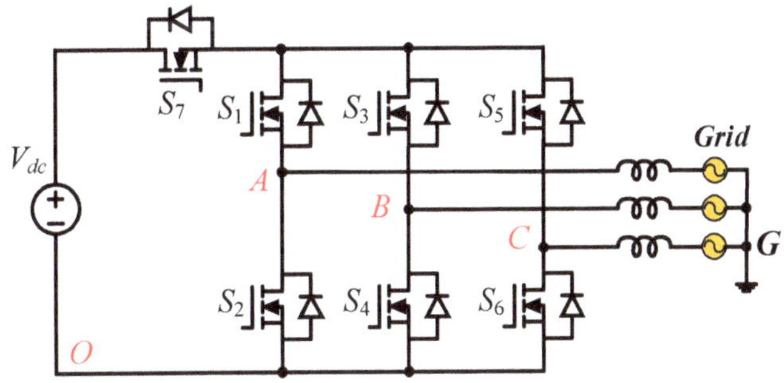

Figure 1. Topology of H7 inverter.

To address the CMV issues associated with conventional PWM inverters, a lot of reduction techniques have been proposed in previous studies. These approaches include modifications to inverter topologies as well as the development of advanced PWM control strategies.

Improved modulation control strategies, such as Active Zero State PWM (AZSPWM), Near State PWM (NSPWM), and Reduced Switching PWM (RSPWM), have been proposed to mitigate CMV variation by avoiding the use of zero-voltage states [8,9]. These techniques reduce the magnitude of the CMV to approximately 33.3% of the input DC voltage. However, this improvement comes at the cost of degraded power quality, a reduced modulation index, and lower efficiency, ultimately affecting overall inverter performance. To further reduce the CMV, hybrid space vector PWM (SVPWM) methods have been explored in [10]. The voltage linearity characteristics of these strategies are similar to those of the AZSPWM method. In addition, the CMV frequency was only equal to three times the fundamental frequency. Nevertheless, hybrid SVPWM techniques present challenges such as increased algorithmic complexity and higher switching losses. A novel model predictive control method is proposed in [11] to suppress common-mode voltage by strategically limiting the number of usable voltage vectors within each sector. As a result, the CMV magnitude is effectively constrained to 33.3% of the input DC voltage. In parallel, [12] introduces a low-common-mode SVPWM technique that divides the voltage plane into 12 sectors, achieving the same CMV magnitude while tripling the CMV switching frequency.

In addition to modulation strategies, various inverter topologies have been proposed to reduce the CMV. As discussed in [13–15], three-phase four-leg inverter structures are commonly employed for CMV suppression. In [13], a near state vector selection-based model for predictive control is introduced to limit the common-mode voltage by using a minimum number of usable vectors in every switching period. This method reduces the magnitude of the CMV to approximately 25% of the input DC voltage. In [14,15], the fourth leg of a three-phase four-leg inverter is actively controlled to compensate for CMV variation, effectively achieving zero common-mode voltage. A three-phase H7 inverter was introduced in [16], in which an additional switch is inserted between the input DC source and the conventional H-bridge inverter. A schematic of the conventional three-phase H7 inverter is shown in Figure 1. By controlling this additional switch to float the inverter from the DC source during zero states, the CMV varies between $V_{dc}/3$ and V_{dc}, as summarized in Table 1. A voltage-clamping H8 topology, discussed in [17], employs two additional active switches to disconnect the power source from the inverter during freewheeling states. This configuration maintains a constant CMV during mode transitions, with the CMV magnitude limited to only 33.3% of the input DC voltage. Alternative H8 topologies were also presented in [18,19], utilizing only two extra switches to constrain CMV variation. In these designs, the CMV fluctuation is similarly restricted to 33.3% of the input DC voltage.

Table 1. CMV value of conventional H7 inverter under different switching states.

Switching State	Bridge States ($S_1S_3S_5$)	State of S_7	V_{CMV}
State 1	100	1	$V_{dc}/3$
State 2	110	1	$2V_{dc}/3$
State 3	010	1	$V_{dc}/3$
State 4	011	1	$2V_{dc}/3$
State 5	001	1	$V_{dc}/3$
State 6	101	1	$2V_{dc}/3$
State 7	111	0	V_{dc}

In this paper, a novel switched-capacitor-based quasi-H7 inverter (SC-qH7 inverter) is proposed. The presented topology combines a conventional three-phase H7 inverter with a voltage multiplier network. Unlike existing solutions, the CMV of the proposed inverter varies only from $V_{PN}/6$ to $V_{PN}/3$, whereas the CMV in the conventional H7 inverter topology fluctuates from $V_{PN}/3$ to V_{PN}. Furthermore, the SC-qH7 inverter supports the full modulation index range without compromising on performance. A switched-capacitor network is proposed to achieve low common-mode voltage and voltage boosting in the proposed SC-qH7 inverter. Similar to conventional switched-capacitor circuits, the proposed SC-qH7 inverter also introduces an impulse charging current in capacitors. This issue of current waveforms can be efficiently mitigated by adding resonant inductors in the switched capacitor loop as presented in [20,21]. The proposed inverter structure is introduced in Section 2, followed by a description of the applied PWM control technique in Section 3. Simulation and experimental validation results are provided in Section 4.

2. Topology and Operating Principle of the SC-qH7 Inverter

Figure 2 illustrates the circuit configuration of the proposed SC-qH7 inverter. In this topology, a switched-capacitor network is added to the conventional H7 inverter.

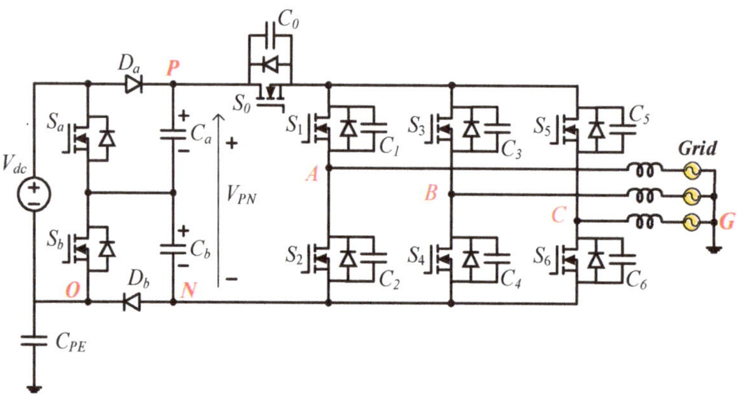

Figure 2. Topology of proposed SC-qH7 inverter.

From Figure 2, the voltage across diode D_b (V_{NO}) is calculated as follows:

$$V_{NO} = \begin{cases} -V_{Cb} & S_b \text{ is on} \\ 0 & S_b \text{ is off} \end{cases} \tag{2}$$

The CMV (V_{CM}) of the proposed SC-qH7 inverter can be rewritten as follows:

$$V_{CM} = V_{NO} + \frac{(V_{AN} + V_{BN} + V_{CN})}{3} \tag{3}$$

2.1. Operating Principle of Proposed SC-qH7 Inverter

The operating principles of the proposed SC-qH7 inverter can be explained through its distinct switching modes, as summarized in Table 2. Similar to the conventional H7 inverter [15], the proposed SC-qH7 inverter has seven switching modes, as shown in Figure 3.

Table 2. CMV value and switching modes of proposed SC-qH7 inverter.

Switching Mode	Switches						Diodes		Output Leg Voltage			V_{CM}
	S_1	S_3	S_5	S_a	S_b	S_0	D_a	D_b	V_{AN}	V_{BN}	V_{CN}	
Mode 1	On	Off	Off	On	Off	On	Off	On	V_{PN}	0	0	$V_{PN}/3$
Mode 2	On	On	Off	Off	On	On	On	Off	V_{PN}	V_{PN}	0	$V_{PN}/6$
Mode 3	Off	On	Off	On	Off	On	Off	On	0	V_{PN}	0	$V_{PN}/3$
Mode 4	Off	On	On	Off	On	On	On	Off	0	V_{PN}	V_{PN}	$V_{PN}/6$
Mode 5	Off	Off	On	On	Off	On	Off	On	0	0	V_{PN}	$V_{PN}/3$
Mode 6	On	Off	On	Off	On	On	On	Off	V_{PN}	0	V_{PN}	$V_{PN}/6$
Mode 7	On	On	On	Off	On	Off	Off	On	0	0	0	$V_{PN}/4$

V_{PN} is the DC-link voltage across the H-bridge circuit.

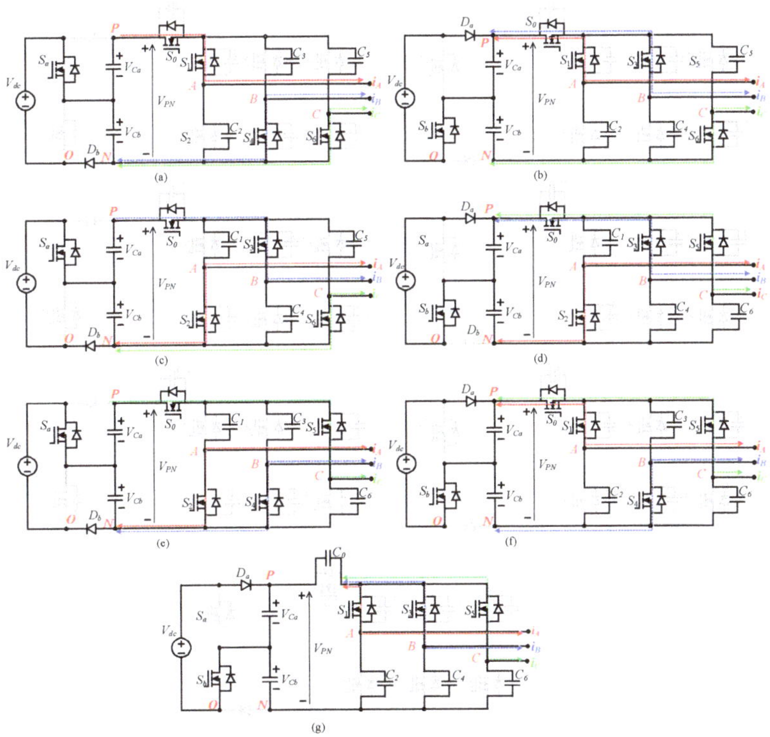

Figure 3. Equivalent circuits of the proposed SC-qH7 inverter. (**a**) mode 1; (**b**) mode 2; (**c**) mode 3; (**d**) mode 4; (**e**) mode 5; (**f**) mode 6; (**g**) mode 7.

Mode 1 (see Figure 3a): five switches S_a, S_0, S_1, S_4, and S_6 are turned on, while four switches S_b, S_2, S_3, and S_5 are turned off. As a result, the voltages between the nodes A-N, B-N, and C-N of the proposed SC-qH7 inverter are V_{PN}, 0 V, and 0 V, respectively. During this mode, diode D_a is blocked, whereas diode D_b is forward-based. Consequently, capacitor C_a discharges, and capacitor C_b charges through the loop V_{dc}-S_a-C_b-D_b. The corresponding CMV during this mode is $V_{PN}/3$.

Mode 2 (see Figure 3b): five switches S_b, S_0, S_1, S_3, and S_6 are turned on, while four switches S_a, S_2, S_4, and S_5 are turned off. As a result, the voltages between the nodes A-N, B-N, and C-N are V_{PN}, V_{PN}, and 0 V, respectively. During this mode, diode D_b is blocked, and diode D_a is forward-based. Capacitor C_b discharges, while capacitor C_a charges through the loop V_{dc}-D_a-C_a-S_b. The corresponding CMV during this mode is $V_{PN}/6$.

Mode 3 (see Figure 3c): five switches S_a, S_0, S_2, S_3, and S_6 are turned on, while four switches S_b, S_1, S_4, and S_5 are turned off. As a result, the voltages between the nodes A-N, B-N, and C-N are 0 V, V_{PN}, and 0 V, respectively. During mode 3, diode D_a is blocked, while diode D_b is forward-based. Capacitor C_a discharges, and capacitor C_b charges through the loop V_{dc}-S_a-C_b-D_b. The corresponding CMV during this mode is $V_{PN}/3$.

Mode 4 (see Figure 3d): switches S_b, S_0, S_2, S_3, and S_5 are turned on, while four switches S_a, S_1, S_4, and S_6 are turned off. As a result, the voltages between the nodes A-N, B-N, and C-N are 0 V, V_{PN}, and V_{PN}, respectively. During mode 4, diode D_b is blocked, while diode D_a is forward-based. Capacitor C_b discharges, and capacitor C_a charges through the loop V_{dc}-D_a -C_a -S_b. The corresponding CMV during this mode is $V_{PN}/6$.

Mode 5 (see Figure 3e): switches S_a, S_0, S_2, S_4, and S_5 are turned on, while four switches S_b, S_1, S_3, and S_6 are turned off. As a result, the voltages between the nodes A-N, B-N, and C-N are 0 V, 0 V, and V_{PN}, respectively. During this mode, diode D_a is blocked, while diode D_b is forward-based. Capacitor C_a discharges, and capacitor C_b charges through the loop V_{dc}-S_a-C_b-D_b. The corresponding CMV during this mode is $V_{PN}/3$.

Mode 6 (see Figure 3f): switches S_b, S_0, S_1, S_4, and S_5 are turned on, while four switches S_a, S_2, S_3, and S_6 are turned off. As a result, the voltages between the nodes A-N, B-N, and C-N are V_{PN}, 0 V, and V_{PN}, respectively. During mode 6, diode D_b is blocked, while diode D_a is forward-based. Capacitor C_b discharges, and capacitor C_a charges through the loop V_{dc}-D_a-C_a-S_b. The corresponding CMV during this mode is $V_{PN}/6$.

Mode 7 (see Figure 3g): four switches S_b, S_1, S_3, and S_5 are turned on, while five switches S_a, S_0, S_2, S_4, and S_6 are turned off. During this mode, diode D_b is blocked, while diode D_a is forward-based. Capacitor C_b discharges, and capacitor C_a charges through the loop V_{dc}-D_a-C_a-S_b.

As shown in Figure 2, the DC-link voltage of the proposed SC-qH7 inverter is:

$$V_{PN} = V_{Ca} + V_{Cb} = 2V_{dc}. \tag{4}$$

For the same input DC voltage, the proposed SC-qH7 inverter achieves a DC-link voltage that is twice that of the conventional H7 inverter. Based on the operating principles described earlier, it is observed that the CMV of the SC-qH7 inverter varies from $\mathrm{V_{PN}}/6$ to $\mathrm{V_{PN}}/3$ (approximately 16.6% of V_{PN}) during powering modes (modes 1 through 6). Notably, during these modes, switch S_0 remains in the ON state, contributing to stable power delivery. When the inverter transitions to mode 7, switch S_0 is turned OFF. Unlike the conventional H7 topology, the proposed topology incorporates seven additional capacitors positioned alongside switches $\mathrm{S_0}$ through $\mathrm{S_6}$, as shown in Figure 2. These capacitors, together with the inherent junction capacitances of the switches, form a coupling path between the input DC source and the inverter's AC output side. Consequently, under the zero-state condition, the CMV is primarily determined by the combined effect of these capacitive elements.

2.2. Capacitor Model During Freewheeling Mode

The transition process from the powering mode to the freewheeling mode is analyzed in this section, with a specific focus on the mode transition from mode 2 to mode 7. The transitions from mode 4 and mode 6 follow a similar process and are therefore omitted for brevity. At the initial time, it is assumed that the system is operating in mode 2, as illustrated in Figure 3b. The initial conditions are formulated as follows:

$$\begin{cases} V_{C0} &= 0V \\ V_{C2} &= V_{PN} \\ V_{C4} &= V_{PN} \\ V_{C6} &= 0V \end{cases} \tag{5}$$

The frequency-domain equivalent of the proposed SC-qH7 inverter is depicted in Figure 4, where the initial conditions of the system are represented by equivalent voltage sources. From Figure 4, the following mathematical equations are obtained:

$$\begin{cases} I_P(s) = I_A(s) + I_B(s) + I_C(s) \\ \frac{I_P(s)}{sC_0} + \frac{I_A(s)}{sC_2} = 0 \\ \frac{I_A(s)}{sC_2} - \frac{I_B(s)}{sC_4} = 0 \\ \frac{I_C(s)}{sC_6} - \frac{I_B(s)}{sC_4} = \frac{V_{PN}}{s} \end{cases} \tag{6}$$

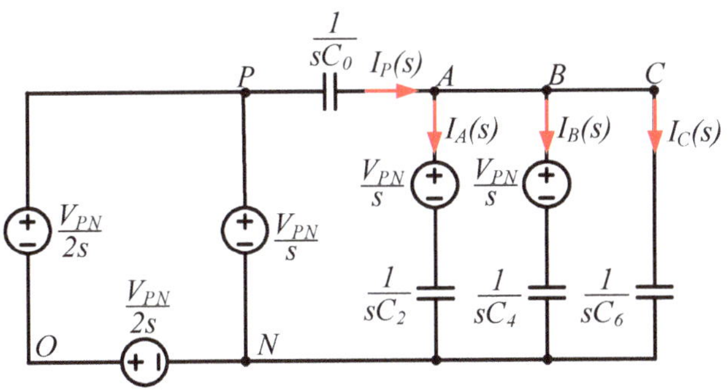

Figure 4. The frequency-domain equivalent of the proposed SC-qH7 inverter when entering mode 7.

With the frequency-domain equivalents, the CMV of the proposed SC-qH7 inverter ($V_{CM}(s)$) when entering mode 7 is expressed as:

$$V_{CM}(s) = -\frac{V_{PN}}{2s} + \frac{1}{3}\left[\left(\frac{V_{PN}}{s} + \frac{I_A(s)}{sC_2}\right) + \left(\frac{V_{PN}}{s} + \frac{I_B(s)}{sC_4}\right) + \left(\frac{I_C(s)}{sC_6}\right)\right] \tag{7}$$

By solving Equations (2) and (3), the CMV of the proposed SC-qH7 inverter ($V_{CM}(s)$) when entering mode 7 can be rewritten as:

$$V_{CMV}(s) = \left(\frac{C_0 + C_2 + C_4 - C_6}{C_0 + C_2 + C_4 + C_6}\right)\frac{V_{PN}}{2s} \tag{8}$$

Taking the inverse Laplace transform, we get:

$$V_{CMV} = \left(\frac{C_0 + C_2 + C_4 - C_6}{C_0 + C_2 + C_4 + C_6}\right)\frac{V_{PN}}{2} \tag{9}$$

When $C_0 = C_2 = C_4 = C_6$, the V_{CMV} under mode 7 is equal to $V_{PN}/4$.

Therefore, the CMV of the proposed SC-qH7 inverter only takes values of $V_{PN}/3$, $V_{PN}/6$, or $V_{PN}/4$. All the switching modes and the corresponding CMV of the proposed SC-qH7 inverter are given in Table 2. As highlighted in the table, the variation range of the CMV in the proposed topology is limited to just 16.6% of the DC-link voltage, representing a significant improvement compared to conventional inverter structures.

3. PWM Control Technique of SC-qH7 Inverter

Figure 5 illustrates the proposed PWM control strategy for the SC-qH7 inverter. Similar to the Discontinuous PWM (DPWM) technique, the proposed method utilizes three types of switching vectors: an odd active vector, an even active vector, and the zero vector V_7. If the final reference voltage is located in sector A_1, as presented in Figure 5a, the switching sequence will be (111)-(110)-(100)-(100)-(110)-(111). The complete switching pattern for each sector under the DPWM strategy for the SC-qH7 inverter is summarized in Table 3. The time durations of the voltage space vectors $\vec{V_1}$, $\vec{V_2}$, $\vec{V_3}$,$\vec{V_4}$, $\vec{V_5}$,$\vec{V_6}$, and $\vec{V_7}$ during a switching period are denoted as T_1, T_2, T_3, T_4, T_5, T_6, and T_7, respectively.

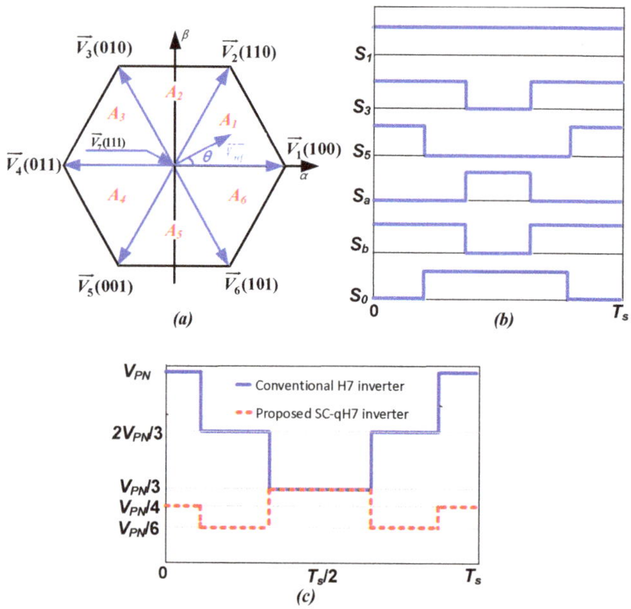

Figure 5. DPWM strategy for the SC-qH7 inverter. (**a**) Space vector and A-type sectors in output of three-phase inverter; (**b**) switching pattern of the proposed SC-qH7 inverter in sector A_1; (**c**) comparison of the CMV between the conventional H7 inverter and the proposed SC-qH7 inverter under the DPWM strategy.

In sector A1, as illustrated in Figure 5 and Table 3, the reference voltage vector ($\vec{V_{ref}}$) is synthesized using the three vectors: $\vec{V_1}$, $\vec{V_2}$, and $\vec{V_7}$. Based on the volt–second balance principle, we derive the following:

$$T_{s\times} \vec{V_{ref}} = T_{1\times} \vec{V_1} + T_{2\times} \vec{V_2} + T_{7\times} \vec{V_7} \tag{10}$$

where T_s is the switching period.

The time durations of the voltage space vectors $\vec{V_1}$, $\vec{V_2}$, and $\vec{V_7}$ can be calculated as:

$$\begin{cases} T_1 = M_i \times T_s \times \sin\left(\frac{\pi}{3} - \theta\right) \\ T_2 = M_i \times T_s \times \sin(\theta) \\ T_7 = T_s - T_1 - T_2 \end{cases} \tag{11}$$

Table 3. Switching pattern of the DPWM strategy for the SC-qH7 inverter.

Sector	Switching Sequence
A1	$\vec{V_7}(111) - \vec{V_2}(110) - \vec{V_1}(100) - \vec{V_1}(100) - \vec{V_2}(110) - \vec{V_7}(111)$
A2	$\vec{V_7}(111) - \vec{V_2}(110) - \vec{V_3}(010) - \vec{V_3}(010) - \vec{V_2}(110) - \vec{V_7}(111)$
A3	$\vec{V_7}(111) - \vec{V_4}(011) - \vec{V_3}(010) - \vec{V_3}(010) - \vec{V_4}(011) - \vec{V_7}(111)$
A4	$\vec{V_7}(111) - \vec{V_4}(011) - \vec{V_5}(001) - \vec{V_5}(001) - \vec{V_4}(011) - \vec{V_7}(111)$
A5	$\vec{V_7}(111) - \vec{V_6}(101) - \vec{V_5}(001) - \vec{V_5}(001) - \vec{V_6}(101) - \vec{V_7}(111)$
A6	$\vec{V_7}(111) - \vec{V_6}(101) - \vec{V_1}(100) - \vec{V_1}(100) - \vec{V_6}(101) - \vec{V_7}(111)$

In the above equations, $M_i = \sqrt{3}\frac{V_m}{V_{PN}}$ represents the modulation index, where V_m is the magnitude of the reference voltage vector and V_{PN} is the DC-link voltage.

Similarly, in sector A2, the reference voltage vector ($\vec{V_{ref}}$) is synthesized using the three vectors $\vec{V_2}$, $\vec{V_3}$, and $\vec{V_7}$. The time durations of the voltage space vectors $\vec{V_2}$, $\vec{V_3}$, and $\vec{V_7}$ can be calculated as:

$$\begin{cases} T_2 = M_i \times T_s \times \sin\left(\theta + \frac{\pi}{3}\right) \\ T_3 = M_i \times T_s \times \sin\left(\theta - \frac{\pi}{3}\right) \\ T_7 = T_s - T_2 - T_3 \end{cases} \tag{12}$$

Similarly, in sector A3, the reference voltage vector ($\vec{V_{ref}}$) is synthesized using the three vectors $\vec{V_3}$, $\vec{V_4}$, and $\vec{V_7}$. The time durations of the voltage space vectors $\vec{V_3}$, $\vec{V_4}$, and $\vec{V_7}$ can be calculated as:

$$\begin{cases} T_3 = M_i \times T_s \times \sin(\theta) \\ T_4 = M_i \times T_s \times \sin\left(\theta - \frac{2\pi}{3}\right) \\ T_7 = T_s - T_3 - T_4 \end{cases} \tag{13}$$

Similarly, in sector A4, the reference voltage vector ($\vec{V_{ref}}$) is synthesized using the three vectors $\vec{V_4}$, $\vec{V_5}$, and $\vec{V_7}$. The time durations of the voltage space vectors $\vec{V_4}$, $\vec{V_5}$, and $\vec{V_7}$ can be calculated as:

$$\begin{cases} T_4 = M_i \times T_s \times \sin\left(\theta - \frac{\pi}{3}\right) \\ T_5 = M_i \times T_s \times \sin(\theta - \pi) \\ T_7 = T_s - T_4 - T_5 \end{cases} \tag{14}$$

Similarly, in sector A5, the reference voltage vector ($\vec{V_{ref}}$) is synthesized using the three vectors $\vec{V_5}$, $\vec{V_6}$, and $\vec{V_7}$. The time durations of the voltage space vectors $\vec{V_5}$, $\vec{V_6}$, and $\vec{V_7}$ can be calculated as:

$$\begin{cases} T_5 = M_i \times T_s \times \sin\left(\theta - \frac{2\pi}{3}\right) \\ T_6 = M_i \times T_s \times \sin\left(\theta + \frac{2\pi}{3}\right) \\ T_7 = T_s - T_5 - T_6 \end{cases} \tag{15}$$

Similarly, in sector A6, the reference voltage vector ($\overrightarrow{V_{ref}}$) is synthesized using the three vectors $\overrightarrow{V_1}$, $\overrightarrow{V_6}$, and $\overrightarrow{V_7}$. The time durations of the voltage space vectors $\overrightarrow{V_1}$, $\overrightarrow{V_6}$, and $\overrightarrow{V_7}$ can be calculated as:

$$\begin{cases} T_1 = M_i \times T_s \times \sin(\theta + \frac{\pi}{3}) \\ T_6 = M_i \times T_s \times \sin(\theta - \pi) \\ T_7 = T_s - T_1 - T_6 \end{cases} \quad (16)$$

The proposed SC-qH7 inverter under the proposed PWM control strategy offers the notable advantage of maintaining linear operation across the full range of the modulation index. A comparison of the CMV between the conventional H7 inverter and the proposed SC-qH7 inverter under the DPWM strategy in sector A_1 is also presented in Figure 5c. As shown in Figure 5b, during even active vectors and the zero vector V_7, switch S_a remains in the OFF state while switch S_b remains ON. Conversely, during odd active vectors, switch S_a is **ON** and switch S_b is OFF. Additionally, switch S_0 remains ON throughout the operation, except during vector V_7, where it is turned OFF. This switching scheme ensures that the CMV of proposed SC-qH7 inverter only takes values of $V_{PN}/3$, $V_{PN}/4$, or $V_{PN}/6$. As a result, the proposed SC-qH7 inverter retains all the inherent advantages of voltage source inverters (VSIs) operated under DPWM, such as reduced switching losses, low total harmonic distortion (THD) in the output current, and full modulation index range ($0 \leq M_i \leq 1$), while also offering the additional benefit of reduced CMV variation.

The power loss of the proposed SC-qH7 inverter topology can be categorized into losses from power switches, power diodes, and capacitors.

The conduction loss of the IGBT is calculated as:

$$P_{con} = V_{CEO} \times \overline{I_C} + I_{crms}^2 \times R_c + V_D \times \overline{I_F} + I_{Frms}^2 \times R_D \quad (17)$$

where V_{CEO}, V_D, R_c, R_D, $\overline{I_C}$, I_{crms}, $\overline{I_F}$, and I_{Frms} represent the voltage drop across the IGBT in saturation mode, the forward voltage of the antiparallel diode, the resistance of the IGBT, the forward conduction resistance of the antiparallel diode, the average value of the collector current of the IGBT, the RMS collector current of the IGBT, the average value of the current through the antiparallel diode, and the RMS antiparallel diode current, respectively.

The switching loss of the IGBT can be calculated as:

$$P_{sw} = (E_{ON} + E_{OFF}) \times f_s \quad (18)$$

where E_{ON}, E_{OFF}, and f_s represent the Turn-ON energy loss of the IGBT, the Turn-OFF energy loss of the IGBT, and the switching frequency, respectively.

The power loss of diode can be calculated as:

$$P_D = Q_{rr} \times V_{Doff} \times f_s + V_{Don} \times \overline{I_{Don}} \quad (19)$$

where V_{Don}, V_{Doff}, $\overline{I_{Don}}$, and Q_{rr} are the forward voltage drop on the diode, the voltage across the diode, the reverse recovery charge of the diode, and the average value of the current through the diode, respectively.

The capacitor loss can be calculated as follows:

$$P_{Cap} = R_{cap} \times I_{caprms}^2 \quad (20)$$

where R_{cap}, I_{caprms}, and f_s represent the equivalent series resistance of the capacitors and the RMS capacitor current, respectively.

4. Simulation and Experiment Results

4.1. Simulation Results

To validate the operating principles and evaluate the performance of the proposed SC-qH7 inverter, simulation studies were conducted using the PSIM 9.0 software environment. The key parameters used in the simulation are given as follows. The switching frequency and the fundamental frequency were 10 kHz and 50 Hz, respectively. The two capacitors C_a and C_b were both selected as 220 µF. The DC input voltage was set to 100 V, and the stray capacitor was 300 nF. The capacitances of the additional capacitors ($C_0 - C_6$), as show in in Figure 2, were each set to 1 nF. An inductive load with R_l = 27 Ω and L_l = 25 mH was used.

Figure 6 presents the simulation waveforms of the proposed SC-qH7 inverter, including the input voltage, the voltages across capacitors C_a and C_b, the DC-link voltage, the voltage stress across switches S_a and S_b, and diodes D_a and D_b, as well as the output line-to-line voltage, output phase voltages, and their corresponding FFT analysis. As highlighted in Figure 6, the DC-link voltage V_{PN} reached a steady-state value of 200 V, which was twice the input DC voltage. The voltage across capacitors C_a and C_b were each maintained at the input DC voltage level. The voltage stress across both switches (S_a, S_b) and both diodes (D_a, D_b) was observed to be equal and limited to the input voltage, confirming the reduced stress advantage of the design. As also shown in Figure 6, the inverter generated a unipolar output line-to-line voltage waveform. The output phase voltage achieved an amplitude of approximately 115 V with a modulation index M = 1. The corresponding output phase currents were sinusoidal and free from noticeable distortion. The amplitude of the output phase current was 4 A, and the THD was measured to be approximately 0.8%, demonstrating excellent power quality performance.

Figure 7 presents the simulation waveforms of the common-mode voltage and the phase voltages with respect to the negative DC terminal, i.e., the voltages between nodes A–O, B–O, and C–O, for the proposed SC-qH7 inverter. The figure also includes a zoomed-in view to clearly illustrate the CMV variation during the different operating modes. As shown in Figure 7b, during the powering modes, the CMV took discrete values of 66.6 V or 33.3 V, while during the freewheeling mode, the CMV stabilized at 50 V. These results confirm that the CMV remained bounded between 33.3 V and 66.6 V, consistent with the theoretical expectations. Consequently, the total variation in CMV was limited to 33.3 V, which corresponds to one-sixth of the DC-link voltage, demonstrating the effectiveness of the proposed SC-qH7 inverter in reducing CMV fluctuations. The waveform of the common-mode voltage across the stray capacitor C_{PE} appeared as highlighted in Figure 8. This low CMV level contributes to a reduced leakage current in the proposed SC-qH7 inverter.

To further evaluate the inverter's performance under varying operating conditions, the input voltage was increased to 160 V and 200 V, while the modulation index was decreased to 0.6 and 0.4, respectively. Figure 8 displays the stray capacitor voltage, corresponding leakage current, and their frequency spectra at different modulation indices. As shown in Figure 8, the higher-order harmonic components of the stray capacitor voltage remained very low across all tested conditions. The rms value of the leakage current was approximately 4 mA when M = 1 and V_{dc} = *100 V* (Figure 8a). When the modulation index was reduced to 0.6 and the input voltage was increased to 160 V, the leakage current increased slightly to 6 mA (Figure 8b). At a further reduced modulation index of 0.4 with an input voltage of 200 V, the leakage current remained modest at 12 mA (Figure 8c). These results confirm that the proposed SC-qH7 inverter can operate across the full modulation index range while maintaining low leakage current levels, demonstrating its suitability for transformerless applications.

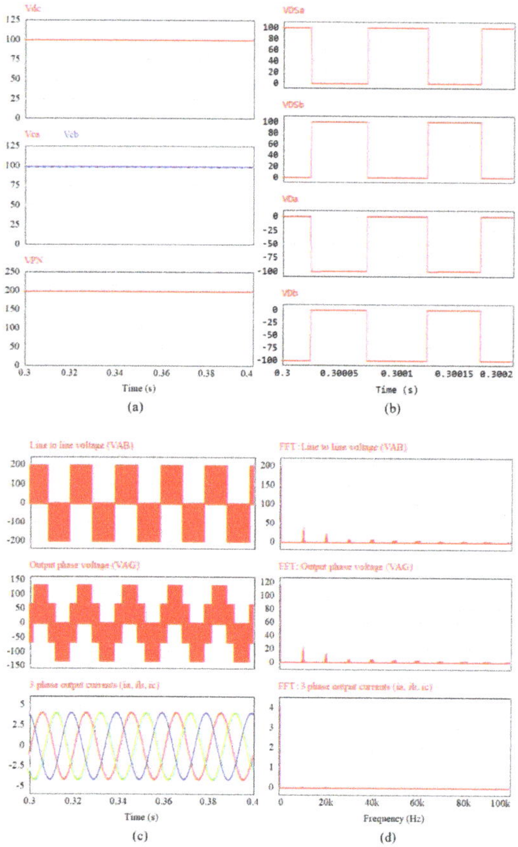

Figure 6. Waveforms of proposed SC-qH7 inverter. From top to bottom: (**a**) input voltage, capacitor C_a voltage, capacitor C_b voltage, and DC-link voltage; (**b**) voltage across switch S_a, voltage across switch S_b, voltage across diode D_a, and voltage across diode D_b; (**c**) line-to-line voltage v_{AB}, phase-A voltage, and 3-phase output currents (Phase A in red, Phase B in green and Phase C in blue); (**d**) FFT of line-to-line voltage v_{AB}, FFT of phase-A voltage v_{AG}, and FFT of 3-phase output currents.

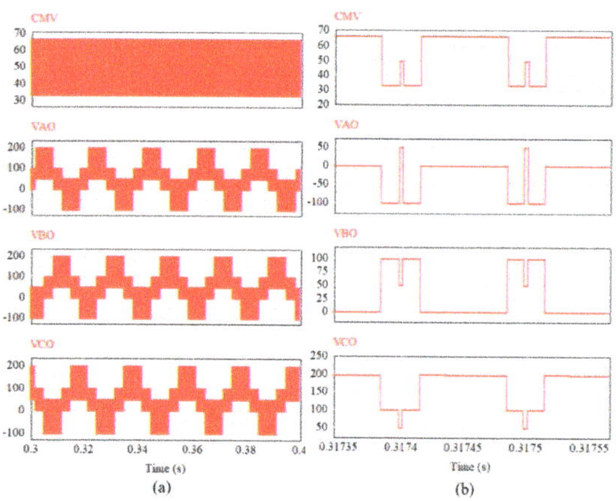

Figure 7. Simulation results of common-mode voltage and the voltages between nodes A-O, B-O, and C-O of the proposed SC-qH7 inverter. (**a**,**b**) From top to bottom: CMV, v_{AO}, v_{BO}, and v_{CO}.

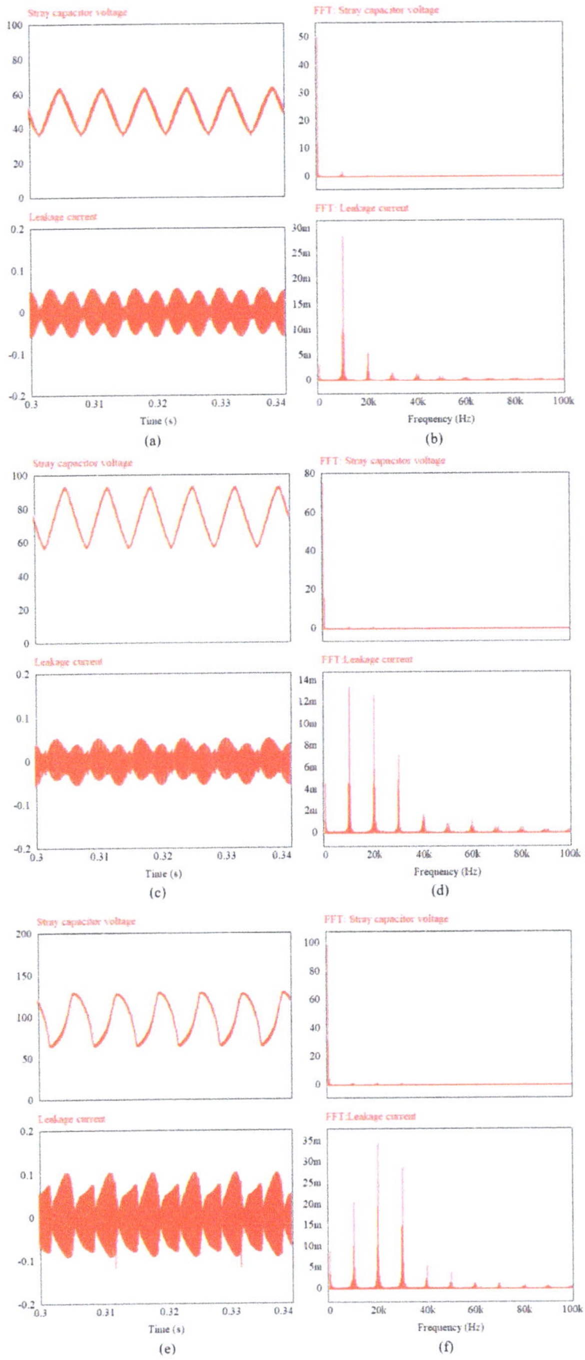

Figure 8. Simulation results of stray capacitor voltage and leakage current for the proposed SC-qH7 inverter at different modulation indices. From top to bottom: (**a**,**c**,**e**) stray capacitor voltage and leakage current; (**b**,**d**,**f**) FFT of stray capacitor voltage and FFT of leakage current. (**a**,**b**) $M = 1$ and $V_{dc} = 100$ V; (**c**,**d**) $M = 0.6$ and $V_{dc} = 100$ V; (**e**,**f**) $M = 0.4$ and $V_{dc} = 100$ V.

4.2. Experimental Results

In order to confirm the effectiveness of the proposed SC-qH7 inverter, a 1 kW laboratory prototype was constructed and tested. The modulation algorithms were implemented using a TMS320F28379D digital signal processor manufactured by Texas Instruments, Dallas, TX, USA, while a DE0-Nano FPGA manufactured by Intel Corporation Santa Clara, CA, USA was employed to handle logic control functions. The parameters used for the experimental setup were identical to those used in the simulation to ensure consistency in performance evaluation. For power switching, two IRFP4668PbF MOSFETs with a voltage rating of 200 V were used for switches S_a and S_b. The six H-bridge switches, along with the additional switch S_0, were implemented using IPW60R045CPA MOSFETs. These eight MOSFETs were manufactured by Infineon Technologies, Neubiberg, Germany. Additionally, two STPS60SM200C Schottky diodes manufactured by STMicroelectronics, Geneva, Switzerland, with low voltage ratings were used for the freewheeling diodes D_a and D_b.

Figure 9 presents the experimental waveforms of the proposed SC-qH7 inverter, including the input voltage, voltages across capacitors C_a and C_b, voltage stress across switches S_a and S_b and diodes D_a and D_b, output line-to-line voltage v_{AB}, output phase-A voltage, three-phase output currents, and the FFT of the output phase-A current. As shown in Figure 9a, the steady-state voltages across capacitors C_a and C_b were 98 V and 95 V, respectively. These values were slightly lower than the input voltage due to parasitic elements present in the experimental hardware. The observed low-frequency peak-to-peak ripple voltages were approximately 4 V for C_a and 6 V for C_b. Figure 9b shows that the voltage stress on switches S_a and S_b was identical at 99 V, while the voltage stresses across diodes D_a and D_b were 98 V and 95 V, respectively. Figure 9c confirms that the inverter generated a unipolar output line-to-line voltage waveform. As shown in Figure 9d, the output phase currents were sinusoidal and free from noticeable distortion. The measured amplitude of the output phase current was 4 A, and the THD was approximately 1.05%, validating the inverter's high-quality output performance. The measured efficiency of the proposed SC-qH7 inverter reached approximately 90.8% when Mi = 1 at a output power of 700 W. However, the efficiency of the proposed SC-qH7 inverter was relatively low due to the use of non-optimal components in the experimental setup.

Figure 10 shows the experimental waveforms of the common-mode voltage and the voltages between nodes A-O, B-O, and C-O, along with their zoomed-in views. As shown in Figure 10a, the CMV during the powering modes took discrete values of approximately 65 V or 32 V, while it stabilized at 50 V during the freewheeling mode. This confirms that the CMV remained constrained within the expected range of 32 V to 65 V.

Figure 11 illustrates the experimental waveform of the CMV across the stray capacitor C_{PE}, along with the corresponding leakage current. The experimental results are in good agreement with the simulation results, validating the accuracy of the proposed inverter model. It can be observed that the SC-qH7 inverter effectively limited the CMV variation, particularly during transitions into and out of mode 7. This suppression of the CMV contributed to a significantly reduced leakage current. As shown in Figure 11, the rms value of the leakage current was approximately 13.5 mA. Therefore, the proposed SC-qH7 inverter effectively limited the leakage current to a level well below 300 mA, as specified in the standard VDE-0126-1-1 [22].

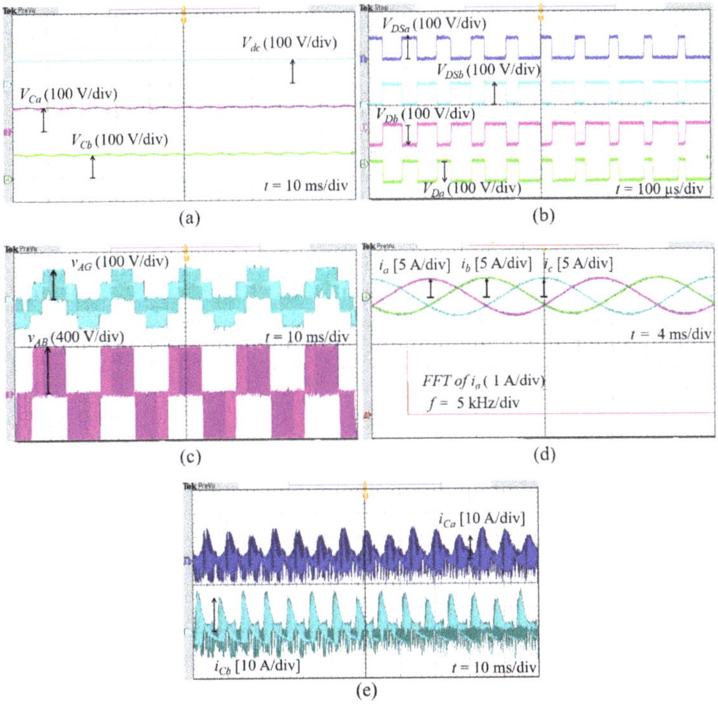

Figure 9. Experimental waveforms of the proposed SC-qH7 inverter. From top to bottom: (**a**) input voltage, voltage across capacitor C_a, and voltage across capacitor C_b; (**b**) voltage stress across switches S_a and S_b and diodes D_b and D_a; (**c**) phase-A voltage and line-to-line voltage v_{AB}; (**d**) three-phase output currents and FFT spectrum of output phase-A current; (**e**) current of capacitors C_a and C_b.

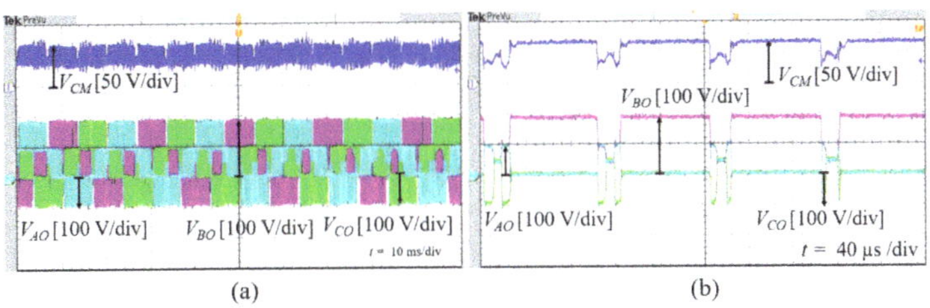

Figure 10. Experimental measurements of common-mode voltage and the voltages between nodes A-O, B-O, and C-O of the proposed SC-qH7 inverter. (**a**,**b**) From top to bottom: common-mode voltage and the voltages between nodes A-O, B-O, and C-O.

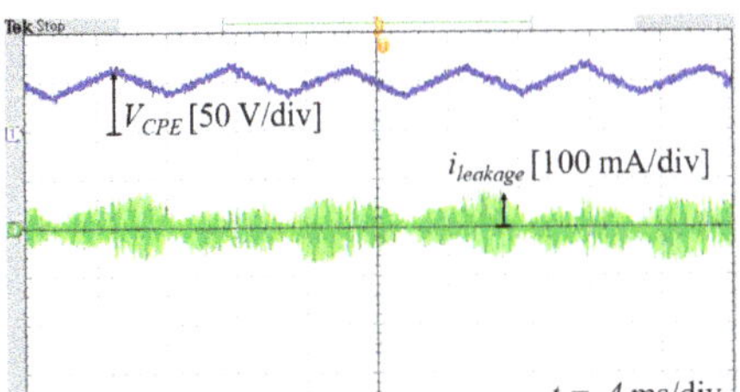

Figure 11. Experimental measurements of stray capacitor voltage (**top**) and leakage current (**bottom**) for the proposed SC-qH7 inverter.

5. Conclusions

In this paper, a novel three-phase two-level DC–AC inverter with low common-mode voltage, CMV, referred to as the SC-qH7 inverter, was proposed, analyzed, simulated, and experimentally validated. By integrating a voltage multiplier network into the conventional H7 inverter topology, the proposed structure achieves a DC-link voltage that is twice the input DC voltage. As a result, the SC-qH7 inverter is capable of operating in both buck and boost modes. The inverter generates a unipolar output line-to-line voltage, which reduces the voltage stress across the output filter inductor. This leads to a more compact filter design and lower power losses. Compared to conventional topologies, the SC-qH7 inverter exhibits a significantly reduced CMV, with its magnitude limited to only 16.6% of the DC-link voltage. Furthermore, the voltage stress across the additional switches in the voltage multiplier network is limited to half of the DC-link voltage, enhancing device reliability. The voltage across the stray capacitor remains low, which effectively suppresses the resulting leakage current. Simulation and experimental results confirm that the leakage current remains below 13.5 mA. Additionally, the output current THD is measured at only 1.05%, demonstrating excellent power quality. These features make the proposed SC-qH7 inverter a promising and efficient solution for three-phase transformerless photovoltaic (PV) systems. Because two capacitors in the switched-capacitor network suffer impulse charging current, the selection of the capacitors needs to meet the requirements of reliability and size of the power converter. The reliability and lifetime of the capacitors are calculated based on the switching frequency and spectral content of the impulse current through the capacitor. The size and type of the capacitor are calculated based on the level of harmonics and the lifetime of the capacitor.

Author Contributions: Conceptualization, T.-T.T. and S.-J.A.; methodology, T.-T.N.N.; software, T.-T.N.N. and M.-D.N.; validation, T.-T.T., M.-D.N. and S.-J.A.; formal analysis, T.-T.T. and S.-J.A.; investigation, S.-J.A.; resources, T.-T.T.; data curation, M.-D.N.; writing—original draft preparation, T.-T.N.N.; writing—review and editing, S.-J.A. and T.-T.T.; visualization, T.-T.T.; supervision, M.-D.N.; project administration, S.-J.A. and M.-D.N.; funding acquisition, S.-J.A. All authors have read and agreed to the published version of the manuscript.

Funding: This work was supported in part by the National Research Foundation of Korea (NRF) through the Korean Government (MSIT) under grant RS-2024-00454464.

Data Availability Statement: The original contributions presented in the study are included in the article.

Acknowledgments: Thanks to TNUT's support in completing the article.

Conflicts of Interest: The authors declare no conflicts of interest.

References

1. Lopez, O.; Freijedo, F.D.; Yepes, A.G.; Comesaña, P.F.; Malvar, J.; Teodorescu, R.; Gandoy, J.D. Eliminating ground current in a transformerless photovoltaic application. *IEEE Trans. Energy Convers.* **2010**, *25*, 140–147. [CrossRef]
2. Yang, B.; Li, W.; Gu, Y.; Cui, W.; He, X. Improved transformerless inverter with common-mode leakage current elimination for a photovoltaic grid-connected power system. *IEEE Trans. Power Electron.* **2012**, *27*, 752–762. [CrossRef]
3. Adabi, J.; Boora, A.A.; Zare, F.; Nami, A.; Ghosh, A.; Blaabjerg, F. Common-mode voltage reduction in a motor drive system with a power factor correction. *IET Power Electron.* **2012**, *5*, 366–375. [CrossRef]
4. Murai, Y.; Kubota, T.; Kawase, Y. Leakage current reduction for a high-frequency carrier inverter feeding an induction motor. *IEEE Trans. Ind. Appl.* **1992**, *28*, 858–863. [CrossRef]
5. Tran, T.T.; Nguyen, M.K.; Duong, T.D.; Choi, J.H.; Lim, Y.C.; Zare, F. A Switched-Capacitor-Voltage-Doubler Based Boost Inverter for Common-Mode Voltage Reduction. *IEEE Access* **2019**, *7*, 98618–98629. [CrossRef]
6. Wang, L.; Shi, Y.; Shi, Y.; Xie, R.; Li, H. Ground leakage current analysis and suppression in a 60-kW 5-level T-type transformerless SiC PV inverter. *IEEE Trans. Power Electron.* **2018**, *33*, 1271–1283. [CrossRef]

7. *IEC 62109-2:2011*; Safety for Power Converters for Use in Photovoltaic Power Systems—Part 2: Particular Requirements for Inverters. International Electrotechnical Commission: Geneva, Switzerland, 2011.
8. Un, E.; Hava, A.M. Performance characteristics of the reduced common mode voltage near state PWM method. In Proceedings of the European Conference on Power Electronics and Applications, Aalborg, Denmark, 2–5 September 2007; pp. 1–10.
9. Un, E.; Hava, A.M. A near-state PWM method with reduced switching losses and reduced common-mode voltage for three-phase voltage source inverters. *IEEE Trans. Ind. Appl.* **2009**, *45*, 782–793. [CrossRef]
10. Janabi, A.; Wang, B. Hybrid SVPWM scheme to minimize the common-mode voltage frequency and amplitude in voltage source inverter drives. *IEEE Trans. Power Electron.* **2019**, *34*, 1595–1610. [CrossRef]
11. Medekhel, L.; Hettiri, M.; Labiod, C.; Srairi, K.; Benbouzid, M. Enhancing the Performance and Efficiency of Two-Level Voltage Source Inverters: A Modified Model Predictive Control Approach for Common-Mode Voltage Suppression. *Energies* **2023**, *16*, 7305. [CrossRef]
12. Zheng, J.; Peng, C.; Zhao, K.; Lyu, M. A Low Common-Mode SVPWM for Two-Level Three-Phase Voltage Source Inverters. *Energies* **2023**, *16*, 7294. [CrossRef]
13. Dadu, A.M.; Mekhilef, S.; Soon, T.K.; Seyedmahmoudian, M.; Horan, B. Near State Vector Selection-Based Model Predictive Control with Common Mode Voltage Mitigation for a Three-Phase Four-Leg Inverter. *Energies* **2017**, *10*, 2129. [CrossRef]
14. Xiaoqiang, G.; Ran, H.; Jiamin, J.; Zhigang, L.; Xiaofeng, S.; Guerrero, J.M. Leakage current elimination of four-leg inverter for transformerless three-phase PV systems. *IEEE Trans. Power Electron.* **2016**, *31*, 1841–1846.
15. Yang, D.; Yang, K.; Gao, F. Topology Optimization Analysis of Leakage Current Suppression in Three-Level Four-Leg Photovoltaic Grid-Connected Inverter. *IEEE Access* **2023**, *11*, 50145–50156. [CrossRef]
16. Freddy, T.K.S.; Rahim, N.A.; Hew, W.-P.; Che, H.S. Modulation Techniques to Reduce Leakage Current in Three-Phase Transformerless H7 Photovoltaic Inverter. *IEEE Trans. Ind. Electron.* **2015**, *62*, 322–331. [CrossRef]
17. Concari, L.; Barater, D.; Buticchi, G.; Concari, C.; Liserre, M. H8 Inverter for common-mode voltage reduction in electric drives. *IEEE Trans. Ind. Appl.* **2016**, *52*, 4010–4019. [CrossRef]
18. Rahimi, R.; Farhangi, S.; Farhangi, B.; Moradi, G.R.; Afshari, E.; Blaabjerg, F. H8 inverter to reduce leakage current in transformerless three-phase grid-connected photovoltaic systems. *IEEE J. Emerg. Sel. Top. Power Electron.* **2018**, *6*, 910–918. [CrossRef]
19. Gupta, A.K.; Agrawal, H.; Agarwal, V. A Novel three-phase transformerless H-8 topology with reduced leakage current for grid-tied solar PV applications. *IEEE Trans. Ind. Appl.* **2019**, *55*, 1765–1774. [CrossRef]
20. Ye, Y.; Cheng, K.W.E. A Family of Single-Stage Switched-Capacitor–Inductor PWM Converters. *IEEE Trans. Power Electron.* **2013**, *28*, 5196–5205. [CrossRef]
21. Rosas-Caro, J.C.; Mayo-Maldonado, J.C.; Valdez-Resendiz, J.E.; Alejo-Reyes, A.; Beltran-Carbajal, F.; López-Santos, O. An Overview of Non-Isolated Hybrid Switched-Capacitor Step-Up DC–DC Converters. *Appl. Sci.* **2022**, *12*, 8554. [CrossRef]
22. *DIN VDE V 0126-1-1 VDE V 0126-1-1:2013-08*; Automatic Disconnection Device between a Generator and the Public Low-Voltage Grid. VDE Publishing House: Berlin, Germany, 2013.

Disclaimer/Publisher's Note: The statements, opinions and data contained in all publications are solely those of the individual author(s) and contributor(s) and not of MDPI and/or the editor(s). MDPI and/or the editor(s) disclaim responsibility for any injury to people or property resulting from any ideas, methods, instructions or products referred to in the content.

MDPI AG
Grosspeteranlage 5
4052 Basel
Switzerland
Tel.: +41 61 683 77 34

Energies Editorial Office
E-mail: energies@mdpi.com
www.mdpi.com/journal/energies

Disclaimer/Publisher's Note: The title and front matter of this reprint are at the discretion of the Guest Editors. The publisher is not responsible for their content or any associated concerns. The statements, opinions and data contained in all individual articles are solely those of the individual Editors and contributors and not of MDPI. MDPI disclaims responsibility for any injury to people or property resulting from any ideas, methods, instructions or products referred to in the content.

www.ingramcontent.com/pod-product-compliance
Lightning Source LLC
Chambersburg PA
CBHW061553210926
51850CB00009B/2087
* 9 7 8 3 7 2 5 8 5 4 8 9 9 *